The Highlands and Islands

A volume in the New Naturalist Series, a survey of British natural history published by Collins and edited by John Gilmour, Sir Julian Huxley, Margaret Davies, and Kenneth Mellanby. The New Naturalist has been described by the *Listener* as 'one of the outstanding feats of publishing since the war', and by *The Times Literary Supplement* as 'a series which has set a new standard in natural history books'. Founded in 1945, it now contains more than sixty volumes, of which the following are available in Fontana:

F. FRASER DARLING
and
J. MORTON BOYD

The Highlands
and Islands

Collins
THE FONTANA NEW NATURALIST

First published in the New Naturalist series 1964
Reprinted with revisions 1969
First issued in Fontana 1969
Tenth Impression May 1979

To
J. R. MATTHEWS
and
C. M. YONGE

Made and printed in Great Britain by
William Collins Sons & Co. Ltd, Glasgow

Contents

Figures

Plates

Preface

By F. Fraser Darling

It is nineteen years since the publication of this book in its first form, a period during which the advance of our knowledge of natural history in the Highlands and Islands has been greater than at any time since Hugh Miller, Archibald Geikie and the compilation of what is called the Harvie-Brown Series of volumes.

I thanked James Fisher, Charles Elton and my late wife Averil Morley for their help in the First Edition, in criticism and suggestion. Nevertheless, the book was very much a one-man effort, written largely during the later war years when I myself was living in island remoteness and when travel to libraries was difficult. Specialist workers were scattered and doing other things. The book bore all the marks of one man covering fields in which his knowledge was not equal to his enthusiasm, but, my back being fairly broad, I am glad I had the temerity to write the book, for it tried to imply the modern ecological view of wholeness and interdependence, and that man was an important part of natural history, a point of view which has continued and been further developed in almost all work in Scottish natural history since then. I went still further myself in *West Highland Survey* in trying to show that man's interference with the processes of natural history in the Highlands and Islands had been a major factor in the development of the human, social and economic problems of the area, and that rehabilitation must work in terms of natural history as well as of administration divorced from ecological realities.

I have become familiar with natural history in two other continents since my work in the Highlands, but the Highlands always seemed to me to be my laboratory and what I saw abroad I would compare with Highland phenomena. So much

ix

was different, but ecological principles are the same everywhere and adjustment was easy. But all this travel from Arctic
to Tropic tended to take me away from the exciting immediacy
of the fast-developing natural history of the Highlands and
Islands and I knew a sound second edition of this book was
beyond me while I was so thinly spread, if I may use a vivid
colloquialism.

Looking around for a possible victim, I did not take long
to come back to the figure I first thought of, my present
collaborator Dr. John Morton Boyd: he was one of the new
generation of Scottish naturalists who could see things in the
round. He had trained under a Master at Glasgow, a tremendous advantage in itself, but not enough for the task ahead
of him if he had not had the initial fire in the belly. This is
what Morton has got in large measure, heating his curiosity,
driving him into many fields and providing great physical
energy to get to remote places. His industry in this present
edition masks the natural contemplative idleness of the senior
author. That industry led Morton to long personal interviews
and correspondence with many individual workers and
corporate groups who have been active in Scottish natural
history in these nineteen years. John Morton Boyd and I
thank these workers for the time they have given and the
completely generous scientific spirit displayed in sharing the
results of their work, especially the following: T. B. Bagenal,
G. P. Black, R. N. Campbell, G. M. Dunnet, F. H. W. Green,
H. R. Hewer, R. A. Innes, D. Jenkins, J. R. Lewis, J. D.
Lockie, V. P. W. Lowe, A. Macdonald, A. T. Macmillan,
D. N. McVean, B. Mitchell, D. Nethersole-Thompson, K. A.
Pyefinch and colleagues, B. B. Rae, J. G. Roger, D. Stephen,
L. K. Stewart, and A. Watson. We are very grateful to W. J.
Eggeling for his criticism of the manuscripts and correction
of proofs, and to Miss Mary Peters for her help with the
bibliography. One name which does not appear is that of the
Publisher, but he should be thanked for his vision, his
courteous patience and the extra pages he has given us.

The Conservation Foundation,   F.D.
New York

Chapter One

Geology and Climate

The scenery of the Highlands and Islands is that of solid rock. The country is essentially a plateau deeply trenched in all directions by the rivers and glaciers of past ages. In the Grampians and Cairngorms, semblance of the high plateau still remains, but to the west the land has been more severely gouged with deep glens, ranges of sharp peaks and narrow ridges. In the north-west the plateau has not merely been trenched; it has been planed down over wide areas of west Sutherland, leaving the mountains as isolated humps on a low spacious plain of rock and peat. The floors of the glens are filled with alluvium grooved by fast flowing rivers and the lower slopes have a skin of acid soil and peat. Platforms such as the Moor of Rannoch and the Outer Hebrides have their rocky faces blanketed in deep peat, but the mountain tops are for the most part stark naked rock. In the interpretation of the scenery the geology cannot be ignored.

Geology linked with climate has determined the nature of the initial vegetation of the Highlands. Man himself has been, and still is, determining the secondary vegetation to some extent through his management of agriculture, forestry, water and sport. There are, however, limits to what he can do by way of alteration of the natural system, in the face of underlying rock. Vegetation in turn has a remarkable effect on the animal life, both in variety and distribution. The rocks, through the relief of the land to which they give rise, also have a definite effect on climate; the presence of mountains, for example, results in cloud formation and heavy rainfall, especially in a warm moist air stream such as prevails in Scotland for the most of the year.

The vegetation influences slightly the local climate.

1

Trees for example reduce wind and provide shade, and the vegetation and climate acting together have their effect on the surface rock and soil; woodlands prevent erosion. We should remember the constant interplay of these dynamic forces as well as life itself in studying the natural history of the Highlands and Islands. It is a noble drama of weather and mountain and sea and plant and animal.

GEOLOGY

From almost every point of view of natural history, the Highland region of Scotland is demarcated by the sharp geological line known as the Highland Boundary Fault which runs north-east—south-west across Scotland from the mouth of the Clyde to Stonehaven (fig. 1). Although a good deal of Scotland's best agricultural land lies along the east coast north of this line, the great mass of the country to the north-west of the Fault is the Highlands, yet the Highlands are not necessarily high ground. The highest point of the Hebridean island of Benbecula is only 409 feet, but Benbecula is unquestionably as Highland in its natural history as in its human cultural relationships. The same can be said for Colonsay with its miniature mountain landscape rising to only 470 feet, and its miniature forest with small feral goats in place of red deer.

Another major fault plays a large part in the topography of the area, for along it the Great Glen of Scotland has been gouged. This great dividing line between the Northern and North-West Highlands on the one hand and the Central and South-West Highlands on the other is marked not only by the Great Glen itself, but by the chain of long freshwater lochs it contains, the most famous of which is Loch Ness, 21¾ miles long and 754 feet deep. The loch drains north-eastwards by the short Ness River to the Moray Firth at Inverness. The very low watershed which crosses the Great Glen, in so far as the flow north-east or south-west is concerned, is above the head of Loch Oich and is no more than 115 feet above the sea level. The next

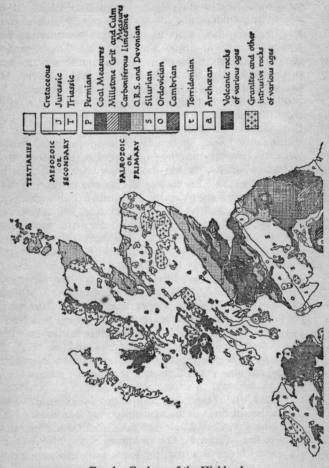

FIG. 1. Geology of the Highlands

The exposures of Tertiary and Cretaceous are not large enough to show on this scale

loch south-westwards is Loch Lochy. The Lochy River
which flows from it runs into the salt water of Loch Linnhe,
and ultimately fans out after the Corran Narrows and becomes the Firth of Lorne, which is such a distinctive feature of the western coastline. At the south-westward end
of the Great Glen and at the head of Loch Linnhe stands
the sentinel-like massif of Ben Nevis, the highest hill in
Scotland, 4,406 feet high. The summit is only four miles
from the sea as the crow flies, so the full sense of height of
this mass can be appreciated by the traveller coming eastwards down Loch Eil. Ben Nevis is formed by a granite
intrusion between the two great areas of the Moine and
Dalradian schists. It has the highest sheer cliff face in
Britain of about 1,500 feet as well as one of the very few
semi-permanent snow patches in Britain. This patch of
snow is untouched by the sun. The actual summit of Ben
Nevis is not of the granite of which the main mass of the
hill is composed. The summit has been carved from a
gigantic cake of basalt with some Old Red Sandstone.

It is worth noting that the two highest mountainous
groups in Scotland are formed of granite—Ben Nevis and
Cairngorms region, 4,300 feet, with Lochnagar, 3,786 feet,
in the eastern Highlands. Topographically, the Cairngorms
viewed from afar may seem as uninteresting as Ben Nevis,
but the physical beauty of these hills is for intimate
observation in the magnificent corries and on the high
plateaux. It is possible to walk a pony on to these high
ridges and plateaux without any trouble, the ground being
good all the way.

There are several other great geological phenomena which
will be less obvious to the casual observer than those of the
Highland Boundary Fault and the Great Glen. First, that
definite line of tectonic tumult known as the Moine Overthrust which emerges from the sea at the south-west corner
of Skye and runs north-north-eastwards to the eastern shore
of Loch Eriboll on the north coast. Here the older rocks
of the Moine schists have been thrust westwards over
younger rocks along a hundred-mile line. Along this line,
like a sandwich filling between Archaean Gneiss and the

overlying Torridonian sandstone on the west side, and the Moine Schists on the east, are beds of Cambrian age including an extremely hard and shiny quartzite which makes a barren strip of country.

Here and there along this line there appears a geological titbit in the shape of an outcrop of the famous Durness Limestone of Cambrian age. The effect of this is to enliven the natural system and change the scenery of this vast geological sandwich. Instead of bareness and blackness of peat we get greenness and soil. It is this limestone that makes the Assynt district a place which no naturalist whether his interests be geological or biological should ignore. A little further south than Assynt, in a black area of bog to the east of Suilven and Canisp, there rises an island of limestone a few hundred acres in extent. The climber on these hills in spring or autumn will experience pleasure and something of a shock to see the townships of Elphin and Knockan on their geological emerald. All this is part of the paradox of the Highlands and of Highland life. They are full of surprises and facts which do not seem to fit in. No sooner does the theorizing type of mind construct a hypothesis which looks neat than some disconcerting fact will create a paradox.

So much for the Moine Thrust and its consequences; the main geological phenomenon is the mixture in the West Highlands of old rocks and new volcanic ones. From Cape Wrath to Applecross the west coast is a wild jumble of the ancient Hebridean gneiss and that old and barren sandstone known as Torridonian. These two formations make for scenery which is exceptionally wild in quite different ways. Then at the foot of Loch Linnhe is the Isle of Mull and, to the north of it, some similar Tertiary volcanic rocks in Morven and Ardnamurchan. Such volcanic rock is a dull grey in colour and amorphous in texture, except where there occur amygdaloid pockets of crystals of much beauty, but it makes some distinctive scenery. These Tertiary rocks as we see them in Mull are the remains of immense beds of lava, possibly about 50 million years of age as against the 1,500 million years or so of the gneiss farther north. The lava is eroded into terrace-like formations which correspond to the actual flows, and on these terraces the soil is found to be brown and rich—a real soil without peat—and the grass grows thick in summer. The country of the Tertiary terraces is cattle country and turns up again in north-central Skye. Sometimes the terraced denudation and landslips give way to a natural castellated architecture such as the Castle Rock of the Treshnish Isles and further still to towers and spires like the Quirang and the Old Man of Storr in Skye. This latter rock is like a natural Tower of Pisa and is visible from many miles away. Sometimes, again, the basalt has solidified in a peculiar way to make those hexagonal columns much visited at Staffa, the small island between Iona and the Treshnish Isles. Such columns may also be found elsewhere, as in Mull, Canna and Skye, on a much less spectacular scale, but probably the most impressive examples in Scotland are those of the northern face of Garbh Eilean of the Shiant Isles. The Shiants are the northern outposts of Scotland's youngest rocks; only four miles away to the west is the east coast of the Hebrides, composed of her oldest rock, the Archaean gneiss.

The Moine Thrust has been mentioned as one of the major geological features of the Highlands. The Moine schists which occur to the east of the Thrust are the most extensive group of rocks found in the Highlands as a whole. They are sometimes called the Undifferentiated Eastern Schist, and stretch in a broad, roughly parallel-sided band, fifty miles wide, from the north coast of Scotland to the foot of the Great Glen where it passes under the basalt of Mull and western Morven. South of the Great Glen more schists and slates, the Dalradian, are the building rocks of a large part of the Central Highlands and reach far into Aberdeenshire. Many of the high tops of the Highlands are of this formation—Ben Lawers 3,984 feet, Craig Meagaidh 3,700, and Carn Eige 3,877 feet. The schists form a great plateau on the western side of the Spey opposite the Cairngorms. This schist plateau has the Gaelic name of Monadliadh—the grey mountain. The

grey is most obvious when it comes up against either the
granite, as on the east side of the Highlands, or against the
reddish-purple Torridonian, as on the west. The Moine
Schists break down more easily than the Lewisian gneisses,
they contain a fair quantity of alumina and often make
good soil, and for example produce a different vegetational
complex from the adjoining Torridonian. The presence
of overmuch peat, of course, may entirely cut out the influence of the slowly-disintegrating rock.

The third geological phenomenon is the Tertiary Volcanoes which consist of huge emplacements of granite beside others of a coarse igneous rock known as gabbro.
The gabbro represents the solidified remains of the sub-
terranean cauldrons in which molten lava was stored
before eruption from volcanoes. The granite most probably represents portions of the earth's crust which were
converted to granite by the heat escaping from the copious
streams of upwelling molten basalt, deep in the roots of
the volcanoes. The erosion of these great masses of rock
of contrasting character by ice, sea and rain has produced
some of the most spectacular scenery not only in the Highlands and Islands, but in the world. A small-scale example
of the gabbro scenery forms Ardnamurchan, the most
westerly point of the mainland of Great Britain. It is wild
but not high enough to be considered grand, in the sense
of Skye. Gabbro and granite are the stuff of the Cuillin
Hills of Skye and away to the north into Trotternish
stretch the lavas from the volcano which once lay above
the Cuillins. The Black Cuillin is gabbro rising to 3,309
feet in Sgurr Alasdair the highest point of the tumultuous
arcuate ridge. Detached to the east is the gabbro massif of
Blaven. By way of contrast, the Red Cuillin to the north is
granite with the elegant pap-like mountains of Beinn
Dearg and Beinn na Caillich. Gabbro is hard and knobbly
with large crystals protruding from the matrix, which
makes it safe for the experienced climber. The Cuillin
range was the centre of an ice-cap in glacial times and the
glaciers working outwards to the perimeter have carved
the great corries and left the sharp ridges in which we de-

light today. With the retreat of the glaciers the shape of the moraines become obvious, and subsequent weathering of the hills has produced some great scree slopes, such as Alasdair's stone-shoot.

In Eastern Rhum there is another ice-carved volcanic root of gabbro similar to that forming the Black Cuillin of Skye. To the west and south of the island, granite and other igneous rocks are present while the northern part consists of Torridonian sandstones into which the volcanic rocks were emplaced. The lavas of this volcano survive only in four small localities in Rhum, but the neighbouring islands of Canna, Muck and Eigg are composed almost entirely of lavas from this source. The Sgurr of Eigg is, however, composed of pitchstone, the origin of which has long been in dispute. The highest hills of Rhum do not go beyond 2,700 feet but in beauty of line they are no less magnificent than the Black Cuillin.

Mull was the centre of great volcanic activity in Tertiary times between 15 million and 75 million years ago. A large volcano occurred over south-east Mull which was the focus of a great swarm of dykes which are beamed south-eastward through mid-Argyll and Cowal and stretch through the Southern Uplands as far as Yorkshire. Portions of the volcano from time to time subsided into the lava cauldron below the cone, and the pressure of the up-welling lava pushed the surrounding rocks back to form arcuate folds. The root of the Mull volcano has been partly exposed by erosion but much of the lava has remained. In consequence the scenery of Mull is dominated by basalt lavas. There are none of the castellated skylines of Skye and Rhum, but massive terraced hills.

The one remaining ring volcano in the area (excluding Arran) is perhaps, in its own way, the most spectacular of all—St. Kilda. The islands of St. Kilda are unique in British scenery and in British natural history. Here the bergs of the ice-ages and the ocean have in their time carved the gabbro and the granite into the most fantastic display of island sculpture in Britain. The scenery of the Cuillins is repeated in a different setting and with a different finish.

Hirta, the main island, is a gigantic harlequin of granite and gabbro. Like the Red Cuillin, the granite hills of Conachair and Oiseval are smooth elegant paps ; like the Black Cuillin the hills of Mullach Bi and the islands of Dun, Soay and Boreray are shattered castellated ridges of gabbro. The cream-coloured granite rises in stupendous walls vertical from the sea to a height of nearly 1,400 feet in the summit of Conachair, and the purple gabbro in a contrasting style of bastions and turreted buttresses rises well over 1,000 feet in Mullach Bi, Soay and Boreray. Stac Lee and Stac an Armin, 544 and 627 feet respectively, lie close to Boreray and the sea-cliff scenery is spell-binding, the more so because of the vast concourse of gannets which nest here in the largest gannetry in the world. Rockall lies 184 miles to the west of St. Kilda ; it is 70 feet high and composed of a variety of granite.

Calcareous sedimentary rocks of Jurassic and Triassic age are of local ecological importance in the West. Like the Durness limestone these give rise to oases of fertile soil and greenery in the widespread heaths. Outcrops occur at Aultbea, Applecross, Raasay, Broadford, Strathaird and Staffin in Skye, Kilchoan in Ardnamurchan, Gribun, Inchkenneth, Loch Don and Carsaig in Mull, and in Morven. Most of these support a combination of well cultivated arable land and pasture, and give a varied coastline with deciduous woodland at Ardtornish, hazel scrub and a rich flora. The exposures at Mingary and at Elgol are a fine stratigraphical display. Near Elgol on the shore there is a flat slab which shows the cracks which the sun made on the estuarine muds over 100 million years ago.

In this first look around the Highlands and Islands we have given a sketch of their geological make-up and how it affects the topography and vegetation ; this is a basic ecological theme which threads its way through this book. Not enough has been said, however, about glaciation as the most tremendous carver of scenery from the matrix of rocks of one kind or another. We have mentioned the glaciation of the entire Highland plateau, with its greatest effects in the west and its least in the Central Highlands. Glaciologists now appear to agree that there were four glacial periods in which the Highlands were involved, though it is improbable that the whole region was covered each time. There were certainly large long-enduring ice-caps in the North-West and Central Highland plateaux from which the glaciers streamed in all directions, carving and planing the terrain. The late glacial history of the area is described in detail by Charlesworth (1955).

From these two or more central ice-caps there are well defined radial courses which have gouged out or deepened many of the glens and polished the summits of some of the lower hills. The country of the Hebridean gneiss in Sutherland and north-west Ross has been heavily scored by glaciers flowing westwards from the inland Torridonian sandstone hills and in the melting of the ice many boulders of this dark red rock have come to rest on the grey gneiss hills. In the Torridonian area we see some fine hanging corries at about 1,750 feet with large areas of boulder moraine below their lips. Where the boulders, long carried in the ice, have been open to the weather, only the scorings and rounding are obvious. In most glens there are lateral, terminal and recessional moraines cut by rivers and streams.

The end of the ice-ages was followed by upward movement of the earth's crust freed from the great superincumbent ice-sheet. This gave rise to the raised beaches which are such a characteristic feature of the coasts of Argyll. Sea-cliffs now appear many hundreds of yards inland and still bear the unmistakable scars of the sea's temper, in caves and detached stacks of rock. These ancient beaches are now fertile, with a well drained veneer of rich loam masking the old shore deposits and yielding early crops of potatoes. Tiree is, for the most part, a raised beach platform, which at the end of the ice-age was no more than extensive reefs ; Islay, Colonsay and Jura have some of the finest raised beaches in Scotland.

The glaciers have not only carved and smoothed the

countryside but have had a profound effect on the subsequent natural history. The rock surfaces, shorn and left bare, have often remained bare or have gathered only a superficial layer of peat, supporting a scanty vegetation. Again, some of the gougings have made saucer-like depressions which will not drain completely and either become lochs or peat-filled bogs with their own flora. The moraines are exceptionally well drained with good grass and heather growing from the rock detritus and a very thin layer of peat ; in many areas they still hold remnants of the native forest. The Central and Eastern Highlands have a lower rainfall and possess much more extensive glacial deposits than the west. For this reason the valleys of the Spey, Dee, Tay and Tummel are probably the best coniferous tree-growing areas in the Highlands.

CLIMATE

The relation of geology and scenery we have taken for granted ; that of geology, climate and vegetation we take almost for granted, and because of that we are too apt, perhaps, to generalize. Old school geography books informed us that the climate of the British Isles was mild and humid and that the laving waters of the northern sweep of the Gulf Stream, which we call the North Atlantic Drift, kept our insular climate equable. Though it is generally true that the south-eastern side of Britain is drier than the north-western, it is for the ecologist to enquire more deeply into such generalizations, for he knows that altitude, position and slope in relation to sun, nearness to sea and so on have considerable local effects on climate. There is a mosaic of climatic effects throughout the Highlands as there is elsewhere, which can be seen on the vast scale of comparison of Western, Northern, Eastern and Southern Highlands, or on the small scale of comparing the weather on different sides of the same mountain.

Weather varies according to latitude, longitude and altitude ; it varies according to the relief of the country, position on the coast or inland, direction and velocity of the prevailing wind, the humidity and temperature of the air, the temperature of land and sea. There are so many variables, which interplay to determine the climate of any one locality, that the differences can be detected over very short distances; the climate can almost be considered as being infinitely diverse.

Loch Tay is 20 miles long; from east to west there is an increase of an inch per year of rainfall for every mile you go. We should rightly judge, therefore, that the natural history of Killin would show differences from that of Aberfeldy on the score of rainfall alone. At Dundonnell at the head of Little Loch Broom the rainfall is about 72 inches a year. Seven miles to the south in Strath na Sheallag beyond the Torridonian cones of An Teallach the rainfall is about 100 inches. Seven miles north of Dundonnell lies Ullapool, which receives an average of 48 inches. Both snow and frost are more severe in Strath na Sheallag than in Ullapool. If climates alter so markedly within a few miles—and that is the rule rather than the exception in the Highlands—so can they alter in a few yards.

The north side of a tree possesses a small micro-climatic region quite distinct from that of the south side, and as a result of this there are definite zones of disposition of mosses and lichens. Similarly, the upper canopy of a tree has a different climate from that at its foot. There is micro-habitat with micro-climates under every stone whether in the sea, in a river, in a loch or on land, also under dung-pats, logs, leaves, sea-weed and a vast miscellany of litter, also in pools in situations from the mountain tops to the sea-shore, in the forks of trees and in water butts beside cottages. We could go to great length in our description of the diversity of the habitat of the Highlands in terms of the micro-climate, but it is beyond the scope of this book. The number of micro-habitats is legion; it is the integration of them all which give the grand wholeness to the Highlands and Islands, of which every naturalist is acutely aware.

In considering climate, let us first of all realize the

FIG. 2a. January Isotherms
(reduced to sea level)

FIG. 2b. July Isotherms
(reduced to sea level)

amount of indentation of the land of the West Highlands,
which allows the sea to enter far into the countryside. The
western ocean, being relatively warm for this latitude of
53-59°N, in itself makes for mildness in the immediate
neighbourhood of the sea lochs. It must be remembered,
however, that altitude far outweighs latitude and distance
from the sea in the matter of climate. West Highland hills
tend to rise steeply from the sea, and as a great deal of
ground lies above the 1,500-foot contour, the overall temperature is low. The coasts, especially in favoured places,
are exceptionally mild, and years may pass without more
than 2°F of frost being recorded. Only the coasts of
western Wales and southern England can exceed the mean
annual warmth of those of the West Highlands. The mean
January temperatures of the West Highland coasts are
between 40° and 42°F compared with 38° and less on the
east coast of Scotland. In July the West Highlands have a
mean summer temperature of 55-57°F compared with
56-58°F on the east coast (fig. 2, above). If we take the
differences between the annual mean summer and winter
temperatures, there is only 14°F of difference on the West
Highland coast, 20° at Dundee and 24° in London, Kent
and East Anglia. These are sea-level temperatures. The

Highlands show a completely different story as soon as you go uphill, and when you go truly inland into the Monadliadh or Cairngorm regions, conditions are much more extreme. Much work has been done on mountain climate in Britain by Dr. Gordon Manley, who has written the volume *Climate and the British Scene* in this series. The twenty years' work on Ben Nevis is available to us in summary and we give these figures as a comparison with conditions which prevail elsewhere. The mean temperature of the hottest month, July, was 41.1°F, and of the coldest month, February, 23.8°F, a range of 17.3°F. The extreme records were, Maximum 66°F (28 June 1902) and Minimum 1°F (6 January 1894).

In 1956 a weather station near the summit of Ben Macdhui (4,120 feet) recorded a minimum of 11°F in February and a maximum of 57°F in September, compared with 7°F in February and 75°F in June at Braemar at 1,120 feet (Baird, 1957). The mean annual temperature at Ben Macdhui was 31.3°F, giving a climate of frequent freeze and thaw.

With rainfall (fig. 3), we see how easy it is to fall into a trap by generalizing that the west coast is wetter than the east. The belt of very high rainfall is not on the coast but a few miles inland, and even then it does not extend uniformly from north to south of the Highlands. Reference to Bartholomew's Atlas of Scotland will show the monthly rainfall distribution in fair detail. Coming south from Cape Wrath, the strip of country with a rainfall over 100 inches a year does not start until Latitude 57.30°N, and then goes southwards and slightly south-eastwards to Latitude 56°N. There is a good area all round this strip with a precipitation of 60-100 inches, but the coastal promontories, especially in the north, and the Hebrides receive only 40-60 inches of rain, a low figure which is not reached elsewhere in the true Highlands until the central area. A very few parts of the eastern Highlands receive as little as 30-40 inches.

These two climatic factors of temperature and rainfall are of immense importance in determining the vegetation

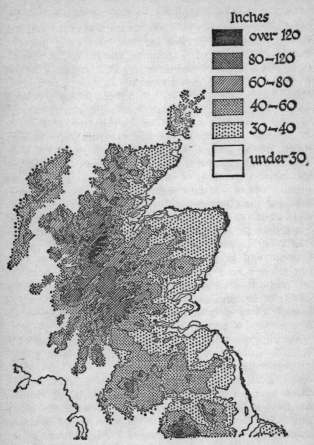

Inches

over 120

80–120

60–80

40–60

30–40

under 30.

FIG. 3. Average annual rainfall in the Highlands

Based on a map prepared in the Meteorological Office and reproduced by permission of Her Majesty's Stationery Office. Crown copyright reserved

of an area, but there are inter-relations of these two which must always be taken into consideration when natural history is being studied. For example, something must be known of the rate of evaporation relative to the precipitation. The ratio between these two in any given situation has a big influence on the types of plants and to a lesser extent on the animals to be found there. There is often a constant variation in the relative humidity during the day and this variation may influence the behaviour of animals. When rainfall exceeds potential evaporation in every month of the year, as it does in most parts of the West Highlands and the Inner Isles, several consequences follow, the most important of which is waterlogging of the ground. Not only is the soil deprived of regular re-oxygenation, but the plants never have a respite from transpiring at full rate; this may sometimes favour vegetative growth, but it does not always favour maturation and regeneration. The West Highlands are in this respect the wettest part of Europe, and great interest therefore attaches to their humid environment. Land-use is much less flexible in this environment than elsewhere in north-west Europe; the effects of burning and overgrazing can be disastrous and a study of these conditions is being made by the Nature Conservancy (Green, 1959).

Sunshine records in the Highlands are highly variable. Broadly speaking, the amount of sunshine is in inverse proportion to the rainfall, so the strip of the Highlands to which allusion has already been made as enduring 100 inches or more of rain enjoys least sun. It is also possible to draw a line bisecting the Outer Hebrides longitudinally, showing an annual total of sunshine of less than 1,200 hours on the west side and 1,200-1,300 hours on the east. One of the surprises of the west is the local area of high sunshine records for the island of Tiree, the outermost of the Inner Hebrides. The island is low (beneath the waves, as Gaeldom described it in the old days) and has but little cloud-stopping or cloud-gathering power. The soil is mostly of shellsand or loam, so its moisture drains or evaporates quickly. It was not without good reason that in past days Tiree was known as the granary of the Isles. Only the south coast of England equals or exceeds Tiree's record.

The West Highlands do not suffer nearly so much snow as the Central and East, or the Southern Uplands or Pennines. Snow comes earlier and stays on the tops of the hills because the factor of altitude is concerned, but the Atlantic mildness pervades much of the lower ground. Nearness to the sea affects how soon the snow will come on a hill face and how long it will stay. Snow as a climatic factor can be of much importance in a region long after it has fallen if the catchment area of the snow is large. The Cairngorm hills provide an excellent example; their greatest accumulation of snow is at the end of April or even in early May, whereafter there is a steady melting which is not complete until August. It is in the dry month of June that the snow held on the spacious tops and plateaux of this region can maintain the water even in rapid-running rivers and affect the fish life down in the Spey and Dee Valleys. As we will later describe, snow has a great effect on plant and animal life in areas where it lies long.

The shore line of the Hebrides and West Highlands is particularly free from snow. It has lain no more than a couple of hours in some winters and then, like as not, it has been at the end of April or in the first week of May. That first week or so of May is regularly a wintry period, and is so well known in the North that it is called the Gab o'May. Of course, a great deal of the winter snow comes on easterly winds which traverse the whole of the Highland plateau before reaching the west coast, bereft of a great deal of their snow burden. In the far west the annual number of days on which snow lay at St. Kilda between 1957 and 1960 was about twelve. During the severe winter of 1962-3 while the Eastern Highlands was snowbound from the coast upwards for months, Benbecula and Tiree had twelve and one day respectively of lying snow.

Frost has a distribution in the Highlands somewhat like that of snow, except for the peculiar conditions which will produce spring or autumn frosts on the floor of a glen and not at a few hundred feet up the hillsides. In calm frosty nights freezing fog develops in the bottoms of the glens and coats the vegetation with thick rime, which is left like a sparkling tide mark on the hillsides when the fog rolls away in the morning. The shore line of the islands off the western seaboard may register no more than 2 Fah. deg. of frost all winter, but in occasional bad years may show up to 14 Fah. deg. of frost. At Braemar in the Cairngorms the average daily minimum for January and February is 29°F and the absolute minimum there in December and January is—7°F.

Wind is a factor of great variability and of exceedingly great importance in the natural history of the area. The topography enters greatly into exposure to wind, and in many places is the salient factor in preventing healthy tree-growth. The temperature and rainfall have been mentioned as important factors in determining the character of the vegetation and the communities of animals dependent on it. Wind is particularly important in tree growth though it also affects the field layer. The wind passing over the surface of the leaves increases the transpiration rate. They soon become parched and are sometimes lost in summer; growth is stunted. It would, however, be wrong to consider the Highlands as windswept; it can be more truly said of the Islands. Though some coastal areas such as Mull of Kintyre, Ross of Mull, Ardnamurchan and Cape Wrath are particularly exposed, others like the shores of Lochs Linnhe, Sunart, Duich, Carron, Torridon and Ewe to mention only a few, are comparatively sheltered. In the inner recesses of the sea-lochs the trees around the big houses have a beautiful symmetry in contrast to the wind-tortured scrub in the more sheltered clefts of the exposed headlands.

The summits of the hills are windswept. Gales above 4,000 feet are worse than on the coast; the meteorological data from the Ben Nevis Observatory period gave an average of 261 gales a year of more than 50 m.p.h. Consequently, as you climb higher and higher on the hillsides there unfolds around you the changing character of the vegetation. You pass from the forest to the dwarf shrub heath and the grass heaths on the naked shoulders to the moss heaths on the alpine summits. The heaths of the forest floor, where wind effect is least, may be waist deep ; those of the mountain tops where the wind is strongest hardly rise over the sole of your boot.

On the outer coasts of the Highlands and in the Islands gusts of well over 100 m.p.h. occur each winter, and in certain places, where the configuration of the hills governs the play of wind, there are freak gusts and updraughts of excessive strength. The outer coasts of the North-West Highlands and Islands are the windiest parts of Britain ; much windier, for example, than the Shetland Isles or the outer Norwegian coast, or Valencia Island off south-west Ireland. The soldiers who have garrisoned St. Kilda since 1957 have recorded gusts of 130 m.p.h. at near sea-level ; what the force was on the summit of Conachair almost 1,400 feet above the sea we can but imagine. During one of the winters, the radar saucer aerial on the summit of Mullach Mor, which was secured by a strong angle-iron gantry, was bent with the gantry through an angle of 90° from the vertical to the horizontal. The senior author had an anemometer on Priest Island in 1936, but the cups blew away. In the Outer Hebrides trees are restricted to native scrub and small plantations of which that around Stornoway Castle is by far the largest ; it is surprising what can be achieved, however, when the force of the westerlies is broken. Raasay, lying in the lee of Skye, is a good example.

The effect of the wind is much less in the interior of the Highlands. The tree line on the coasts may be no more than 200 feet—assuming that trees will grow at all— whereas it is 1,800 feet on the western side of the Cairngorms. The prevailing south-west winds come off relatively warm waters of the North Atlantic Drift and are laden with moisture. The weather is rarely cold during the time they blow.

On the western seaboard the view to the west is of a vast

sky and seascape upon which the weather forecast is generally written in bold outline. The development of the frontal systems, with the cirrus-stratus grading westward into the alto-stratus, heralds the storm. The watery sun in the alto-stratus casts a sinister light over the sea and mountain before the rain and wind strikes. Following the warm front comes the cold front, with its squally finale to the storm; the land is refreshed, the sun shines on the white crested sea with renewed brilliance and the rivers and mountain burns gush.

In times of north-westerly and northerly winds the sky may change in a few hours from being cloudless to being filled with great billowing cumulus and cumulo-nimbus clouds—the thunderheads with the spreading anvil-shaped crowns. These bring sudden heavy showers of rain and hail, and may be so severe as to be termed a "cloud-burst". A flash-flood may result with trees and bridges swept away and deep erosion scars carved on the mountain sides. The airstream from the west often brings in little puffy fair-weather cumulus from the ocean which, on reaching the sun-warmed mountain massifs, mushroom into great thunderheads which disgorge their rain in the Central and Eastern Highlands.

So much for the moist oceanic air, some of which comes to us from the Azores and some from Iceland; mention must also be made of the dry continental air which comes on easterlies from central Europe or from Scandinavia and Siberia. This is the weather which brings long periods of heat and sunshine to the Highlands in summer and long periods of frost and sunshine in winter. When the continental anti-cyclones develop troughs, the frosty weather is punctuated by blizzards, which may be followed by a thaw. If the troughs are strong enough they dislodge the continental airstream from Britain and admit a flow of warm rain-bearing air from the Atlantic. North-west Britain is an area of constant advance and retreat of the oceanic and continental air. It is this which makes its climate so changeable, and when modified by the relief of the mountain country, so very diverse.

This chapter is a first brief look around at the form of the Highlands and Islands in the basic elements of geology and climate. We have mentioned briefly how these are related to the cloak of soil, vegetation and animal life with which, in calm and tempest, Nature has clothed the naked rock. Before going on, however, to dissect the habitats of the area into various major and minor components by stages from the floor of the coastal seas to the summits of the mountains, it is important to appreciate the ecological interpretation of the scenery.

Chapter Two

The Highlands

Let us look at a physical map of the Highlands and Islands
(fig. 4) and make a tour as observers of country rather
than as naturalists making detailed studies of habitats. It
will be convenient to divide the areas into five zones (fig. 5);
three of which are essentially Highland and described in
this chapter. The remaining two zones appertain mostly
to the Islands and are discussed in the next chapter. The
zones are as follows:

1 The southern and eastern fringe, which is in effect a
  frontier zone between Highlands and Lowlands.
2 The Central Highlands, which may be likened to a
  continental or alpine zone.
3 The Northern Highlands, a zone with sub-arctic or
  boreal affinities.
4 The Inner Hebrides and West Highlands south of Skye,
  which may be called the Atlantic or Lusitanian zone.
5 The Outer Hebrides and islands of Canna, Coll, Tiree,
  and such small islands as the St. Kilda group, the
  Treshnish group, the Flannans, North Rona and Sula
  Sgeir: an oceanic zone.

THE SOUTHERN AND EASTERN HIGHLAND FRINGE

This zone follows the line of the Highland Boundary Fault
from Helensburgh almost to Stonehaven, and then turns at
right angles north-westwards to include the middle Dee.
The Lochnagar massif, 3,786 feet, may properly belong
to the Central Highland zone, but its long southern slopes
all drain into the eastern plain below the Highland Boundary Fault. From Lochnagar we can cross to Pitlochry and
thence to Loch Tay and south-westwards to the head of

Loch Lomond and to the sea at the head of Loch Fyne.

The land to the south and east of this zone is highly productive agricultural ground which shows some of the best farming in Scotland. The zone itself is largely occupied by sheep farms, but the farther north-eastwards we go from Cowal to the Glens of Angus the better are the sheep, and the hills on which they graze become easier and better grouse moors. That part of the zone east of the Tay Valley from Stormont through the hills of Angus to the Moor of Kerloch has a very high value as grouse moors for they are among the best in the country. There are also deer forests in the following areas: east of Loch Lomond where cattle and sheep are also grazed, the Forest of Glenartney south of Loch Earn and east of Loch Lubnaig, Invermark Forest south of Lochnagar and in the upper reaches of the Glens of Angus.

The changing nature of this zone within historical time may be gathered from such names on the maps as Forest of Alyth and Forest of Clunie. There would be a large number of trees there hundreds of years ago, but the word forest would be given in the particular connotation of a large uncultivated tract, a usage of the word with which we are more familiar in the Highlands where a deer forest may be practically treeless. The Forests of Clunie and Alyth are now rearing farms for cattle and sheep, though, of course, there are still large areas of grouse moor. The golden eagle has gone from here, no longer tolerated by grouse-shooters and the farmers, and the country is not rough enough to give it sanctuary. But in Invermark Forest at the head of the Angus Glens the eagle is still present. One might say that the red deer have gone from the forests of Clunie and Alyth, and so they have as full residents. This, however, is a frontier zone by our definition, and in hard winter weather the stags come down the long glens of Isla, Fernait and those in Atholl. A fair amount of coniferous timber is grown in the frontier zone, particularly in Cowal, the Trossachs, Strathyre, Strathtay and on the Braes of Angus.

Dunkeld is one of the gateways to the Highlands proper,

B

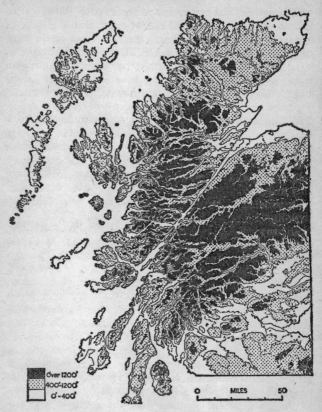

Over 1200'
400'-1200'
0'-400'

0 MILES 50

FIG. 4. Generalized relief features of the Highlands
By courtesy of the Land Utilisation Survey of Great Britain

at the foot of Strathtay. From Dunkeld to Pitlochry we are in a valley made famous by an earlier Duke of Atholl in his zeal for planting. Larch became one of our most important conifers after the Duke had planted it so extensively during the 18th century. It is interesting to note, also, that it was in this afforested country that the hardy, canker-immune hybrid between the European and Japanese larch occurred by a fortunate accident. The Japanese larch, which the hybrid is now superseding, is reddish in colour when the leaves are shed and is a faster grower than the European variety which is straw-coloured when bare of leaves.

West of Dunkeld we are into Strath Bran, still timber country, grouse moors and rearing farms. The fauna of Highland hills are constantly pressing down into this frontier zone and are as surely being scotched before the plain of Strathmore and the Straths of Tay, Earn and Forth are reached. Peregrine falcons, wild cats, eagles, foxes, red deer—all these come through and rarely return. This frontier zone generally carries a big stock of roe deer; despite their unpopularity with the forester, they happily persist, apparently as numerous as ever. There are no high tops in Strath Bran until the head of Glen Almond where the summit of Ben Chonzie, 3,048 feet, dominates everything else in the district; yet there is big country here which the relative smoothness of the hill faces tends to emphasize. The streams have good brown trout and the valleys are always well wooded among the numerous farms. The bird population is rich and varied.

West again, we come into the Forest of Glenartney with its two sharp peaks of Ben Vorlich, 3,224 feet, and Stuc a' Chroin, 3,189 feet, which are visible from Arthur's Seat, Edinburgh. Glenartney is the most southerly of the deer forests proper and though the high country of the two peaks is very suitable for red deer, the winter trek of the animals makes the forest harder and harder to maintain in an age when the voice of agriculture is clamant. Glenartney carries a sheep stock.

The country to the west is now getting much wilder and

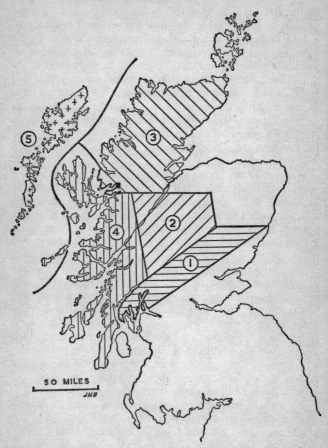

FIG. 5. The zones into which the Highlands and Islands have been
sectioned for the purposes of this book, see p. 22

the easily walked slopes of good heather are giving way to some bare rock faces, to the wetter hills and birch woods. Such is the country on either side of Loch Lubnaig where the Forestry Commission has changed the face of the hillsides. The varied scheme of planting here and farther west in Glen Croe can serve as a model to confound those who hold that forestry spoils scenery. There is the same kind of country in the Trossachs round Loch Katrine. Here in the forests of Rowardennan and Loch Ard the Forestry Commission have created the Queen Elizabeth Forest Park. Plantations of conifers covering an area of 13,000 acres have been established with nearly 20,000 acres of lochs and hill ground available for public recreation.

We now come to Loch Lomond, beginning at the foot of Glen Falloch as a narrow and quite uninteresting loch ; it widens and becomes more impressive farther south. The shores are fringed with birches and oaks, and on the west bank particularly there are some fine groups of trees. In this superb landscape Ben Lomond looks splendid seen across the loch from Tarbet or Luss ; a still finer view can be obtained from the other side where there is no road except that from Loch Katrine to Inversnaid Lodge, which can be reached also from Aberfoyle. The view westwards from above Inversnaid looks to the Loch Sloy power station and includes a group of "Munros" draining to Loch Sloy—Ben Vorlich (another of the name) with its two peaks, Ben Vane and Ben Ime. This is the catchment area of the Loch Sloy reservoir and a road has been made into the area which allows more people to see the fine scenery than have been able heretofore.

A "Munro", by the way, is a hill 3,000 feet or over, separated from another by a dip of 500 feet or more ; the name is from the Scottish mountaineer H. T. Munro, who listed them. There are 276 Munros, and 543 tops over 3,000 feet.

This group of hills in the Arrochar district is fairly and squarely in the west of Scotland and has a high rainfall. The most southerly of the group is Ben Arthur (the Cobbler), 2,891 feet, where there is much bare rock and excellent climbing. At the foot of this hill we are on the west coast of Loch Long. The frontier zone is practically lost here, for there is not the rich agricultural land immediately to the south.

Glasgow is fortunate in its landowners to the north. On both sides of Loch Lomond access is given to all, and a National Nature Reserve has been set up to preserve the fine natural type woodlands on the islands of Clairinsh, Inchcailloch and Torrinch and the wildfowl grounds at the mouth of the River Endrick. The Queen Elizabeth Forest Park links up with the Argyll National Forest Park established by the Forestry Commission south-west of Arrochar in 1936, which includes the privately-given peninsula between Loch Goil and Loch Long. It includes the forests of Ardgartan, Ardgoil, Glenbranter, Loch Eck, Glen Finart and Ben More, with 25,000 acres afforested and 33,000 acres available to grazing and public recreation. This extremely broken stretch of Highland country is within a few miles of the busy industrial Clyde. The establishment of a national park, and the faithful implementation of the Town and Country Planning Act should ensure to Glasgow an area of wild beauty with a rich natural history, much of which yet awaits patient investigation.

THE CENTRAL HIGHLAND ZONE

This area gives the nearest approach to continental and alpine conditions that we have in Scotland. The southern boundary is a line drawn from Lochnagar to the head of Loch Lomond, including the high hills on the north side of Loch Tay. The western boundary would be a line from Loch Lomond through Ben Nevis to Carn Eige and Mam Soul on the borders of Inverness and Ross, thence almost due east across the Great Glen at a point just south of Urquhart Castle. This northern line would continue from that point to Tomintoul, one of the highest inhabited villages in Scotland, at 1,160 feet; and the line from Tomintoul to Lochnagar forms the short eastern boundary. The south-western and north-western corners of this arbitrarily delimited zone are the least typical, in that they lose the plateau-like quality of the Central Highlands proper, but we should not like to include the peaks round the head of Glen Lyon in the West Highland zone, nor do we think the triangle of country north of the Great Glen may rightly be said to have an ecological complex similar to that of the Northern Highlands. Between 80 and 90 per cent of the ground in this central zone is above the 1,000 foot contour. Arable farming is scarcely practised except in Glen Moriston, Cromdale, Boat of Garten and in the narrow straths. The slopes of the hills are mainly of dwarf shrub and grass heaths and after 2,500 feet become alpine in character.

The Central Highlands have their particular interest for naturalists who may be specialists. There is the botanical field of the high tops, among which Ben Lawers, 3,984 feet, has always held a special place. The schist of which the hill is composed breaks down easily, and there are exposures of other rocks as well, providing soil which allows a greater variety of alpine plants to grow than on most other summits.

The Cairngorms' tops with snow bunting, dotterel and ptarmigan are our most considerable arctic relic. The ancient pine forests at the eastern and north-western side of the Cairngorms, also a relic of a past age, contain the Scottish crested tit and the Scottish crossbill. The central Highland area contains some of the biggest deer forests in Scotland, such as Blackmount in Breadalbane (about 108,000 acres) some of which also carries sheep, the Forest of Mar, which is almost as large, and the wonderful deer country between Loch Ericht and Loch Laggan, which includes Ben Alder, 3,757 feet.

Our central zone holds the upper reaches of three large river systems—the Dee which flows eastwards from the Cairngorms and the Grampians; the Spey which rises from tiny Loch Spey in the Corrieyairick Forest north of the high top of Creag Meagaidh above Loch Laggan; and the Rivers Garry, Tummel and Tay flowing southwards, joining and continuing as the Tay outside the central alpine zone. The much shorter River Spean which flows westward from Loch Laggan has now all but disappeared because of the erection of a hydro-electric dam and aqueducts at the west end of Loch Laggan. The Spey gathers its waters from the Monadliadh hills, the Grampians and the Cairngorms, the largest area of long-snow-lying country in the Highlands, and flows north-east 120 miles to the Moray Firth. The River Truim, the Spey's first large tributary, runs through Badenoch, one of the barest parts of the Highlands. Badenoch has the appearance of a devastated countryside, an appearance partly due to nature and partly to the destructive hand of man several hundred years ago. This area was fought over many a time and bands of broken men were burnt out of their retreats just as the last wolves were a century or two later. To our minds, the Forests of Drumochter and Gaick, a little to the east, are the most depressing part of the Highlands. The hills are big humps without the individuality of crag and forest.

West of the road, in the upper Spey Valley region and south of Loch Laggan, the hills become sharper and more shapely and there is a good deal of natural birch, among which are many stands of coniferous timber which in no way spoil the landscape. There is still plenty of natural birch and juniper scrub as far as Aviemore and beyond. We are in a very beautiful area which is one of the most popular holiday resorts in Scotland for those who like a mixture of woodland and high hill and a sharp healthy climate of low summer rainfall and plenty of snow and crisp conditions for winter sports. At Aviemore, the valley of the Spey widens, and if the observer climbs the wooded hillock of Craigellachie south-west of the village, now a National Nature Reserve, he will see the old Scots pine forests of Rothiemurchus and Glenmore as the floor of a great basin formed by the Cairngorms and the little range of hills to the north which culminates in Meall a' Bhuachaille, 2,654 feet. Loch Morlich lies in the middle of the basin and its bright sandy shores at the eastern end are visible. The dark green of the pine forest stretches through the pass or *bealach* at the foot of Meall a' Bhuachaille into the old and much cut forest of Abernethy. Rothiemurchus and Abernethy are still very beautiful, however, with birch, juniper and good natural regeneration of young pine taking away the grim formality of the solid stands of planted timber.

The Cairngorms, which form the heart and have the most extreme alpine conditions of our central zone, are fairly easily reached from Aviemore by means of the track and pass known as the Lairig Ghru. The Lairig splits the granite massif of the Cairngorms into two halves at a height of 2,750 feet, and is the most spectacular part of the Cairngorms seen from Aviemore or farther west of the Spey. Ben Macdhui, 4,300 feet, is on the east side and Braeriach and Cairntoul on the west side of the pass. Just south of the summit of the Lairig are the very small lochans known as the Pools of Dee. The water is extremely clear and probably originates from springs. The source of the Dee is at Wells of Dee, at 4,000 feet on the plateau of Braeriach, and in 12 miles it is a considerable river at the Chest of Dee. Salmon ascend the Dee as far as Glen Geusachan and up Glen Dee to Corrour. By the time the Linn of Dee is reached we are into the forest again, mostly planted Scots pine until we get below Braemar, where Ballochbuie still holds a fine show of old pines. These are part of the Royal property at Balmoral. The plantations are restricted to the main Dee Valley but the tributary glens of Derry, Luibeg, Lui and Quoich hold some of the finest and oldest natural pine forest in Scotland. In the heart of the mountains the Shelter Stone in Glen Avon is ringed by an array of 800 feet precipices and is the haunt of mountaineers and naturalists. One of the best routes into the Cairngorm mountains is up Glen Tilt from Blair Atholl, past the Falls of Tarf. It is a long and arduous U-shaped glacial valley for most of the way until the Bynack shieling is reached at 1,500 feet. After that there is the sense of height and space, and the high hills of the Cairngorms lie ahead in a much more picturesque group than when seen from the west. This time it is the noble Glen Dee which splits the massif rather than the sharp nick of the Lairig Ghru. Trees are few, though the narrow dens which cut down to the Tarf from round about Fealar have plenty of small birches, and curiously enough there are a few well-grown spruces at Bynack.

The Grampian hills south of the Cairngorms give a sense of vastness. Ben Iurtharn, 3,424 feet ; Glas Thulachan 3,445 feet ; and the tops of Beinn a'Ghlo, 3,671 feet ; all these and many another 3,000-footer can be easily climbed on a pony, and once on those clean, smooth summits the pony can be let out to a gallop, so different are they from the sharp peaks of the West. The snow lies long on these high remote hills, but in summer there is a wealth of excellent grazing for deer, sheep and cattle. We have found patches of beautiful brown soil as high as 1,800 feet.

South-west of the Grampians lies the Moor of Rannoch, a broad undulating waste of bog, heath-clad hillock and loch. It stretches eastwards from the Black Mount and narrows like a wedge between the massifs of Ben Alder (3,757 feet) and Beinn Udlamain (3,306 feet) in the north and the Glen Lyon hills, dominated by Carn Mairg (3,419 feet) and the beautiful cone of Schiehallion (3,547 feet) in the south. The Moor of Rannoch was the site of a great reservoir of ice in glacial times, fed by the glaciers streaming from the ice-caps on the Black Mount and the other surrounding massifs. The ice piling up in the spacious saucer overspilled eastwards along the line of Lochs Rannoch, Dunalastair and Tummel, westwards through Glen Etive and Glencoe and southwards through Glen Orchy. Moor of Rannoch ice also streamed northwards into the heart of Badenoch through the long glen now filled by the narrow finger of Loch Ericht. The paths of the ice are marked by belts of moraines and the courses of streams laid down in glacial times.

On the moraines, trees followed the ice and the Moor may have been extensively wooded as recently as Roman times. Surviving fragments of this old forest can still be seen in the Black Wood of Rannoch, Crannach Wood and at Loch Tulla ; in the blanket bogs for which the Moor is famous, there are the roots of the ancient forest embedded in the peat. The moraine systems of the Moor are probably still capable of growing trees and the Forestry Commission have planted a dry area at over 1,000 feet beside the main road to the north of Loch Ba, the islands of which have a natural scrub. The entire Moor of Rannoch is of out-standing scientific interest, and the Nature Conservancy have purchased over 3,000 acres of the country between Loch Laidon and the railway as a reserve containing a fine example of blanket bog with moraines. The Moor is but a stage in the pageant of the ages of ice, forest and mankind.

THE NORTHERN HIGHLANDS, A ZONE WITH SUB-ARCTIC AFFINITIES

The rocks of the Northern Highlands on their western side are mostly very hard, and poor in such minerals as make good soil; they are Lewisian gneiss, Torridonian sandstone and Cambrian quartzite; these three have little either of calcium or of fine particles which will become clay and contribute to the soil picture. Furthermore, where the bedrock itself is not showing through (and often it is over 50 per cent of the landscape) the ground is covered with peat which has no bottom of shell-sand or clay which, on disintegration or removal of the peat, might become productive soil. Sand dunes occur on the coast at only a few places such as Gairloch, Gruinard Bay, Achna-haird on the north coast of the Coigach peninsula, across Rhu Stoer, at Achmelvich, and at Sandwood Bay a few miles south of Cape Wrath. None of these has sands of very high shell content, as found in the Hebrides.

It is a rocky coast to which run a multitude of short, rapid rivers—the Laxford from Loch Stack and Loch Mor into Loch Laxford; the Inver from Loch Assynt into Enard Bay; the Kirkaig out of the lochs below Suilven; the Polly, the Kannaird, the Broom and the Dundonnell Rivers; the superb Gruinard River which is only six miles

long on its run from Loch na Sheallag ; the Little Gruinard,
even shorter, coming from the Fionn Loch which is one
of the most famous trout lochs in the North ; the River
Ewe, only two miles long after it leaves Loch Maree, but
very broad ; the Kerry River running into Gairloch, famed
in the past for its pearls ; and the Applecross River which
drains much of the peninsula of that name. As things stand
at the moment the rivers of this region, so variable in their
flow from day to day, make up in wealth of salmon and
sea-trout for the poverty of the land for afforestation and
agriculture.

The boreal or sub-arctic affinities of the northern zone
are most marked on the gneiss and the sandstone. Each
rock has its very distinctive form and each contributes to
some of the wildest scenery in Scotland. But here in the
interplay of gneiss, sandstone and quartzite the naturalist
may walk for a week or more and see no human habitation
other than an occasional stalker's cottage. So rough and
wild is most of the country that habitations unconnected
with sport are difficult to find away from the sea's edge.

The limestone outcrops running south just to the west
of the line of the Moine Overthrust are green oases of high
ecological interest. At Borralie near Durness there is an
outstanding development of Dryas heaths on limestone
pavement similar to those in the Burren district of Ireland.
At Inchnadamph the limestone pavements and scarps have
a scrub vegetation rare in Scotland, though common in
Scandinavia. At Rassal near the head of Loch Kishorn and
at Tokavaig in Sleat, Skye, the limestone supports a rich
brown soil and natural woodlands of ash and hazel in
otherwise bare country.

The Lewisian gneiss of the mainland rises to greater
heights in the general run of the country than it does in
the Hebrides, except in Harris and at one place in South
Uist. Also, it is not hidden under such a blanket of peat
as in Lewis. The gneiss country of Sutherland and Ross is
one of a myriad little hills of great steepness, with little
glens running hither and thither among them. The lochans
are seemingly countless and most of them have a floor of

peat. The gneiss hills themselves are like rock buns, looking as if they had risen in some giant oven and set into their rough shapes. This ground holds up the water in pockets in the rock and allows the formation of cotton-sedge bogs and such very shallow lochans as grow water lobelia and water lilies. When these lochans are near the sea and grow reeds the bird life is rich. Greenshanks are common ; about twice as many are found on the gneiss as on the adjoining Torridonian sandstone. Good heather is not common on the gneiss ; the complex is one of bog myrtle and deer's hair sedge and poor grasses. Also, this type of vegetation does not appreciably alter in the altitudinal range of the gneiss. For example, we could find no major difference in sample patches in the Gruinard Forest at the foot of Carn nam Buailtean at 600 feet, and at the top of Creag Mheall Mor in the Fisherfield Forest at over 2,000 feet. Loch Laxford, a sea loch, shows some typical low gneiss country with coastal crofts at Foindlemore and Fanagmore.

The gneiss tends to get higher the farther it goes inland. A'Mhaighdean (the maiden) reaches 2,850 feet above the Dubh Loch in Ross, ten miles from the sea as the crow flies. It has a high cliff face of notable grandeur, a rare thing for the formation on the mainland. The sea cliffs of gneiss are nowhere impressive but near Cape Wrath are up to 450 feet high. Even the Torridonian, a formation which one might expect to make magnificent cliffs, does not provide these in any quantity at the sea's edge. The island of Handa, near Scourie and opposite Fanagmore at the mouth of Loch Laxford, is a splendid exception. Handa, now a bird sanctuary, is one of the few places on the Torridonian sandstone which provide true sea-bird cliffs. The rock is stratified horizontally and the galleries make nesting ledges for sea birds such as guillemots, razorbills and kittiwakes. There are sheer cliffs of nearly 400 feet, and in the little screes of earth among these, now covered with fescue and scurvy grass, there are large colonies of puffins and fulmar. The white-tailed sea eagle nested here until the second half of the 19th century. No other place on the Torridonian formation can compare with Handa for numbers of auks, except perhaps Clo Mor, about four miles east of Cape Wrath, where there is a cliff of over 600 feet.

The splendour of the Torridonian is in the peaks it makes inland. Some are fantastic and others superb. There is only one Suilven and it is undoubtedly the most fantastic hill in Scotland. It rises to 2,309 feet out of an undulating platform of low gneiss. Seen from north and south it has a distinctive shape of a very steep frontal cliff and rounded top called Casteal Liath (the grey castle), then a dip and a lesser knob before a more gentle slope down to the east. But when seen from west or east the extreme thinness of the hill is apparent. Suilven means the pillar which is a good name for the hill seen from the west. It is often likened to a sugar loaf, also. There are greyish-white quartzite boulders sprinkled on the top, yet there is a little alp of grass up there and an occasional bed of *Rhacomitrium* moss.

One of the striking things about the Torridonian peaks of the far north-west is their isolation, caused by the vast denudation which has taken place, leaving these few outliers of sedimentary rock protruding above the wilderness of gneiss hillocks and innumerable lochans. North of Suilven and Loch Assynt is the massif of Quinag, five conical peaks capped with quartzite, with a fine rampart of cliff and scree on the west side, which is nearly three miles long. South of Suilven there is Cul Mor, 2,786 feet, surrounded on three sides by great precipices ; and Stac Polly, 2,009 feet, the narrow ridge of which is like one of those fairy castles of childhood tales perched on the top of steep slopes. Ben More Coigach rises to over 2,000 feet in under a mile from the sea as the crow flies. The air of this sub-arctic countryside with its lower rainfall is generally much clearer than farther south in the Highlands.

The group culminating in Foinaven (2,980 feet) is without doubt the barest range in Scotland, and composed of that unyielding white rock, the Cambrian quartzite. The northern part is like a giant E, the crossbars being ridges peppered heavily with boulders which form screes below the shoulders: the hollows of the E are fine corries on the slopes of which the snow bunting has bred. The southern part is a horseshoe-shaped ridge of which Ben Arkle, 2,580 feet, is the western rampart. This hill of Cambrian quartzite with its banding of white scree may be viewed to perfection from the highroad on the shores of Loch Stack; but for the greatest glory of this range a six-mile trek must be made to reach the vast horseshoe corrie and Loch an Easain Uaine, the loch of the green falls.

Ben More Assynt itself is a mass of Lewisian gneiss resting on Cambrian quartzite. There has been a series of geological overthrusts in the region, which have resulted in areas of limestone coming to the surface. This limestone makes its appearance in Assynt, causing a wealth of plant and animal life in the waters affected by the lime content and the temperature of the water emerging suddenly from the rock. Water-worn caves occur in this district in which have gathered soil and bones of animals of earlier times. The Lewisian and Torridonian rocks were laid down before the advent of determinable life in the area and have no fossils; these rocks are of pre-Cambrian age and approximately 1,500 million and 800 million years old respectively. The Lower Cambrian which lies on top of the Torridonian in gentle unconformity consists of quartzites, dolomitic mudstones, serpulite grits and Durness limestone and contains what are believed to be casts of small soft-bodied worms which left no other durable traces along with sparse remains of molluscs and arthropods. In the mudstones soft-bodied animals have been flattened and give the impression of fossil fucoids. In the quartzite the casts are locally well displayed in what is known as "pipe-rock". All other remains of life are recent, probably within the last 10,000 years.

The comparatively low ground of all this northern region of the gneiss, so difficult of access and so plentifully strewn with lochs, is also a place where sub-arctic birch scrub is common and there is a certain amount of hazel. There are large stretches of birch in Inverpolly Forest in the vicinity and on the islands of Loch Sionascaig which is now within the Inverpolly National Nature Reserve (26,800 acres). There is more birch round Lochinver, below the north face of Quinag, in Strath Beg at the head of Loch Eriboll, and many a stretch may be found on the hills far from the roads, in places which are almost unknown to the naturalist. Some day soon we may find the redwing building in these woods, for this bird has been heard singing here from time to time in April, found breeding elsewhere in the north-west, and is essentially a native of the sub-arctic birch wood. In remote Cape Wrath the sub-arctic vegetation occurs at a low level and ptarmigan breed nearer the sea than anywhere else in Scotland.

The eastern side of the extreme Northern Highlands is given up to extensive sheep-farming. The hills are of no great height and are of easy slope. The herbage is sweet and good. The largest sheep farms in Great Britain are here, some having upwards of 10,000 ewes. The breed kept is the Cheviot of the distinctive lustrous-woolled Sutherland type. The lambs are sold annually at the great sales at Lairg. No man was more responsible for the development of Cheviot sheep-farming in the north than Sir John Sinclair of Ulbster, in the 1790's. The influence of sheep-farming on the natural history has been profound.

South of Loch Broom, the Torridonian hills are more thickly grouped and reach their highest peaks. Their spiry form and the high corries facing to the east are distinctive. The quality of herbage is generally poor and the terraces formed in the lower reaches of the Torridonian hold up the heavy rainfall so that it is often quite impossible to get about dryshod. How different is the nature of the ground from those smooth dry slopes of glacial sand and gravel which are such a marked feature of the Central Highlands! The differences brought about in the vegetational complex have not been sufficiently stressed by plant ecologists in the past, but the monograph by McVean and Ratcliffe (1962) has done much to correct this. The ground had not been well walked through and ecologically dissected until the 1950's. The establishment of the Nature Reserves in the Loch Maree basin, at Inverpolly and in Sutherland, and their potential as open-air laboratories should further help to improve our inadequate knowledge.

Two high hills of the Torridonian have north-eastern corries quite the most magnificent of their kind—and few who have seen them both can decide which is the better. We allude to An Teallach of Dundonnell, 3,485 feet, and Beinn Eighe, 3,309 feet, between Kinlochewe and Loch Torridon. Each of these hills has three corries facing to N.N.W. to N.E. Coire Mhic Fearchair is the most westerly of the corries of Beinn Eighe, and the Toll Lochan corrie of An Teallach is the most easterly one of that range. Some of the buttresses in Coire Mhic Fearchair are exceptionally fine and the corrie makes an almost perfect horseshoe. In the Toll Lochan corrie, the cliff at the head of the lochan is nearly 1,800 feet and of superb architecture.

Between An Teallach and another corried Torridonian peak, Beinn Dearg Mor, 2,934 feet, is the broad amphi-theatre known as Strath na Sheallag. This strath is all deer forest, and though so remote it draws cattle, sheep and ponies to it from far away. Liathach, 3,456 feet, Beinn Alligin, 3,232 feet, and Slioch, 3,217 feet, these are just three more of these splendid Torridonian peaks—clear of peat from 1,750 feet upwards—and some of the tops have a white cap of quartzite. The sudden change from wet peat-laden terraces to the upper slopes of bare rock or thin covering of brash and alpine vegetation, results in a sharp snow line in winter which gives these hills a special seasonal beauty. This sudden cessation of the peat immedi-ately allows a different flora, one of plants which can with-stand droughts and sudden changes of humidity, and which prefer sweeter conditions than are possible on peat. Here and there among the alpine grasses and sedges are straggling plants of dwarf juniper, clinging close to the rock. Sea pink and thyme are also to be found on the gravel. Eagle, peregrine falcon, pine marten and wild cat are common in this country.

The glens of the Torridonian area of the north are often

well wooded. They have been owned by people with a fair (or perhaps unfair!) measure of worldly riches, who have been able to spend a good deal of money on planting for amenity. Take Dundonnell for example, at the head of Little Loch Broom: the loch side is bare of trees and is given up to crofting townships, but soon after the head of the loch is reached one is into a fine wooded glen. There are natural Scots pines, birches, alders, oaks, rowans and hazels, but all round the cultivated strath and the house which was built in 1769, there are signs of planting for beauty: limes, many fine beeches, sycamores, ashes, elms, oaks, chestnuts, big old geans, and until a few years ago many acres of fine larches. The wild life of such a glen is obviously profuse and varied. We have these men of a past age to thank for planting that which we now enjoy, just as we may blame those of a century earlier who were denuding the Highlands of timber.

Loch Maree is another place where there are some very fine woods, but here the sub-arctic quality of the northern zone is being lost and replaced by the complex of sub-alpine vegetation. Near where the Ewe River from Loch Maree goes into the sea in Loch Ewe there is the famous garden of Inverewe planted a century ago by Osgood MacKenzie and further cared for and developed by his daughter until the end of the second world war: belonging now to the National Trust for Scotland, it grows a great variety of rhododendrons and azaleas and many sub-tropical plants and plants from Oceania. This is just another facet of the Highland paradox: the garden at Inverewe lying between the stark precipices of Ben Airidh Charr and the bare windswept slabs of Greenstone Point where the sea is never still.

Chapter Three

The Islands

South of Skye the Outer Hebrides do not mask the influence of the Atlantic as they do to the north. Its effect on plant growth in the Inner Hebrides is both direct and inhibitory, and indirect and encouraging. The island of Islay, for example, changes character completely between its western and eastern halves. On the Atlantic side there is the lack of trees and shrubs, and the presence of short sweet herbage salted by the spray from innumerable south-westerly gales, whereas there are beautiful gardens, yuccas and some tall woodlands on the south and east. The Rhinns of Islay on the Atlantic coast are not heavily covered with peat as is a good deal of the eastern half. Islay is an island of many good arable farms, and it has several square miles of limestone country.

The waters of the North Atlantic Drift cast up on these Atlantic shores pieces of wood and beans of West Indian origin. Plants such as the pale butterwort (*Pinguicula lusitanica*) and the moss *Myurium hebridarum,* which occur again on British coasts only in the south-west, are present here in fair numbers. The pale butterwort occurs also in the bogs of Portugal and western Spain, and on the west coast of France ; Myurium moss is found in the Azores, the Canaries and St Helena, as well as in our Outer Isles. Crabbe (*cit.* Campbell, 1945) found another Lusitanian plant in Stornoway Castle park, namely *Sibthorpia europaea*. It has been suggested, and a certain amount of evidence has been brought forward in support of the suggestions, that some of these western cliff edges escaped the last glaciation so that their Pleistocene flora was not exterminated. The opposite view of recent introduction is also held.

Jura is not so well served with the rich quality of vegetation that we find in Islay or even in small Colonsay and in Mull. It is composed of quartzite, which is poor stuff. Jura is also heavily covered with peat and is an island of high hills; the Paps rise to 2,571 feet and are quite rough going. It was on them that Dr. Walker of Edinburgh in 1812 conducted his classic experiment on the boiling-point of water at sea level and high elevations. Jura has a very small population of human beings on over 90,000 acres. The island is so poor that its long history of being a deer forest will probably continue. It was on Jura during the latter part of the 19th century that Henry Evans conducted careful studies on Scottish red deer. His were the first researches of a scientific character on these animals. A book on Jura has recently been published (Budge, 1960).

The high and rocky island of Scarba (3,676 acres) lies north of Jura. The Gulf of Corryvreckan is the narrow sound between the two islands. This celebrated whirlpool with overfalls is caused by the strong tide from the Atlantic being funnelled through a strait, the floor of which is extremely uneven. The Gulf is quiet at the slack of the tide but is hazardous to small craft when the tide is running. The largest whirlpool is on the Scarba side of the sound, but there is a spectacular backwash on to the Jura coast, considered very dangerous in the days of sailing boats. The maximum current is probably about $8\frac{1}{2}$ knots which is fast for a large bulk of water. No ordinary motor fishing-boat or "puffer" (small coaster) could hope to make headway against such a current.

The West Highland zone has what the North Minch lacks, a number of sizeable islands which are not big enough to have lost their oceanic quality, and not so small that they are utterly windswept. Colonsay and Oronsay, west of Jura, are an excellent example of islands which have the best of almost all worlds. Naturalists may be glad that these islands are in the possession of one who recognizes its value and beauty in the natural history of the west.

Most of the islands is of Torridonian sandstone of a different type from that farther north. Here and there on the islands there are good soils, particularly on the 100-foot terraces. There are sand dunes, cliffs and rocky beaches where several rare maritime plants are to be found. There are freshwater lochs with water lilies and royal fern in profusion. Natural woods of birch, oak, aspen, rowan, hazel, willow and holly also occur. Beech, sycamore, alder, poplar and sea-buckthorn have been planted and some are regenerating naturally. The sight of these, so near the Atlantic and its gales, may be imagined from this passage from Loder's exhaustive book (1935): "The woods are being rejuvenated by young plantations of Birch and Aspen, which are springing up naturally and contending for supremacy with an annual luxuriant growth of bracken. The Woodbine twines over the trees, and festoons along the edges of the numerous rocky gullies that cut up these slopes. Ivy has climbed up and formed pretty evergreens of the more stunted of the forest trees. The Prickly-Toothed Buckler Fern grows in profusion, and the little Filmy Fern is also to be seen under mossy banks."

But in gazing on these woods now and noting Colonsay's wealth of small birds, we should remember the effort entailed in establishing these conditions. Loder says: "When planting in the island first began, the trees made so little headway that it was considered amply satisfactory if they formed good cover. For the first ten years or so they made little progress, and many places had to be planted over and over again. Protection from animal and weather was provided in the first instance by dry-stone dykes, 5 feet high."

The trunks of trees in these Atlantic places tend to become covered with lichens such as *Lobaria pulmona* and *Usnea* spp. and mosses such as *Eurhynchium myosuroides* (on birch), *Ulota phyllantha*, *Hypnum cupressiforme* and *Brachythecium rutabulum*. Native trees seem to be much more affected by the humid climate than such exotics as *Escallonia, Ceanothus, Verbena* and wattle

(*Acacia*) which grows luxuriantly. This is one aspect of Colonsay, but there are also its sedgy and heathery moors and at the southern tip the grey seal breeds in fair numbers. Elsewhere, on the cliffs, kittiwakes, razorbills and guillemots breed; and three species of tern, arctic, common and little, nest on the island. In winter, several hundred barnacle geese frequent Oronsay, finding sanctuary on the grassy Eilean nan Ron at the southern tip, and the owner has a fine flock of feral Canadian geese.

Colonsay and Oronsay together might well be looked upon as an epitome of the West Highland world in its full range of Atlantic exposure and sheltered mildness.

Farther to the north-west are Coll and Tiree, two more islands which receive practically the full force of the Atlantic, but which show decided differences in natural history. Tiree is very low indeed. The rocky portion of the island, of Lewisian gneiss, reaches its highest point in Ben Hynish, 460 feet, but by far the greater part of Tiree is about 25 feet above sea level and composed of raised beach and blown sand resting on the gneiss platform. The island is one of good-sized arable crofts and in this respect differs from its neighbour Coll, which is divided into farms. The sandy pastures of Tiree are deficient in cobalt. This is an induced deficiency caused by the plants being unable to assimilate the cobalt, because of the very high alkalinity of the soil. The cobalt deficiency results in a disease in sheep called "pine", and to prevent this the sheep in Tiree are given a cobalt pellet.

The island is of particular interest to the bird-watcher. It gets passages of migrants in spring and autumn, but the pattern of the migration is peculiar because of the geographical position of Tiree on the very edge of the country. It also has fine assemblies of breeding birds in summer. In winter large concourses which have been ousted by a freeze-up on the West Highland mainland enjoy the milder climate of Tiree, where the soils are open and there is food for all. Loch a' Phuill and Loch Bhasapoll are famous for ducks and swans. The vast beaches encourage certain waders, including the bar-tailed godwit, sanderling and greenshank. In the past the snipe-shooting was reckoned the best in Europe. Happily, there is less shooting now. Tiree has an interesting outcrop of marble on the shore at Balephetrish and the Ringing Stone in the same locality is a huge rounded boulder which probably came in the ice from Rhum. The stone is marked by many ringed hollows on its surface.

The interior of Coll rises to 339 feet and reminds one of the low gneiss country of Sutherland. The island presents a uniform rocky appearance when seen from a distance on the east. On the west are miles of shell-sand dunes, a feature which tends to be characteristic of many of the islands which meet the full force of the Atlantic and are low enough to have allowed the sand preliminary lodgment. Coll has always had the reputation of being a good place for cheese and sound dairy cattle. This island is important for the student of plant distribution in relation to the last glaciation and associated changed ocean levels. The *machair* of Coll seethed with rabbits before myxomatosis swept the island in the late 1950's, but in Tiree there are no rabbits and it is unlikely that there ever have been, despite an old record of an introduction that failed.

Coll has breeding arctic skuas, red-throated divers (plate xv) and red grouse which Tiree does not possess. A few pairs of grey lag geese may still breed there. The low sandy islet of Gunna lies between Coll and Tiree and in summer has fine colonies of arctic and little terns and three common species of large gull. Sandwich terns may breed there too. Shelduck are plentiful generally on these islands, especially on the rabbit-pitted sand hills of South Coll. In recent winters about 300 barnacle geese have found sanctuary on Gunna, roosting there at night and ranging by day along the north coasts of Tiree and Coll. Flocks of white-fronted and grey lag geese are also present in winter, usually based on the Reef, Tiree. In autumn, Gunna is a grey seal nursery with over 50 calves born annually and all year round it is an attractive little sanctuary.

The Treshnish Isles lie between Coll and Mull and are composed of Tertiary basalt; they are the eroded outliers of a lava plateau and have the characteristically terraced architecture of basalt flows. The most southerly island has a rounded cone 284 feet high, which is not a volcanic cone but is all that remains of upper flows after ice, sea and weather have had their play. Its shape gives the island its name, Dutchman's Cap. The middle island of Lunga is a pile of eroded flows rising 337 feet, but the other islands are all flat-topped with sheer sides of uniform basalt resting on a lava platform which is washed by the sea. This platform is of great importance to the grey seals which use it as a breeding ground. The Treshnish group, especially the Harp Rock of Lunga, is a nesting place of kittiwakes, auks and fulmars. Storm petrels nest in the Treshnish also, and the Manx shearwater on Lunga at least. Lunga has rabbits and the excellent grass on these islands attracts barnacle geese in winter. The islands are at present used as sheep grazings and are a haunt of summer yachtsmen. The main islands have probably been in continuous use for over a century as seasonal grazing for cattle, sheep and horses. Harvie-Brown (1892) found them heavily grazed by both domestic stock and wild geese. "The whole sweet pasturage," he wrote, "foot by foot, yard by yard, is manured by both quadruped and bird"; the same could probably be said of the Treshnish Isles today.

The Cruachan of Lunga will be a good place to rest for a few moments and look at the topography of Mull. The eye is first struck by the shapely peak of Ben More, 3,169 feet. This is the highest point reached by the Tertiary basalt in Scotland. The cone itself is the result of great weathering, and the various beds of this lava are evident now in the truncated edges of the lower slopes of the hill. For sheer hard going, the descent from the summit to Loch Scridain takes a lot of beating, for the traveller is constantly having to make his way round these faces of rock which are not readily obvious to him as he comes down the hill.

The cliffs of the Gribun are one of the most striking

features in Mull, for both scenery and geology. The 1,000 feet cliffs have Mesozoic sedimentary rocks trapped beneath the Tertiary lavas, and the ascending succession of rocks can be seen along the course of the Allt na Teanngaidh at Balmeanach. On the shore, Moine schist is overlain in unconformity by red Triassic sandstone and conglomerate. The section along the stream shows clean exposures of the ascending succession from Triassic through Rhaetic to Cretaceous rocks above which the stream cuts a gorge in the overlying basalt. The nearby island of Inchkenneth, the burial place of old Scottish kings and chieftains, is part of the Triassic conglomerates and limestones, which has been eroded to give fine physiographical variation and fertile soils.

The whole of the north end of Mull consists of green even terraces with occasional gullies. The islands of Ulva and Gometra are flat terraced cones. The ground is porous and does not form basins for freshwater lochs ; peat is absent and bracken grows rampant. Trees of many kinds grow well in the sheltered parts of Mull on this base-rich soil from the volcanic rock and reach extraordinary luxuriance and beauty at Carsaig Bay on the south coast of the island, where there are also outcrops of calcareous sandstone of the Lias. In this locality a xenolithic and sapphire-bearing sill has been intruded into the sediments at Rudh' a' Chromain.

There are two interesting fossil sites in south-west Mull. At Ardtun near Bunessan, there are leaf beds consisting of mudstone underlying columnar basalt, and on the shore at Ardmeanach there is a fossil tree embedded in the lava. This tree, first brought to notice by MacCulloch (1824), is now represented only by the empty column from which the petrified trunk has been eroded or chipped away by collectors. The south-east end of the island is dominated by a mass of gabbro which culminates in Sgurr Bhuidhe and Creach Bheinn (2,352 and 2,344 feet). From them we may look down on the north side to the long, bare, impressive valley of Glen More and on the south to the tree-lined waters of Loch Uisge and Loch Spelve. The southern peninsula of Laggan is extremely rough and rocky, with plenty of scrub birch.

The islands of Muck and Canna are both almost entirely of basalt with an erosion platform of lava that looks like clinker. Their soil is so good and their position in the Atlantic so favoured that these islands can grow what are probably the earliest potatoes in Scotland, i.e., May 31. The sheep and cattle of these islands do extremely well and come to the mainland in very good order. The wealth of plants and animals is much greater than would be found on the Torridonian or gneiss formations, provided, of course, the situation has not been improved by blown shell-sand. The Glasgow University Expedition to Canna in 1936 published a full report of their extensive finds. Muck and Canna both offer the right kind of cliffs for sea-birds, and Canna is also a breeding station for the Manx shearwater. John Lorne Campbell, the owner of Canna, has made an outstanding contribution to the entomology of the Small Isles. The noctuid moth "the grey", (*Hadena caesia*) was first reported in Scotland from Canna and the convolvulus hawkmoth (*H. convolvuli*), the death's head hawkmoth (*Acherontia atropos*) and the clouded yellow butterfly (*Colias edusa*) have also been found. Most of the trees on Canna have been planted; they include Corsican and Austrian pine, Japanese larch, Sitka spruce, ash, elm, sycamore, oak, birch, elder and willow.

Eigg is not now a big shearwater station; in 1964 there were probably less than 50 pairs. The island is rich in mosses and possesses remnant hazel shrub. The Sgurr (1,280 feet) is the obvious physical feature of Eigg and by far the island's most interesting natural phenomenon. The Sgurr itself is of pitchstone lava, resting on a thin river bed of conglomerate which contains fossil pieces of driftwood from some far distant time. Beneath this is the basalt plateau in which the river valley was incised. The pitchstone shows columnar jointing in places, a character which is still more strongly marked on Oidhsgeir, 18 miles away to the W.N.W. This low islet of pitchstone is considered to be part of the same sheet as the Sgurr of Eigg. The musical sands of Camus Sgiotag, a small bay on the north side of the island, are of partially rounded quartz grains of similar size, which make a shrill sound as one treads upon them.

To return for a moment to the few acres of Oidhsgeir, an islet which does not reach higher than 38 feet above sea level. Here on the top of the pitchstone columns which are 8 inches or so across the top are found the nests of kittiwakes in the season. There are also great numbers of common and arctic terns and eider ducks. The deep-cut channels among the pitchstone columns are also a playground for the grey seal. One channel on the south side runs up into a pool where a boat may lie in perfect safety.

The island of Rhum is now owned by the Nature Conservancy and was declared a National Nature Reserve in April, 1957. More than any other Nature Reserve in Scotland, this 26,400 acre island is a research reserve; it is an open-air laboratory offering great opportunity for research directed at improving the degraded land characteristic of so much of the Highlands and Islands, by restoring woody cover and increasing biological turnover and for elucidating the effects of a varied range of climate and soil on plants and animals. The diverse geological character of Rhum has already been mentioned, and this is mirrored in a wide variety of soils, vegetation and ecological conditions for herbivorous animals such as red deer.

The work on the red deer now in progress on Rhum is referred to later; suffice it is to say here that conditions on the island are good for a classical study of population dynamics, movements, behaviour, food, mineral requirements, parasites and diseases, and of statistical and field techniques directed at the scientific and economic management of deer. There is opportunity also of studying the ecology of the golden eagle, large mountain-top colonies of Manx shearwaters and the establishment and effect of shelterbelts of trees. This work can only proceed, however, if there is a viable community on the island working as a team, and the Conservancy realise how vitally important is the welfare of the estate staff and research workers.

Rhum was for long known as the "forbidden island", but this no longer applies. Day visitors are free to land at any time at the harbour in Loch Scresort where there are facilities for picnicking and enjoying the scenery. Anyone, including mountaineers and naturalists, wishing to stay overnight must, however, obtain permission from the Nature Conservancy in advance, before going to the island. The restrictions are the minimum required to achieve the objects for which the Reserve was established.

A list of the breeding birds of Rhum is given in Chapter 10. The mammals include the Rhum mouse (*Apodemus hebridensis hamiltoni*), brown rat, pygmy shrew, pipistrelle bat, grey seal, common seal, otter. Besides the red deer, there are feral goats and ponies. The common lizard and palmate newt are present ; salmon are rare, brown and sea-trout abundant, and eels common. Two of the island's moths deserve special mention: the belted beauty (*Nyssia zonaria*) and the transparent burnet (*Zygaena purpuralis*).

Skye has suffered human depopulation like many another Highland area, but it is still one of the most heavily crofted areas of the West. Preservation of game has practically ceased and almost all the hill ground is now crofters' grazing. Topographically, Skye is magnificent, with its Cuillins and its Quirang, but from the point of view of wild life it is somewhat disappointing. Harvie-Brown pointed out seventy years ago that the whole area facing the Minch is faunistically poor, and it is so today.

What Skye lacks in attraction for the biologist it makes up for in geological interest. The Cuillin Hills have already been mentioned, but there are other outstanding sites. The Quirang in Trotternish is a most spectacular example of land-slipping along the massive escarpment of Tertiary lava flows. The erosion of this complex has resulted in an amazing terrain with pinnacles, hills and valleys.

The Table, a wide recessed platform high on the cliffs, has green sward and has been used as a shinty pitch. The Storr near Portree has fine basalt pinnacles and cliffs with rare mountain plants. The country between Broadford and Loch Slapin has diverse geological interest with a wide variety of Torridonian, Cambrian, Ordovician, Triassic and Jurassic sediments and Tertiary igneous rock. The Elgol coast in Strathaird has impressive outcrops of Jurassic limestone exposed immediately below the crofts and showing an exquisite form of honeycomb weathering. In Sleat, there is the ashwood at Tokavaig growing from the most southerly large outcrop of Durness limestone, and the classical sections of Tarskavaig Moine schists and the Tarskavaig Thrust.

The island of Raasay between Skye and the mainland is ecologically diverse. The diversity arises from a complex and varied geology with Torridonian sandstones, Lewisian gneiss, Tertiary granite, Jurassic limestones and other sediments. This has a profound effect on the distribution of plants and animals in all departments of the habitat. The island has been further diversified by the coniferous plantations of the Raasay Forest. There are patches of natural scrub woodland and numerous lochans and streams on bedrocks of contrasting character. The community of small birds is rich and the Raasay vole is distinct from that on the mainland. Although the island has attracted many naturalists, its outstanding potential for comprehensive field studies has not yet been exploited.

The mainland coast of the Atlantic zone running north from the Mull of Kintyre to the Kyle of Lochalsh is the most indented part of the British coast. The influence of the sea is felt deep among the mountains. Take the tidal backwaters of Loch Fyne, Loch Etive, Loch Eil, Loch Sunart, Loch Hourn and Loch Duich; little or nothing of the ocean surge is felt in these fjord-like recesses and the seaboard, fingering its way into the massif, pervades the area with a maritime mildness. It is a coastline of sharp contrasts and great variety of marine and terrestrial habitats.

The districts of Kintyre, Knapdale, Cowal and mid-Argyll are composed of a complex of Dalradian schists and quartzites. This is country with multiple land-use in sheep, timber and cattle. Red deer are present but little of the ground is managed as deer forest, as happens in Jura. The area is transgressed by a swarm of dykes from the Mull volcano and the scenery is that of low broken country with a mosaic of coniferous forest, green maritime grasslands and rolling moorlands of heather and mixed grass heaths.

In the present climatic period before the advent of man these districts were probably clothed with a mixed forest of ash, oak, birch, hazel and rowan. On the northern slopes of the higher hills there was probably some pine, but this was essentially an oak forest which stretched northwards across the ancient lavas of the Lorne plateau and into the coastal strips of the West Highlands. These were the woods which over the space of several centuries, were fed to the Lorne furnace at Bonawe and the Argyll furnace on Loch Fyne. The entire forest appears to have been consumed or cleared for agriculture, but in some areas small oak-woods have regenerated from coppice, natural seeding and planting.

The shores of Loch Etive are cattle country with fine herds of the Highland and Galloway breeds. The fold of Highlanders at Achnacloich is spectacular on the lochside pastures with the fine woods and gardens of the house, and the distant hills of Benderloch in the background. Across the loch the Bonawe Quarry from which the setts for the streets of Glasgow were taken, still thrives. The granite, which is available in various shades of pink or grey, is now used mostly as road metal and pre-cast concrete blocks which are shipped by small coasters through the narrow navigation channel in the Falls of Lora. Farther north, behind the decaying cottages of the old village and the council houses of the new, the slate quarries at Ballachulish survive only as a monument to bygone prosperity. The galleried faces of the quarries plummet into deep dead ultramarine waters, and cascades and dripping seepage echo and re-echo in the deserted amphitheatre.

This hard macabre savagery of bare rock, twisted metal and brilliant yet evil-looking tarns is in sharp contrast to the soft gentleness of the woods of Appin and the green pastures of Lismore. To walk the length of Loch Sunart, ten miles out of the twenty through oakwoods and young conifer plantation, is an aesthetic experience which is the antithesis of the Ballachulish Quarry. The scenery of the distance is as beautiful as the redstarts among the oaks and hazels near at hand. Sanna Bay on the northward tip of Ardnamurchan is one of the most beautiful shell-sand bays of the West. The peak of Ben Resipol, 2,777 feet, dominates the landscape and the traveller can hardly miss seeing Ben Iadain, 1,873 feet, on the other side of the loch in Morven. It is a little cap of Tertiary basalt perched on the Moine schist. Morven, more so than Kingairloch and Ardgour, is fine sheep and deer country from which the oak was stripped for the Lorne furnace and is now replaced by the fine coniferous stands of the Fiunary Forest. At Lochaline the seam of white Cretaceous sandstone is quarried and milled for the glass industry.

Moidart, Morar and Knoydart are Moine schist, but this time more mountainous than the Dalradian schists of south Argyll. This is mixed deer and sheep country with little forestry and it is the wettest part of the Highlands. Claish Moss at the west end of Loch Shiel is one of the finest examples of West Highland raised bog. When viewed from the summit of Ben Resipol, the curved alignment of the pool and hummock system of the bog gives the impression that the bog is "flowing" like a glacier. The backbone of the West Highlands runs through the middle of Ardgour, Morar and Knoydart; to the west spreads the old mixed deciduous forest and to the east the old pine forest which continued in the Great Glen. Remnants of old oak still survive in Glen Beasdale in South Morar and Coille Mhialairidh on Loch Hourn, and of the old pine forest above the shores of Loch Arkaig and in Glen Loy. The islands of Loch Morar are heavily wooded and very beautiful with oak, ash, bird cherry, yew and juniper.

Kinlochhourn is one of the most out-of-the-way places in Scotland. To get there you must either drive along Glen Garry past Tomdoun and the Loch Quoich reservoir or go by boat from Mallaig to Inverie. The car run takes you from the Great Glen, a country of forests, sheep, cattle and deer, progressively westwards. First the trees dwindle, then the cattle, then the sheep until when you reach the great fastness of Knoydart, Ladhar Bheinn (3,343 feet), only deer remain. When the road ends at Kinlochhourn there is yet one step further which you can take into this wild country: along the fjord to the remote community at Glen Barrisdale overshadowed by Ladhar Bheinn with its magnificent eastern corrie. The plantations around Kinlochhourn Lodge contain eucalyptus and the glens of Knoydart have birch woodlands with pines and oaks in places.

The Rattagan and Eilanreach Forests occupy great tracts of land to the east and south of Glenelg respectively. Glen Beag with its fine brochs is part of Eilanreach, and Glen More is part of Rattagan. The only natural woodland worthy of note in this district is a thick scrub of birch, ash, alder, hazel and rowan on the south side of Glen More. We are now into the gneiss country, deeply gouged by glaciers of bygone times, descending from the elegant peaks of Kintail and the high plateau around Glomach into the basins of Loch Duich and Loch Long. The lower slopes now carry the coniferous stands of the Inverinate Forest and the remainder of the country is managed in sheep and deer with cattle and arable land in the bottoms of the glens.

The key-note of the Atlantic zone which we have just described is the diversity of its ecology which possesses the full stratification from the floor and depths of the Sea of the Hebrides to the summits of Ben Cruachan (3,689 feet) and Carn Eige (3,877 feet). No other zone has a wider interest for the naturalist. The treatment we have given here can be no more than a sketch, but some of the detail will be added in subsequent chapters. Yet this is not an encyclopaedia of natural history loaded with minutiae; rather is it the putting together of a connected story in which detail and bold outline are used at will to achieve the wholeness of the Highlands and Islands. It is in the

Atlantic zone that this wholeness of sea and land is most readily apparent.

THE OUTER HEBRIDES OR OCEANIC ZONE

The Outer Hebrides are the most westerly portion of Scotland on the seventh degree of west longitude. The great majority of people on the Long Island, as the whole group is called, is fairly densely packed on to the western fringe, with concentrations at Stornoway and on the Eye Peninsula in east Lewis. One might ask why the people are so densely grouped on the west side where harbours are fewer and where the force of the Atlantic Ocean is unbroken. To obtain the full answer one must see the Outer Hebrides for oneself and preferably from the air. It is on the west that the grim blanket of peat ends and the ocean has thrown up an immense weight of shell-sand.

A flight on the service 'plane between Benbecula and Stornoway shows the contrast between the green western fringe and the browns and greys of the peat-covered interior. The white surf-girt shores of the west coast intensify the contrast, for the east coast is dark, rocky and usually tranquil. Barra, Benbecula and the Uists comprise the southern half of the Outer Hebridean gneiss platform. The Barra Isles and South Uist are hilly and of elegant shape. The main ridge stretches for about ten miles north of Lochboisdale and is dominated by Beinn Mhor (1,994 feet) and Hecla (1,988 feet). In Benbecula and North Uist the platform flattens into a low hillocky land with a maze of lochs and ramifying inlets of the sea. From the low terrain the peak of Eaval (1,138 feet) in North Uist rises like an island in a rough sea of rock, bog and water.

This can all be seen in the brief moment which it takes the aircraft to climb from the runway and turn on to its course north. The panorama on the starboard side is that of the barren land with roads winding tortuously over hillocks and around lochs, connecting solitary crofts. On the port side there is a great sweep of sand, the sand-

smothered islands of Monach, and the croftlands of
Balranald heavily dotted with houses. In minutes the aircraft is flying over the east end of the Sound of Harris
with all its little islands. At a glance, there are the seal
islands of Haskeir, Shillay and Coppay; the low cone of
Pabbay and the green platforms of Bernera, Ensay and
Killegray; beyond, on the horizon, is the St. Kilda group
with Boreray standing out well to the right of Hirta. Below
can be seen the shoals of sand and reefs which make the
Sound of Harris so treacherous to the mariner.

The aircraft is flying along the east coast of Harris and
out of the port windows can be seen a cross-section view
of Harris; the scenery has now suddenly become hilly. The
Outer Hebridean platform of gneiss contains an emplace-
ment of foliated granite which occupies most of south and
west Harris. Bordering on the granite the gneiss is seen in
a mass of grey smooth hills of which Clisham (2,622 feet)
is the highest. The red deer which live in these fastness are
small, but have well-shaped heads.

Directly below the 'plane can be seen Scalpay in East
Loch Tarbert. The bare inhospitable ground is thick with
houses, and the sheltered bays are busy with boats. Here
is seen most vividly expressed the Hebridean's love for his
remote and difficult land. Scalpay itself is unyielding, but
there are fish in the sea and the Scalpachs working abroad
return home for part of the year to plough their earnings
into their homes and their boats. Scalpay possesses several
large fishing boats working to Stornoway and mainland
ports, and it is the only community in the west still with
coasters which trade all round Britain.

Loch Seaforth is a long bent finger of the sea pointing
at the heart of Lewis: a patchwork of heather, bog and
loch not unlike Benbecula. But the twenty minute flight is
almost over and as the aircraft turns to land at Stornoway,
the rotating scene shows the rolling heather moorlands of
north Lewis with its highest point in Muirnag (808 feet),
the great sandy strand round Broad Bay, the skinned lands
and crofts of Point (Eye Peninsula) and the prosperous-

looking little town with its superb harbour and interesting woodlands.

As the dunes have stabilised through the millennia and the stiff marram grass has given way to kinder herbage, a light lime-rich soil has formed. There are miles and miles of the white sand on the Atlantic shore, and above it the undulating *machair* of sweet grass on which are reared great numbers of cross-bred cattle. The prevailing south-westerlies continue to blow winter and summer, year after year, century after century. The tangle from the shallows of the ocean, the various Laminarias of the marine botanist, is torn from its bed and washed up on the beaches. Ponies and carts have gone and man now comes down with his tractors and trailers and takes up some of it to spread on ploughed portions of the *machair*. In many areas this pratice is dying out in favour of farmyard manure and chemical fertilisers. All these things are helping to make soil, and the sand itself in these gales, especially if the winds are dry, is being blown up towards the blanket of peat. The sand sweetens the peat and causes its barren organic matter to be unlocked and become fruitful of herbage for man's beasts, dung from which still further ameliorates the peat. Such is the constant process, in which the storm is a necessary and beneficent factor in allowing and maintaining fertility. Throughout the Outer Hebrides, the crofters are transporting thousands of tons of shell-sand to the moorlands and bringing fertility to localities which it was thought had been lost for ever (plate II).

But once the coastal strip is crossed the peat reigns supreme. Its blanket must have increased about ten feet in thickness since early man came to the Outer Isles, for only the tops of the fine Megalithic stones at Callernish, Lewis, were showing when Sir James Matheson of the Lews undertook their excavation. The landscape in the bog is shortly described—a low undulating plateau of peat, bare grey rock of gnarled shape, and thousands of small and large lochans of brown acid water. If we wander through these areas of peat we shall come upon drier knolls, and here we shall find green turf and the ruined shielings. They are the summer dwellings of a pastoral people taking advantage, for their cattle and sheep, of the short spell when the peat grows its thin crop of sedge and drawmoss. The people lived on the little knolls as on islands, bringing their cattle up to them twice a day for the milking, throwing out their household waste. Lewis still has its shielings in the modern idiom: huts of wood, felt and corrugated iron used by peat-cutters. A study of the peat, which is still widely used as a domestic fuel in the Outer Hebrides, is interesting not only for what it grows and harbours now, but for the history to be deduced from a deep profile. The requirements for peat formation are a high ratio of precipitation to evaporation and a general coldness of atmosphere in the growing season sufficient to inhibit bacterial activity in the waterlogged soil, but not cold enough to prevent growth of certain plants. Acidic rocks also help. A vegetation complex of sour bog plants, such as sphagnum moss, sedges and grasses of various kinds and cross-leaved heather, soon occupies the ground to the exclusion of all those plants which need a well aerated soil and a supply of basic compounds. The rain impoverishes the original soil by washing out plant foods and then, by creating waterlogged and therefore anaerobic conditions, prevents the action of normal soil bacteria in breaking down the dead vegetation into humus. Such necessary decomposition does not keep pace with vegetational production by the plants, so that a gradually thickening layer of peat forms. The peat, thus composed of organic matter without lime, is highly acid in character, which is a still further check to bacterial action. The normal water content of peat as it lies in the bog is as high as 93.5 per cent.

Peat varies in consistency from being highly fibrous to the state of a black amorphous substance, depending on age and the type of vegetation. Cotton-sedge peat is tough and fibrous and can be set up on end to dry in pyramids of four bricks and handled later with very little loss. Lower, older, amorphous peat is very brittle and cannot be set up.

The ages of the peat deposits have been tentatively fixed

as beginning about 7000 BC at the close of the Boreal period. The warmish dry climate which grew forests of pine, birch and hazel became warmish and wet, bringing about destruction of the scrub hazel vegetation by moss. The Atlantic period closed between 5000 and 4000 BC and a cooler and somewhat drier sub-Boreal period set in with a rapid development of peat. This continued until near our era which is cold and wet and called sub-Atlantic. The peat today is still making in some places as on the main bog of Lewis, and receding in others, as on many burned and overgrazed watersheds in the Central and Eastern Highlands. In many localities from the Moor of Rannoch to the high moors to the east of the Cairngorms, the stumps of ancient forest trees are coming forth as the peat crumbles away.

In 1949 the Scottish Peat Committee was set up following the pioneer research of the late G. K. Fraser in Scottish peat deposits, to investigate the problems of exploiting peat in Scotland. This led to the survey of peat bogs, many of them in the Highlands and Islands, research into the use of peat in the generation of electricity, and the general exploitation of peat. In 1962 in its second report the Committee state: "the problems of exploitation arising not only from inherent characteristics of the peat itself, but no less from the changing economic environment, difficult as they are, present a challenge to our scientific and technical skills in Scotland." One of the encouraging advances has been the development of peat lands for agriculture and forestry.

By cutting peats the islander is doing two jobs—providing the wherewithal for comfort at the fire, and removing some of the great pervading blanket. He does not come upon bedrock at the foot of the peat banks, but on a layer of boulder clay which, when mixed with the top thin layer of sedge and peat, will shortly turn into soil providing much better grazing than anything from the top of the peat. If modern mechanical tools such as the scraper and bulldozer were brought into operation on what is commonly called the skinned land, the agricultural scientist would make much good land throughout the Long Island, particularly in Lewis. The management of the peat land is now being taken a stage further in the reclamation schemes already mentioned, with the heavy dressing of the peat-cuttings and bog surfaces with lime, compound fertiliser and seeds of grasses and clover.

The Outer Hebrides are often described as being treeless, but the term is relative. The people who write about them are usually those who have a considerable experience of trees and tend to take them for granted. The Outer Hebrides are not treeless, nor need they continue to be so desperately short of trees as they are. The grounds of Stornoway Castle on the east side of Lewis are famous These are Lady Matheson's legacy to the Hebrides. She planted another piece with larch and other conifers half-way across Lewis, near Achmore, and these made good trees, but were blown down by a terrific gale on March 16, 1921. There are Corsican pines of hers at the head of Little Loch Roag, growing quite straight to 35 feet high. There is another plantation of deciduous and coniferous trees at Grimersta on the Atlantic coast of Lewis. A plantation of conifers sheltering a house, Scalisgro, on the east side of Little Loch Roag, is about fifty years old and in the 1930's several acres of conifers were planted in Glen Valtos in the Uig district of Lewis. Another plantation has been established near Balallan, and much more experimental planting of this sort is required with a view to the establishment of shelterbelts in the Outer Hebrides. Several good sycamores are to be seen at Tarbert, Harris, and at Borve on the west side of Harris there are several acres of stunted mountain pines. More trees are to be found about Newton Hotel, North Uist, and at Grogarry Lodge in South Uist, and there are plantations in defiles at Northbay, Ard Mhor and Breivig in Barra. There is a birch wood complete with bluebells and wood sorrel on the slopes of the Allt Volagir, South Uist. Many of the islands in the lochs are covered with dwarf scrub and those in Loch Druidibeg, South Uist, are outstanding (see also Spence, 1961). That the Outer Hebrides were once wooded may be deduced from pollen analysis and the remains of wild cat and blackbird in an Iron-age midden (Baden-Powell and Elton, 1936-37). Even St. Kilda is likely to have had birch-hazel scrub at one time (McVean, 1961) and pollen from its peat indicates that pine, alder, elm and oak were probably widespread in the Outer Hebrides, their pollen having reached St. Kilda in the aerial plankton.

The sea-cliffs are important in the natural history of the isles. The great ocean pounds against them and is gradually wearing them away, but the rock is the old gneiss and holds remarkably well. Sir Archibald Geikie in his *Scenery of Scotland* calls to mind the measurement of the effect of waves made at the Atlantic rock of Skerryvore before the lighthouse was begun in 1845. The summer average weight of pounding was 611 lb. per square foot; in winter, 2,086 lb. with 6,083 lb. in a very heavy south-westerly gale. Even when it is water alone that strikes the rock, the wearing effect is far from negligible, but when loose rock is moved by the water and pounded against the cliff, even in our short lifetimes we may be able to notice the denuding effect of wave action.

The force of the sea on the north point of North Rona has been an eye-opener to us on our respective visits to the island. These cliffs are perhaps forty feet high but sheer, and going into deep water. The top is irregular with occasional ten-foot gullies a few yards wide in which are some very big boulders eight to ten feet thick, and a lot more smaller ones too heavy for a man to lift. These boulders and the walls of the gullies which contain them are peppered with scars and chips caused by the boulders rolling to and fro under the impulse of a sea which fills the gully about thirty-five feet above its normal level.

Some of the cliffs of the Outer Hebridean coasts become crowded haunts of ledge-breeding sea-birds. None are more spectacular than those at St. Kilda. In the south Barra Isles, however, the precipices of Berneray and Mingulay provide a fine spectacle . The precipice of Aonaig in Mingulay is 793 feet high and the stacks of

Arnamull and Lianamull are very fine. These islands form one of the main breeding stations for guillemots and kitti-wakes in the Long Island and its outliers ; remote, they pro-vide a first-class goal for expeditions of young naturalists.

The land-making influence of the sea in times of storm on the sandy western coast of the Hebrides has already been mentioned. The islands in the Sound of Harris prob-ably change shape through the years, sand being laid down in one place and taken away in another. Pabbay, for example, was once the granary of Harris but the sand has encroached over the south-east end and gone from the west end, showing the remains of trees at low spring tides. This submerged forest is probably the result of Holocene sinkings, but nevertheless the shell-sand beaches have cer-tainly advanced within historical times. Peat is also exposed at low tide at Borve in Benbecula. The minister in Harris who was responsible for the account of that parish in the Old Statistical Account of 1794 remarks that certain lands had been lost to the plough within living memory, and that when a sand hill became breached by some agency, and was eventually worn away, good loam was sometimes found beneath—and even the ruins of houses and churches. Whatever we may have lost in the Holocene sinkings, it may be remarked that the last three thousand years have seen more rising than sinking along Highland coasts. The covering of the ruins of wheel-house dwellings in the sand hills of South Uist, and the more recent transgression of sand over the ruins of Mingulay bear testimony to this.

None of the sounds through the Long Island is navig-able by large ships. The major through routes between the ocean and the Hebridean seas are by the Sounds of Harris and Barra, and both are tricky. The north deep-water channel in the Sound of Harris takes one on numerous headings to and fro, at one place coming close to the shore at Leverburgh and at another, breathtakingly near a reef. In the narrow straits the tide runs at about 5 knots. These sounds have been the scene of many wrecks notably that of the *Politician* in the Barra Sound, which gave the background to Sir Compton Mackenzie's lampoon *Whisky Galore*. Both these sounds are a haunt of common seals. In the Harris Sound they frequent the calmer reaches in the centre and at the eastern end, while the grey seal is found in strength in the surf and swell of the ocean at the western end. The minor sounds separating Benbecula from North and South Uist are shallow tidal sand-silted passages across which bridges and causeways have been built to carry the main road through the islands.

The east side of the Outer Isles is entirely different from the populous and spacious west. Admittedly, north of Stornoway there are the sandy lands of Gress, Coll, Back and Tolsta, and the Eye Peninsula, supporting many crofts, but south of these the land is peat-laden and comes to abrupt cliffs at the sea's edge. Long arms of the sea, like Loch Seaforth and Loch Erisort, Loch Maddy and Loch Eport, run far into the interior ; indeed, the last two named almost reach the west coast. This east coast of the Hebrides is uninviting and curiously dead, but the coastal sea is far from dead. There is a winter fishery for herring in the east coast lochs and large numbers of basking sharks appear off-shore in summer. There are plenty of seals and otters and diving gannets, and breeding divers flight to and from the inlets of the east coast. The interior and east coast districts hold golden eagle, buzzard, hen harrier, short-eared owl, raven, grey lag, heron and skua, but pairs are well scattered and there is nothing to compare in spectacle to the breeding concourses of auks, gulls and fulmars on the western sea-cliffs, supported by the terns, plovers and oystercatchers on the beaches. Dunlin, ducks and a few pairs of red-necked phalaropes are on the *machairs* and their lochs.

The east side of Harris from East Loch Tarbert to Rodel is well worth a visit to see what man can do in the shape of difficult cultivation. Take for example the township of Manish where the ground rises at a steep slope from the sea. It is in reality a rough face of rock devoid of soil but holding the peat here and there. The lobster fishers of

Manish have actually built the soil of their crofts by creating lazy-beds or *feannagan* with seaweed and peat. Tracts of the moorland interior of the Long Island have been cultivated in this way at some time in the past. By building up these little patches varying from the size of a small dining-table to an irregular strip several yards long, the inhabitants have overcome the difficulty of drainage. The seaweed is carried up to the lazy-beds in creels, and cultivation is with the spade. Two crops only are grown, potatoes and oats, and the oats are *Avena strigosa*. The industry of the people of east Harris and their steadfast persistence with a thousand-year-old style of husbandry are remarkable. The potato is the only new thing, brought to the Outer Isles in 1752.

The Outer Isles also have their outliers which are described in detail in Chapter 10. St. Kilda, Monach Isles, Haskeir, Gasker, Flannan Isles, Sula Sgeir and North Rona are but small spots on the map, but their size is out of all proportion to their importance in the natural history of the British Isles. Within their small compass there is a welter of life and unsurpassed interest to the naturalist.

We have come to the end of our arm-chair tour of Highland country, from the frontier zone of Perthshire and Angus to the zone of the Atlantic outliers. We have given but a glimpse of what is without doubt one of the finest scenic and faunistic areas in Europe. Whether it survives as such depends very much on the good will and care of British people. Any area of natural beauty and interest adjacent to a highly populous industrial region is in peril from that very proximity, but there is always the hope that men's minds will become awakened to natural beauty and the right of wild life to existence for its own sake. Then, human proximity may be to the advantage of wild life and the wild places, for no country sparrows or moorhens are as tame and safe as those in St. James's Park.

Chapter Four

The Human Factor, and Populations of Animals

It is of the very nature of humanity to alter the complex of living things wherever man is found. Man must be considered as part of the natural history of the earth's surface, however unnatural he may be. Of course, all animals influence the rest of the complex of living things in some way or other, but none does it with reflective intention as man does, and we might add, none does it with much less reflective intention or regard for consequences. He can make the desert bloom, or ultimately fill an off-shore island with the beauty of bird song, and equally he makes deserts as spectacularly as any horde of locusts.

What has man done to the Highlands and Islands and what is he doing? The late Professor James Ritchie wrote a large volume entitled *The Influence of Man on Animal Life in Scotland*. It is an interesting and often depressing story, but Professor Ritchie would have been the last to suggest that he had told the complete tale. The senior author has carried the story rather farther in *West Highland Survey* (1955). There is much of it we do not yet know or are only just learning how to infer and deduce. And our methods of recording are never complete enough to mark down, for future minds to work upon, the doings of the present generation of men.

Scotland, and the Highlands particularly, have nothing like such a long human history as England. The last glacial epoch prevented that, for Scotland was under the ice thousands of years after man had inhabited the south of England, and it is possible that such areas as west Sutherland and the north-west corner of Ross-shire did not know man until two or three thousand years ago. When man first came to the Highlands he was a hunter and fisher and knew

65

no arts of husbandry. One wonders what large effects early man can have had, because a small population of hunters taking life only for its own subsistence, and not for any export, would hardly bring to extinction many of the animals we know were present at that time. It is probable that natural causes were much more important in changing the natural history of the Highlands in those days. A few degrees' change of temperature for a period of years, for example, whether up or down, would work very great changes in the tree line and the specific constitution of the vegetation. The growth-rate of sphagnum moss under optimum condition has, in the deduced history of the Highlands, felled forests as surely as the fires and the axes of mankind.

The biggest effect man has exerted on the history of the Highlands has been in the destruction of the ancient forest —the great Wood of Caledon. This has happened within historic time, partly between AD 800 and 1100 and then from the 15th and 16th centuries till the end of the 18th. Even our own day cannot be exempt from this vast tale of almost wanton destruction, for the calls of the two German wars have been ruthless. Even after the war, in the name of scrub-clearance, native oakwood has been cleared.

Many of the priceless remnants in Strathspey, and at Rothiemurchus and Kinlochewe were felled, mostly for ammunition boxes, and the old pines of Locheil Old Forest went up in smoke during Commando training. These facts should never be forgotten as one of the consequences of war, and now that National and Forest Nature Reserves (plates VI, IX *and* X) are a reality in the Scottish Highlands, it is good to know that the Forestry Commission and the Nature Conservancy have instituted long-term programmes of research into the natural regeneration of the true Scots pine and its forest associates. The bottle-green is distinctive, and so is the redness of its boughs; the needles are very short and the shape of the mature tree is often much more like that of an unhindered hardwood than the commonly accepted notion of a pine.

The old forest consisted of oak with an understorey of holly and hazel at the lower levels on the sunnier slopes and on the best soils. The remnants of the oak forest such as those at Letterewe in Wester Ross, Ariundle in Sunart (plate x) and Clais Dhearg in Lorne, usually occupy south-facing slopes and are rooted in rich brown earths. There were alders along the watercourses and alders and willows in the soft places. Pines and birches with a little aspen and a good sprinkling of rowan occupied the rest of the forested ground reaching to the tree line. Fire and hurricane occurred at long intervals of possibly a hundred or two hundred years. Birch is a pioneer on the burnt forest floor and is a natural nurse crop preparing the ground for the infiltration of pine once again. Excellent examples of the opportunism of the birch can be seen in many places where the old forest has been felled and the ground left unplanted and lightly grazed. The felling lines are clearly marked with pine above and birch below, the opposite to what would be found in nature. The oak forest has nearly all gone but substantial oakwoods, growing on ground which has been continuously under trees since the time of the old forest, and more or less modified by man, are still present in Ross-shire, Inverness-shire, and Argyll. Surprisingly enough some of the finest trees lie on the islands of Loch Lomond close to the industrial heart of the country. A fragment of old oakwood is to be found at Dinnet in Aberdeenshire, and on the shores of Loch Tay in Perthshire.

On the shore of Loch Maree in Ross-shire there is one of the finest collections of woodland relics in Scotland. On the south-facing slope and the brown earths of Letterewe there is a fine oakwood while on the north-facing slopes and on the islands there are impressive stands of native pine rooted in wet peaty podsols. In the pine woods there is evidence from charcoal layers in the soil sections of great forest fires, perhaps at the time of the Viking occupation. The oakwoods on the other side of the loch were exploited in the 17th century for the bloomeries at Letterewe, Poolewe and Kinlochewe, where local bog iron was smelted with charcoal for the manufacture of cannons among other things. Later, when the woods and the bogs were exhausted, the bog excavations became the arable land of cottars who settled in the devastated woodland through which, at that time, ran the main track joining the Mackenzie lands in Kintail and Gairloch.

The Durness limestone outcrops among the Torridonian and Lewisian rocks along the Moine Thrust, and provides bright green oases in the dark mountain country. There are scrub woodlands on the two most southern of these outcrops at Kishorn and Sleat. The Rassal wood at Kishorn is composed almost entirely of ash growing out of the cracks in the limestone and rooting into the pockets of rich brown earth. The ash may be a colonization following the felling of the old forest, as may be the alder-ash wood at Carnoch, Glencoe, also sitting on limestone. But those and many other equally notable woodlands in the Highlands are doomed by man's use of them as grazing and shelter for sheep and deer. Since the inception of the Nature Conservancy in 1949, some important remnants of the old forest have been declared National Nature Reserves and steps are being taken to rehabilitate and extend them where possible. The lead given by the Conservancy in recent years has been taken up by a few landowners who, in their wisdom, are now aware of the vital role played by scrub woodlands in the conservation of the soil and the long-term management of the land. On the debit side, the destruction of the oak forest since the second German war has been quite irresponsible. Oak scrub has been given to timber merchants as a perquisite when felling plantations of conifers and it is not without difficulty that the Conservancy has negotiated a little of what is left as reserves.

Nairn (1890) says that the great Caledonian Forest extended "from Glen Lyon and Rannoch to Strathspey and Strathglass and from Glencoe eastwards to the Braes of Mar". The imagination of a naturalist can conjure up a picture of what the great forest was like: the present writers are inclined to look upon it as their idea of heaven and to feel a little rueful that they were born too late to "go native" in its recesses.

The main trouble between A.D. 800 and 1100 was the Vikings, whether Danes or Norwegians. The tradition of the burning by "Danes", or "Norwegians" still exists in legends which may be heard in the north-west Highlands today. The West Highlands were also a source of boat-building timber for the Norsemen in Orkney and Iceland (Brögger, 1929), but the wanton burning of the western portions of the forest would doubtless be eased after Somerled's Lordship of the Isles became established in the 11th century. Even as late as 1549, Dean Monro speaks of the wooded character of Isle Ewe and Gruinard Island in Ross-shire affording good hiding for thieves and desperate men. Today the wooded islands of many of the lochs stand as living monuments to the old forest which has completely vanished from the surrounding moorlands. The islands though accessible to deer have escaped the ravages of fire and sheep and still have scrub woodland of native species. Notable examples are those in Lochs Sionascaig and Maree in Ross-shire, Morar and Arkaig in Inverness-shire, Ba and Awe in Argyll and Druidibeg in South Uist.

The woods of the Central Highlands were destroyed from the south-east. Gentlemen like the Wolf of Bade-noch (*floruit*, 1380) who was a brother of King Robert of Scotland, wandered through the country with large armed bands bent on plunder. Setting light to the forest was an easy way of smoking out or finishing off anyone who resisted. The forests about Inveraray were destroyed by Bruce in an expedition against Comyn.

All these causes of destruction considered, we are still brought back to what we believe is a fundamental factor in the relation of man to the wild life around him, whether animal or vegetable. Man does not seem to extirpate a feature of his environment as long as that natural resource is concerned only with man's everyday life: but as soon as he looks upon it as having some value for export, there is real danger. The forests of the Highlands were *dis-covered* (this word was used at the period) by the Lowland Scots and the English at the beginning of the 16th century. Queen Elizabeth of England prohibited iron smelting in Sussex in 1556, and in the Furness district of Lancashire in 1563, because of the devastation caused to English woodlands. The smelters had to move farther north. The Scottish Parliament saw to what this would lead and passed an Act (1609) prohibiting anyone "to tak upoun hand to woork and mak ony issue with wod or tymmer under payne of confiscatioun of the haill yrne". We can see exactly how this Act would work from the operations of black markets in Britain during the second German war. The game was so profitable that an occasional heavy fine was accepted as a normal tax on trade.

At this time thieves and rebels hid in the woods and wolves bred therein. It seems that infestation of the forests with these two forms of predatory fauna was so bad that it could be endured no longer. Menteith in *The Forester's Guide* quotes an order by General Monk, dated 1654, to cut down woods round Aberfoyle as they were "great shelters to the rebels and mossers". Ritchie, in giving an account of the extinction of the wolf in Scotland, mentions local tradition and definite record of woods being destroyed in the districts of Rannoch, Atholl, Lochaber and Loch Awe for this very purpose.

The suppression of the first Jacobite rebellion of 1715 gave an impetus to destruction. English business enterprises such as the York Building Company purchased forfeited estates and quite unashamedly set out to exploit them. But for the obstructive tactics of the Highlanders themselves it is probable that every vestige of pine forest would have gone at this time. Even after this period between the rebellions, the higher standard of living, which was more or less imposed on Highland proprietors by their taking up the English way of life, caused them to sell large areas of forests for smelting purposes. Ritchie says: "The destruction wrought by these later and larger furnaces was irreplaceable. In 1728, 60,000 trees were purchased for £7,000 from the Strathspey forest of Sir James Grant. ... About 1786 the Duke of Gordon sold his Glenmore

Forest to an English company for £10,000 ; and the Rothie-murchus Forest for many years yielded large returns to its proprietor, the profit being sometimes about £20,000 in one year."

The last of the felling and smelting with charcoal seems to have been as late as 1813. The brothers Stuart, 1848, mention twelve miles of pine, oak and birch being burned in Strathfarrar to improve the sheep pasture.

The effects of the normal spread of arable cultivation with a rising population may be taken for granted, but this does not by any means round off the story of the changed face of the Highlands through the destruction of the pine and oak forests. The passing of the forests heralded another biological phenomenon of great significance for the natural history of the Highlands, and which was also brought about by man's agency. This was "The Coming of the Sheep". The old husbandry of the Highlands and Islands was in cattle, a well-covered sequence of rearing in the islands and of feeding in the mainland glens before the strong store beasts were driven away south to the great fairs such as Crieff and Falkirk. The Highlands were a country unto themselves into which Lowlanders ventured with some wariness. The collapse of the second Jacobite rising in 1746 allowed flockmasters from the Southern Uplands to think about the exploitation of the new expanses of grazing in the North.

The end of the rising of 1745 meant an end of internecine warfare among the clans, which in turn favoured the survival of more men. The human population of the Highlands rose considerably during the second half of the 18th century, this we know from Dr. Alexander Webster's census in 1755 ; the population story has been brought up to date in *West Highland Survey*. Yet the extension of sheep-farming on the ranching system of the southern uplands meant a way of life in which fewer men were needed ; also, the new sheep farms needed the crofting ground of the glens for winter pasture. The Highland gentry at this time varied greatly in achievement of the aristocratic ideal. Some had little thought at all for the clansfolk in the glens now that they had no further military significance, and others, finding themselves drawn into English metropolitan life, needed ready money—and a lot of it. The flockmaster offered high rents which the new clean ground amply repaid.

The old sheep of the West Highlands and Islands were the so-called "Hebridean sheep" which were sometimes four-horned and dark fleeced, and were the descendants of sheep akin to the Soays of St. Kilda. The sheep coming north with the Border men were blackfaces, which had been bred in the south since the 16th-17th centuries. The Scottish blackface had its origin in the Pyrenees and possibly before that somewhere in central Asia. The sheep were crossing the Highland Line into Dunbarton-shire before 1760; by 1790 the occupation was complete in most of Argyll and in Perthshire and the sheep were plentiful in Mull and Inverness-shire. The first sheep farm in Ross-shire was settled in 1782 and many others followed at the turn of the century. Cheviot sheep-farming in Sutherland and Caithness was begun largely through the energy of Sir John Sinclair of Ulbster in the early years of the 19th century. Extension continued until 1850. Profits were large for both landlord and farmer, but the poor folk found themselves in a bad way. Their husbandry was relatively intensive, the ground being made into lazy-beds (*feannagan*) yielding good crops of barley and oats, and later of potatoes. Fencing was relatively unimportant for they possessed few sheep and the cattle were tended and kept out of the arable ground by the old men and children. The arrival of a heavy stocking of sheep on the hill made the position of these people untenable. They were cleared by the landlords and many thousands chose to emigrate. The folk who remained were pushed to the coasts where their crofting townships are today.

Sometimes these coastal townships were places of such extreme exposure and poverty of soil that after a hundred years of hand-to-mouth existence the crofts have gone empty. The sight of such derelict townships as those on St. Kilda, Mingulay and Monach Isles and so many decrepit ones elsewhere, is a most saddening and disturbing thing. It does not present the ruin of a civilization by sack or natural catastrophe, but the quiet failure of simple folk to obtain subsistence from their environment. In other places, the shift to the coast has proved almost a salvation, for the people have found a mild, sheltered and early climate, and natural resources in fish and seaweed which have enabled them to live much better than they could have in the inland glens. These coastal crofting communities vary greatly in habits and thus in their influence on local natural history. Some have a shore from which they can fish and gather tangles, others turn their energies inland to breeding sheep and cattle and to weaving tweed. It is unfortunate that the small quantity of arable land in many townships is being neglected, rush and sedge creeping into both unoccupied and occupied crofts. The Crofters' Commission has tried to stimulate the economy of crofting in the last ten years but there is little that can be done when there is no younger generation to take over; the unworked crofts are taken over by the most prosperous families in the township in a gradual evolution from smaller to larger holdings which are more correctly described as farms.

The coming of the sheep finished the process of changing the face of the old Highlands of the time of the forests. Large areas of birch were burned and birch bark was exported for tanning sails and rope. Shepherds were occasionally paid in part with the value of birch bark which they themselves had to cut and peel while they were on the hill. The flockmaster's firestick was a destroyer of ground cover over hundreds of thousands of acres. Every spring some patch of heather or purple moor grass would be burnt and seedling trees would suffer. Much birch was cleared in the 19th century by the bobbin-makers working for the cotton mills of the Lowlands and Lancashire. The pirn mill at Salen in Sunart was the principal reason for the establishment of that settlement.

The sheep themselves are destroyers of the scrub habitat of which birch, willow, rowan, hazel and gorse are a part, but even they have not been quite the last straw in man's despoliation of natural Highland forests, because his railways have happened to run through some of the last expanses. The old Highland Line running through the Grampians and Strathspey has been the cause of burning a good many acres of the ancient pine woods. The same has happened on the West Highland Line where the old pine wood of Crannach on the Black Mount has suffered chronically from locomotive sparks. Now that diesels are replacing steam locomotives on these routes this damage is likely to die out. Every year or two there are fires in Strathspey caused often by cigarette-ends or picnic fires, which take away more and more of these beautiful trees. Some of these last remnants have become the property or trust of the nation and others are in the hands of landowners who appreciate them as much from the point of being part of Scottish heritage as from their economic value. Each one of us, however, must be conscious of a personal responsibility in preserving not only them, but the young saplings which they have seeded.

The destruction of the forests meant the end of a habitat and the extinction or near extinction of many animals. Others were able to change their habits and became adapted to the moorland habitats which replaced the forest. The purpose of this chapter is to mention some of the more startling events in vertebrate natural history such as actual extinctions, retrogressions, resurgences, and introductions of new species within the area of the Highlands.

The losses of the last 200 years are large in proportion to those of the previous 10,000 years. Changing climate is an immensely important mover of species and when climate changes in a relatively small island such as Britain, extermination is often the fate of land mammals which cannot readily adapt themselves. Again, if man is present and the animal of fair size, he may speed the influence of climate. The lemming (*Dicrostonyx torquatus*) and the northern rat-vole (*Microtus ratticeps*) are probably examples of changing climate being the dominant factor in extermination in the country as a whole; they probably disappeared with the advent of the warmer Atlantic climate and the extension of forest growth. Vestigial arctic climates such as that of the 4,000-foot plateaux of the Cairngorms have been insufficient to maintain the lemming, which occurs in similar country in Norway.

The giant Irish elk (*Megaceros hibernicus*) disappeared in prehistoric days also, probably before the advent of man. Climate was an active factor, but the organism itself was heading for disaster. The evolution of antler form and weight had no particular relation to function or survival value, but was a concomitant of increasing body size and followed a different growth rate. The great annual drain on the constitution of the Irish elk, of growing 80-90 pounds of antler, was too much in an age which was deteriorating from that of the rich pasturage of the Pleistocene. Whereas the red deer grew smaller in every way, and thus adapted itself, the giant deer apparently died in all its glory. It is thought that the northern lynx (*Lynx lynx*) persisted in the Northern Highlands until man came, but soon afterwards it became extinct. The species was probably in decline with the rise of the warm Atlantic climate, but was given the final push into extinction by Neolithic man. Bones of the northern lynx were found near the hearths in the limestone cave of Allt nan Uamh near Inchnadamph, Sutherland.

The assumption of the disappearance of the brown bear (*Ursus arctos*) in the 9th-10th centuries means that man must have been responsible, for climatic change had long ceased in its more violent form and the destruction of the forests had scarcely begun. More bones of the bear have been obtained from a small, recently discovered cave at Creag nan Uamh by Dr. A. S. Clarke of the Royal Scottish Museum who excavated a few miscellaneous parts of at least one adult and a juvenile. The cave could well have served for hibernation of the bear. A few isolated teeth of the Arctic lemming were also found in the cave. The reindeer (*Rangifer tarandus*) inhabited the Northern Highlands well into the historic period, and had it not been for man's influence it might have survived in Sutherland and Caithness. The destruction of the forest floor by fire must have greatly restricted its range and finally its extermination was probably due to direct hunting. The Orkneyinga Saga mentions the hunting of the reindeer in Orkney about the middle of the 12th century, but the species was extant later than this. Leopold and Darling (1953) have shown how the caribou in sub-Arctic Alaska declined rapidly after burning of tundras and forest floors had removed the lichen growth. An imported herd has been reintroduced to the Glenmore district of Speyside by the Reindeer Company Limited. The herd was brought from Sweden by Mikel Utsi who, with his wife, has been largely responsible for the whole project.

The elk or moose (*Alces alces*) persisted in the north until rather later than this period of the brochs, defensive stone towers which were built during the Norse raids and after they developed into conquest about AD 1000-1100. Man, by direct hunting and the indirect means of destruction of forest which had then begun, was the cause of its disappearance by about AD 1300. The virtual disappearance of the willows must have meant the end of the elk, hunting or no.

The beaver (*Castor fiber*) was found in the Highlands until the 15th-16th centuries, Hector Boece mentioning its existence about Loch Ness, and its being hunted for its skin.

We consider next a group of diverse creatures which are extinct as wild animals of the type they were in the 13th, 14th and 15th centuries when they disappeared; but which still lived on in domesticated forms or crosses with other domesticated stocks. The wild boar (*Sus scrofa*) would be found wherever there were oak woods and by its delving would prevent the establishment of grasses and bracken so characteristic of the non-regenerative woodland floors of today. Though the boar consumed huge quantities of acorns and damaged the tree roots it probably paved the way for oak regeneration. After giving the soil a thorough going-over the herds probably moved on, leaving the brown earth ready for the next fall of acorns and burying at least a few to the right depth. Its domestic descendants persisted in the West Highlands and Islands until the middle of the 19th century, when swine ceased to be kept as a general practice. The conversion of the people to an extreme type of Presbyterianism engendered a Judaic attitude to the pig, and numbers fell away rapidly after the 19th century conversions. The great wild ox or Urus (*Bos taurus primigenius*), surely the most magnificent member of the northern fauna, also disappeared through hunting and the clearing of the forest. The Highlands also had their wild ponies. Cossar Ewart has pointed out that these ponies lacked callosities on the hind legs. Hector Boece mentions the ponies in the same passage as that in which he records the beavers of Loch Ness. The Scottish wild horse received crosses of Norse blood, and later of Arab, so that the Highland pony of today has at least some claims to represent the indigenous stock.

The white cattle with black hoofs, muzzles, eyes and ears remain to us today in a few herds in large parks. They are rather poor creatures, having been greatly inbred through lack of numbers. None of them is in the Highlands. Cattle of this colouring arise from time to time, and it would probably not be difficult to build up a herd of strong-coated white cattle with black points from the existing cattle stocks of the Highlands and Islands. Similarly with the ponies, we could find a few Celtic ponies (*Equus caballus celticus*) cropping up as segregates from the Hebridean herds, and build up a stud of them.

The story of the wolf (*Canis lupus*) in the Highlands is important because this animal was responsible for a good deal of the later history of the destruction of the forests. Clearance of the forest by burning was doubtless the easiest way of restricting the wolf's range. The last wolf of Scotland is said to have been killed by one Macqueen on the lands of Mackintosh of Mackintosh, Inverness-shire, in 1743. Passage through the Northern and Central Highlands in the 16th century was hazardous enough for hospices or "spittals" to be set up where the benighted traveller could rest in safety. All the same, we are dubious of all stories attributing man-hunting to the wolf, although wolves were

plentiful and hungry enough to cause people in the High-
land areas to bury their dead on islands off-shore.
Examples of such islands for which this tradition exists are
Handa, Sutherland; Tanera, N.W. Ross; Inishail, Loch
Awe; and Eilean Munde, Loch Leven, Argyll. A detailed
account of the wolf in Scotland may be found in
Harting's *British Animals Extinct within Historic Times*
(1880).

From the animals and dates mentioned so far in this
chapter, we gather that several mammals disappeared
between the years AD 1000 and AD 1743. Birds were more
fortunate, but their turn was to come with the improvement and lightening of the fowling-piece, the rise of game
preservation and the spread of land reclamation. Almost
as the last wolf howled in the Highlands, extermination of
certain birds began. The first were the crane and the bittern
which went in the 18th century, partly by direct hunting
for feathers and food, but mainly through draining the
marshes for land reclamation. It may be said, incidentally,
that this characteristic 18th century movement for draining also contributed to the extirpation of malaria from
Scotland.

The absolute extinction of the great auk (*Alca impennis*)
is a story so well known that there is no need to recount
more than its gradual diminution on St. Kilda as a breeding species during the 17th and 18th centuries. By 1840,
when the last great auk was caught and killed on St. Kilda,
the captors were unaware of its identity, and the bird was
actually killed because of their fear of it.

All the other extinctions are of the raptorial tribe.
The decline and disappearance of the osprey (*Pandion
haliaetus*) in the 19th and 20th centuries and its subsequent
reappearance in recent years is mentioned in Chapter 12.
The goshawk (*Astur gentilis*) still bred in the 19th century;
Nethersole-Thompson (pers. comm.) states that a pair of
goshawks nested in Inverness-shire in 1960. The kite
(*Milvus milvus*) was finished but a few years afterwards.
The Harvie-Brown *Vertebrate Fauna* Series are good
sources of information on the last haunts of all these

raptors. The white-tailed or sea-eagle (*Haliaetus albicilla*) has been the last to go. Shetland had its last breeding record in 1911. The West Highland and Hebridean coasts, being nearer to extensive sheep-farming interests, lost their sea-eagles rather earlier. By 1879 they had gone from Mull, Jura and Eigg. The species finally ceased to breed in Skye, the Shiant Isles and the north-west mainland about 1890. It is a dismal story. Unlike the osprey no sea-eagles have ever tried to recolonize Scotland; if they did they would find many of their old haunts in the possession of the golden eagle. In any event the vested interest in game preservation, of a decrepit hill sheep-farming industry in the West Highlands and Islands, the pressure of egg collectors and irresponsible sportsmen, are heavy odds which a rare predator must face when staging a comeback in this country.

The history of the golden sea-eagle (*Aquila chrysaëtos*) (plate v) is inextricable from that of the sea-eagle until about the middle of last century. With the coming of the sheep both species were probably dealt with as equals, though the sea-eagle was a more powerful bird and capable of doing much more damage to flocks. The golden eagle population proved the more resilient possibly because of its greater numbers and wider range. By the middle of the 19th century it was a much rarer bird than it had been previously, and with the increase in game protection the outlook was probably black, especially in country destined to become grouse moor. In the deer forests, however, the eagle was less of a nuisance and a rare part of the grand mountain country; it was probably out of sentiment rather than from any appreciation of a useful purpose which the bird served in the forest, that some protection was afforded it.

In the past century the eagle has been hunted remorselessly not only by those who were managing the land in the Highlands and Islands but also by the collectors of skins and eggs. The story is the same as with sea-eagle, osprey and kite with the exception that the golden eagle has, through it all, survived. Its survival when all the others

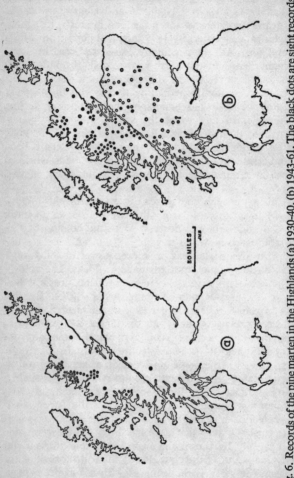

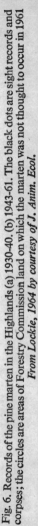

Fig. 6. Records of the pine marten in the Highlands (a) 1930-40. (b) 1943-61. The black dots are sight records and corpses; the circles are areas of Forestry Commission land on which the marten was not thought to occur in 1961 *From Lockie, 1964 by courtesy of J. Anim. Ecol.*

went under has made it an emblem of the serenity and grandeur of the Highlands. It is now protected by law, but is secretly persecuted. Having become gradually more numerous throughout this century, with long respites from killing during the wars and having been made the subject of a bounty scheme since the second world war by the Royal Society for the Protection of Birds, it is now well distributed.

It is thought by some that the local extinction of the ptarmigan (*Lagopus mutus*) in the Hebrides is due to the shrinkage of its range due to the climate becoming gradually warmer and wetter. Watson (in Bannerman, 1963 and 1966) says that numbers fluctuate greatly from year to year, and since the available habitat is very small on isolated peaks in the Hebrides or on peaks in the eastern Highlands such as Sgor More (2,666 feet) and Mount Battock (2,555 feet), local extinctions are likely to occur there in bad years. These may be recolonized in good years ; one has been seen recently on Rhum for instance and others on Scarba. Their last haunt in the Outer Hebrides was on Clisham, Harris, the highest hill in the Long Island. They still persist in small numbers in the Cuillins of Skye.

The polecat (*Putorius putorius*) was thought to have become extinct in Scotland but may still survive in the remote North-West Highlands. Some or all of the reports may refer however to feral ferrets, which closely resemble the polecat, and have been intentionally introduced to keep down rabbits, and now fill the niche of the wild polecat. The intensity of game preservation and the skill of the Scottish gamekeepers in killing so-called "vermin" is doubtless responsible for the decline in numbers to the verge of extinction.

Reafforestation has played a part in the spread of the pine marten (*Martes martes*) (plate IV) in recent years. It persisted in the Highlands only by an extraordinary adaptation to a different way of life. It was primarily and by preference a woodland animal, a beast of the old pine forest, living on squirrels, small rodents, birds and their eggs, frogs, newts and lizards, fish and a wide variety of insects and berries.

Steady persecution and the destruction of the forests reduced its numbers to very few, but it retreated to the cairns of the treeless hills, and in north-west Sutherland was protected in its last stronghold by a group of landowners and sportsmen led by the late Lt.-Cmdr. E. J. Fergusson.

The geographical distribution and food of the pine marten have been studied in the last ten years by Dr. J. D. Lockie (1961 and 1964) of Edinburgh University. Towards the close of the 19th century the marten was rapidly exterminated from many parts of the Highlands due to the use of the gin trap for rabbits and in game preservation. By 1900 they were extremely scarce and confined mostly to the North-West Highlands. They remained scarce till the late 1920's when an increase both in numbers and range was noticed in north-west Sutherland. This was probably due to a relaxation in game preservation during the 1914-18 war. The increase and spread continued during the 1930's in Wester Ross and Sutherland and in 1935 a pair was introduced to the Loch Laggan district. The increase continued throughout the 1939-45 war with a spread east and south from the stronghold in the far north-west. The Caledonian Canal seems to have been a partial barrier to the spread. Nevertheless, between 1943 and 1961 there were twelve known records south of Loch Ness, which was more than had been recorded from 1900 to 1943 (fig. 6). Since 1945 numbers of martens have fluctuated locally ; one such fluctuation lasted seven years at the Beinn Eighe Reserve, and this did not appear to be caused by fluctuations in the principal food—the short-tailed vole.

There are no pine martens in the Hebrides, though they were present in Harris until the 1870's. The late Osgood Mackenzie of Inverewe, who saw much of the old wild Highlands in their last days before the rabbit and the sheep destroyed what was left by the woodcutter's axe, told a story of finding a pine marten and a big wether lying dead together in the Forest of Harris. The marten had evidently got its canines into the throat of the wether and that animal, dashing downhill in fright and pain, had struck a rock and killed the marten; then the sheep had dropped and bled to death. In his book *A Hundred Years in the Highlands*, Osgood Mackenzie left one of the finest accounts of the butchery of Highland wildlife ever written: in this, to some extent, he took an active part.

David Jenkins (1962), who has recently carried out an enquiry into the status of the wild cat (*Felis sylvestris*) in Scotland, found them widely distributed in small numbers. They were numerous in Inverness-shire and on higher moors in Aberdeenshire, but less so in the north and west and were uncommon in Perthshire. There have been remarkable changes in numbers in recent years, in parts of Angus and Moray for example, and this is probably due to the general increase in woodland providing better cover and probably more food, and to myxomatosis which deprived the cats of their staple food and forced them into the open. The effects of myxomatosis would, of course, vary from district to district depending on how dependent the cat population was upon rabbits for food. This is perhaps one of the reasons for the present scarcity in the north-west, compared with the ptarmigan, blue hare and grouse country farther south.

Feral cats are rare on high ground in Scotland, but wild ones do visit farms and there is bound to be some cross-breeding. When the last family left the Monach Isles in the 1940's they left the cats behind, and there is now a thriving population of feral cats there.

In 1956 the Nature Conservancy made a survey of badger (*Meles meles*) distribution, sending questionnaires to Forestry Commission foresters. The badger was present then throughout the mainland areas of the Highlands but on none of the islands. Of 162 replies, 76 reported that it was not known in the forest. About half of those who had badgers in their forests noted that they were increasing and only four that they were decreasing. The increase up till 1956 was most apparent in the south-west; two-thirds of the foresters who had the badger in the country from Lochaber south through Lorne, Knapdale, Kintyre and Cowal noted an increase. In the north-east from Loch Ness to the Moray and Buchan coasts, and in the south-east including Perthshire, Stirlingshire and Angus about half of those who knew the badger to be present said that it was on the increase. In the north-west Highlands from Sunart to Sutherland only a quarter reported an increase. All this goes to show that after a couple of centuries of persecution the badger is well on its way back.

The red deer (*Cervus elaphus*) have had to undergo a considerable change of habits in order to survive. From being creatures of the open forested ground they have had to become animals of the bare hillside, now living the year through without much shelter. They have achieved this survival by a dimunition in their size and in the elaborateness of antler growth The Highlands of today would not carry a stag of the size and magnificence of antlers seen in the Highlands of 500 years ago. The red deer is the subject of chapter 6.

The roe deer (*Capreolus capreolus thotti*) (plate VII) has been a native of Scotland since prehistoric times, and has suffered the same environmental change as the red deer. The destruction of the old forests reduced its range, but unlike the red deer it did not greatly change its habits, body size or antler growth. Roe deer are fairly common on the bare deer forests, but are much more so in woodlands and the new era of reafforestation is bringing about an increase in numbers throughout the Highlands. Roe deer are more difficult to exclude from forest enclosures than red deer and do a great deal of damage in nipping the tops from the young conifers as they protrude green and attractive above the snow. Efforts are made to keep the numbers to a minimum, but this is difficult with such a creature which so skilfully uses the forest cover and can outrun most dogs across the forest floor. The ecology of roe deer is mentioned in chapter 7.

The history of the red squirrel (*Sciurus vulgaris*) in Scotland is one of remarkable interest and was studied in detail by Harvie-Brown (1880). Remembering Dr. Johnson's facetious remarks about the value in Scotland of such a piece of timber as a walking-stick, it may well be understood that the extreme course of deforestation in the greater part of Scotland must have restricted the range of the squirrel until it was nearing extermination. It had long gone from the Lowlands by the time Dr. Johnson was making epigrams, and was in a much reduced state in the Highlands where some pine and oak forest still existed. Birch wood, of course, is of little attraction to a squirrel. Harvie-Brown says that some severe winters killed off most of the Highland stock. Some were left in Rothiemurchus and apparently a few persisted in south-east Sutherland until 1795.

From this date onwards the species began to increase in the Highlands, both from the resurgence of the Rothiemurchus stock and from introductions. The Southern Highlands were colonized in the first half of the 19th century from the stock liberated by the Duchess of Buccleuch at Dalkeith, Midlothian, in 1772. The Eastern Highlands were colonized from the Duke of Atholl's importation from Scandinavia in 1790, turned down at Dunkeld. There was another introduction at Minard on Loch Fyne in 1847 which allowed a spread over the South-West Highland area and northwards to Dalmally and Glen Dochart. Finally, and of great importance, there was the introduction by Lady Lovat in 1844 to Beaufort Castle, Beauly, Inverness-shire. The Beaufort Castle stock spread in all directions, recolonizing Sutherland, crossing from the Oykell to Loch Broom and Little Loch Broom, down the Great Glen and into north Argyll. Harvie-Brown in the *Vertebrate Fauna of Argyll and Inner Hebrides* (1892) says that the squirrels had not yet reached Strontian, but now they are well down Ardnamurchan. There are none in the Coille na Glas Leitire in the Beinn Eighe Reserve or in the Letterewe Oakwoods in Wester Ross and the presence of the pine marten there may prevent colonization.

The squirrel has become a forestry pest of importance. Squirrel-shooting clubs have been active in many districts, but the species remains plentiful and will doubtless hold its own in this era of reafforestation. What the effect of the squirrel has been on the small woodland bird population of the Highlands can never be known, but on such species as the Scottish crossbill it must have been considerable and depressing. The squirrel does not occur on any of the islands except Arran.

The American grey squirrel (*Sciurus carolinensis*) is an introduction to Scotland which has not got far into the Highlands. Here also shooting has had little effect in controlling the population. A single pair was released at Finnart, Loch Long, in 1890 (Ritchie, 1920) and their offspring have spread over to the west bank of Loch Lomond and thus come within our area.

The variable, blue or alpine hare (*Lepus timidus*) is an indigenous member of the mainland Highland fauna. Raymond Hewson (1954, 1955) says that in 1951 it was rarer north of the Great Glen than elsewhere in the Highlands, and most numerous in the eastern Highlands. Numbers were then increasing throughout the Highlands generally except in Wester Ross and Argyll where it has died out in several places. Blue hares were introduced to Harris in 1859 and spread rapidly to Lewis, and were still present in 1951. Those introduced to North Uist between 1890 and 1913 were less successful and there the species is extremely rare. It does not occur in Benbecula, South Uist and Barra. In the Inner Hebrides it occurs in Skye, Mull, Raasay, Scalpay, but though introduced to Eigg in the 1890's it is now extinct there. The hares in Jura are of the Scottish race, but those in Islay are of the Irish.

The blue hare population is subject to considerable fluctuations, the minimum bags being about a tenth of the maximum over a period of 30 years. The best bags obtained from the Islands, in the Loch Seaforth-Loch Erisort district of Lewis, are small compared with those from similar country on the mainland. Numbers of blue hares appear to fluctuate in parallel with those of grouse; there was a decline in both during the 1930's. When one of us was working the Dundonnell, Gruinard and Letterewe Forests in 1934-36, a blue hare was seen on two occasions only, each time on a peat-free gravelly plateau at an alti-

PLATE I: Glen Affric, Inverness-shire. Scene of hydro-electric and forestry development and of great natural beauty

PLATE II: *Above*, Crested Tit (*Parus cristatus scoticus*) in Speyside. *Below* clover-filled grassland recently reclaimed from the peat moorland seen in the background at Barvas, Lewis

PLATE III: Birds of the pine forest *left*, Siskin (*Carduelis spinus*)

Below, Capercaillie (*Tetrao urogallus*)

PLATE IV: Mammals of forest and moor. *Above*, The Highland Fox (*Vulpes vulpes*) at Kinlochewe, Ross-shire. *Below*, Pine Marten (*Martes martes*)

PLATE V: *Above*, Golden Eagle (*Aquila chrysaëtos*) in Ross-shire, the largest raptorial bird in Scotland *Below*, Merlin (*Falco columbarius*) the smallest raptorial bird of Highland moors

PLATE VI: *Opposite*, Winter at Loch an Eilean in the Rothiemurchus section of the Cairngorms National Nature Reserve

PLATE VII: *Above*, a yearling Roebuck (*Capreolus capreolus*) showing forked antlers in velvet, in early March *Below*, a day-old Red Deer (*Cervus elephus*) calf on Rhum with ear-tags

PLATE VIII: *Above*, with the Cairngorms in the grip of winter, a hungry Red Deer (*Cervus elephus*) stag is attracted to a lure in the snow

Right, a Red Deer hind finds a winter bite by the edge of a pool, in the Kinlochewe Forest

tude of 2,250 feet. The ecology of the blue hare is discussed further in chapter 8.

The rabbit (*Oryctolagus cuniculus*) has had an effect on the Highland landscape complementary to that of the sheep, of checking regeneration of trees. Today they are slowly recovering from the plague of myxomatosis which swept through Scotland in 1953-54, and are once more becoming a pest throughout the Highlands wherever ground is suitable for them. The basalt soils of Mull, Morven and Skye are perfect for the rabbit, and the sandy *machairs* are as paradise. Many thousands have been exported annually but these merely represented a healthy crop. Recovery from myxomatosis has been faster in some colonies than in others, which seem to have suffered from recurring spasms of the disease. It is the responsibility of all land-users to keep numbers low ; they are a great capital liability.

The rabbit was introduced to Scotland possibly in medieval times. Dean Munro (1549) wrote that they were present in great numbers on Mull and Inchkenneth. Yet many mainland districts may not have had rabbits until much later. Osgood Mackenzie, for example, writes in *A Hundred Years in the Highlands* of the care taken to acclimatize them to Gairloch about 1850. Now almost all the major islands in the Hebrides and many of the minor ones have rabbits. A notable exception of the larger ones is Tiree, and of the small ones Bernera (Harris), St. Kilda and North Rona. Lighthouse-keepers have introduced rabbits into many islands and promontories, notably the Flannan Isles.

The influence of myxomatosis on rabbits in the Highlands has been less decisive than elsewhere in Britain. The reasons are probably two-fold ; first, the discontinuity in distribution and second, the density of the flea population carrying the infection. Perhaps the doctoring of the ground with infected rabbit carcasses was less systematic in the Highlands than elsewhere, though a great deal of this was done both on the mainland and on the islands.

One good thing the rabbit has done ; it has acted as a

D

buffer species providing food in plenty for predators which we might otherwise have lost by now—wild cat, polecat, pine marten and buzzard. This last has increased greatly in the Highlands and Islands and is now widely accepted as a useful species. When the rabbit population was struck down by myxomatosis many of these predators, which may themselves have played a part in spreading the plague, went hungry, but most predators probably had reserves of other food on which to turn, principally carrion. The buzzard and the rabbit are inseparably linked in the mind of the naturalist as predator and prey, yet this bird has successfully reared young on the rabbit-free island of Tiree for many years. The diet there includes brown hare leverets and the young of a large number of birds—gulls, lapwings and other waders. In winter they take carrion as do wild cats and martens in the mainland forests. From about 1800 till 1915 the buzzard decreased under game preservation, though it never was a serious predator of game. Later it increased when game preservation lapsed somewhat during and between the two world wars. Myxomatosis resulted in a striking decrease in breeding activity of the bird, but was less important than game preservation in affecting the distribution and numbers.

The brown rat (*Rattus norvegicus*) is an introduction to the Highlands as it is to the rest of the country. Rats are found in remote glens and on most islands where there is human habitation. Their worst disturbances in the natural history of the Highlands are when they invade small off-shore islands either permanently or seasonally. Their staple food is shell fish and littoral detritus, but in early summer they play havoc in tern and gull colonies, sucking the eggs and devouring the young. Gunna and Eilean Ghaoideamal off Oronsay are excellent examples of this. We know of several small islands being cleared of terns by rats and the quick change of fortune in many tern colonies on large islands may also be due to the depredations of rats.

Rats often inhabit shore caves, foraging widely along the rocky shores. The rat population at Ceann a'Mhara in

Tiree, for example, is centred in a cave which is used as a roost for most of the starlings in the island and a large number of rock-doves. The floor is piled high with manure from the birds and into this the rats burrow. The Shiant Isles and Ailsa Craig are both heavily infested with brown rats which live in the talus slopes at the foot of the cliffs and feed in summer on the large sea-bird colonies. There are no rats on St. Kilda.

The rat is said to have cleared North Rona of its human population of thirty, soon after 1685. The species, in this case, was probably the ship rat, *Rattus rattus,* since *R. norvegicus* was scarcely known in Western Europe, and quite unknown in Britain, at this date. The rats came ashore from a wrecked ship, and ate up the barley meal stored in sheep skins and the people eventually starved to death, the steward of St. Kilda finding the last woman lying dead on the rocks with her child on her breast. The rats also starved thereafter, probably because, as ship rats, they were ill-fitted for life on North Rona and because the immense swell on the island prevents any hunting of the rock surfaces of the intertidal zone. One also cannot help wondering if the people of North Rona really starved to death, or did the rats bring plague?

Populations of rats like those of other rodents are subject to large-scale fluctuations which appear independent of the constant predatory pressure exerted by man. Migration probably accounts for some of these fluctuations in numbers. In the Highlands and Islands where human habitations are far apart, and where in the course of the years more and more dwellings are becoming derelict, the rat population must be kept on the move. Every crofthouse and farm has its quota of dogs and cats, and the population of these through the area must be vast and out of all proportion to the threat of rats and mice. When ashore, the ship rat is an inhabitant of the wharfside and warehouse; they may have come ashore from shipwrecks in the past and at such ports as Campbeltown, Oban, Mallaig and Stornoway, but none has been reported so far.

Turning to happier mammalian introductions in the

Highlands and Islands there are fallow deer (*Cervus dama*) and Japanese sika deer (*Cervus sika*). In a recent article G. Kenneth Whitehead (1962) has summarized the history and present distribution of those two species in Scotland. Fallow may have been present since about AD 900 introduced perhaps by the monks at Kildalton in Islay. Mention is made of them in records of the 13th century from Stirling and the 16th from Inchmurrin (Loch Lomond) and Atholl when they were certainly living in a wild state. At present fallow deer are still to be found on Loch Lomonside, near Inveraray, near Dunkeld, on Bolfracks estate and elsewhere in Perthshire, on Speyside, at Balmacaan west of Loch Ness near Garve, near Rosehall in Sutherland and Berriedale in Caithness. Other reports of fallow come from Mull, and from Arisaig, Inverness-shire.

The sika deer were introduced to the Highlands at the Achanalt Forest in Ross-shire about 1889 and were enclosed there until 1915. Thereafter they were liberated and have since spread north but not south. The Loch Rosque Forest near Achnasheen is probably still the head-quarters of this species in the Highlands. Those introduced to the Black Isle at the beginning of the century are probably now extinct, but another introduction of sika to Carradale in Kintyre, in 1803, was most successful. They are still there and have now spread north into Knapdale. The domestic goat, (*Capra hircus hircus*) shows a ready tendency to go feral, and there are now "wild" goats in many parts of the Highlands and Islands, from Wester Ross to Dunbartonshire. They also occur on several of the islands including Rhum, Eigg, South Rona, Jura, Harris, Arran and Holy Island in the Firth of Clyde. Those on the slopes of Ben Lomond are reputed to have a history going back at least to King Robert the Bruce. The distribution was summarized by Boyd Watt (1937) and since then the goat has become even more widespread, but is local and occurs in small numbers. The feral goats of the Highlands have become similar to the wild type. They inhabit the hills above the peat line where conditions are drier for them and but rarely descend to the low ground. Their young are born in January and February, so the criterion of selection these animals undergo in the Highlands is a severe one. The habits of feral goats in the Highlands have been described by Darling in a supplement to Boyd Watt's paper.

The two British seals, the grey seal (*Halichoerus grypus*) and the common or brown seal (*Phoca vitulina*) have always been a part of the economy of human communities on the western seaboard of Scotland. More detailed descriptions of both species are given in chapters 11 and 9 respectively. From the earliest times seals have provided man with skins, meat and oil which were used for a variety of purposes. The grey seal has, however, always borne the brunt of the hunting when ashore on the breeding islands and helpless when surprised by men with clubs. The common seal, which does not venture above highwater mark at any time of its life, was much less vulnerable until the rifle came into the possession of the crofter-fishermen. But neither the clubbing of the grey seal nor the shooting of the common has succeeded in bringing either species to the verge of extinction. The common seal on the Scottish coasts is part of a world population of some 3,000,000 (Scheffer, 1958). The grey seal on the other hand is one of the rarer seals ; about 46,000 are distributed on the temperate coasts of the North Atlantic and the Baltic (Smith, 1963). In the 19th century the grey seal was faced with the danger which comes to a comparatively rare species when the toll taken is for export and not for the limited needs of a resident population.

Late last century skins of grey seals slaughtered at North Rona and Sula Sgeir were being bought and resold by the Danish Consul in Stornoway. The fishery was wasteful in the extreme and quite unorganised. Then came a remarkable relief for the seals in cheap rubber boots for the fishermen, and simultaneously, almost, the arrival of cheap and clean paraffin for lamps ; only those whose have tried the smoky flame of seal oil can fully appreciate the boon of paraffin. The seals got some respite except for the fact that the hunting had become a traditional social occasion which had to be gradually broken down.

The common seal is not protected by law. The grey seal on the other hand has been protected since 1914 and the increase in the species throughout the last century combined with the lack of commercial exploitation in the last 50 years, has resulted in numbers which it cannot have known for centuries. It is spreading to islands and mainland coasts from which it had long disappeared. This outstanding story of the survival of the grey seal is an example of the importance of human ecology in relation to the animal, and the latest chapter is now being written concerning the impact of the seals on the mid-twentieth-century coastal fishing industry in Scotland. This we discuss later and would finish with seals for the present by saying what requires to be said again and again about the grey seal —no British mammal could be more easily exterminated.

There have been extinctions, introductions and remarkable changes in the status of several Highlands birds. First, the capercaillie (*Tetrao urogallus*) (plate III) became extinct after the destruction of the old pine woods and its end was hastened by direct hunting. The last capercaillie of the old stock disappeared in about 1771 from the Glenmoriston area. The gap that followed was one of 67 years, though the Earl of Fife attempted an introduction at Mar Lodge, Aberdeenshire, in 1827-9. It was unsuccessful. Lord Breadalbane did better at Taymouth Castle in 1837-8. He got 13 cocks and 19 hens (Ritchie, 1920) and turned them loose in the woods. There were over 1,000 birds present on the estate 25 years later. Harvie-Brown traced the spread of the capercaillie over Scotland from this point up to 1879, in his *Capercaillie in Scotland*, 1880. A successful introduction of caper to Strathnairn just east of Inverness in 1894 is doubtless responsible for the later spread into Easter Ross. Darling was told by an old stalker that he had seen caper in the pine woods at the head of Loch Shiel before these were felled. I. D. Pennie (1950 and in Bannerman, 1963) gives the latest account of the status and distribution of the capercaillie in Scotland.

The general spread from Taymouth was north-eastwards and south-westwards along the river valleys where there were successions of coniferous plantations. The maximum spread was attained by the first world war. The bird is now well established in suitable woods all over north-east Scotland from Tayside to Easter Ross and south-east Sutherland, and in Cowal, Dunbarton, and south-west Stirling. The two major wars were setbacks for the capercaillie with the extensive felling of coniferous woodlands and the purging of the bird by the woodsmen.

Dr. Pennie remarks that the habitat preference of the capercaillie may be changing with the changing scene in Scotland. More are occupying younger plantations and deciduous woodlands (no breeding record in the latter so far), visits to standing corn have been recorded and stubbles are now widely used. When the capercaillie is extending its range the hen birds move first and the cocks follow in a year or two. It has been found that under these circumstances the hen caper will mate with blackcock and pheasant and produce hybrids. They are breeding in the Cairngorms Nature Reserve, both in Speyside and Deeside. Their winter diet of pine, larch and spruce shoots makes them an enemy of single-minded foresters.

The pheasant (*Phasianus colchicus*) has been constantly imported into the Highlands, but those days are probably over. The West Highlands at least are not pheasant country, and unless replenished, stocks are not persistent. Good stocks are still maintained by prosperous estates, particularly in the Eastern Highlands, which possess extensive broad-leaved woodlands and adjoining arable land. They have been introduced to many of the islands including Islay, Gigha, Jura, Colonsay, Coll, Tiree, Mull, Eigg, Rhum and Raasay and still persist on some of those. In Islay, where the woodlands, scrub and farmland provide excellent cover and feeding, there are fine stocks.

Two more birds still present in Scotland have had their ranges severely restricted by the destruction of the forests. These are the crested tit (*Parus cristatus scoticus*) (plate II) and the Scottish crossbill (*Loxia curvirostra scotica*). Although many of their best nesting grounds in the upper Spey Valley were cut down during two world wars, crested tits still nest there in good numbers. They have colonised some of the new forests and have increased in others in the Findhorn Valley and at Culbin. In one of the forests Nethersole-Thompson has seen them nesting in nest-boxes. Crested tits probably now nest in south-east Suther-land; they have certainly nested regularly within two miles of the Sutherland border in an old pine wood in Ross. Although the Cairngorms have apparently prevented the spread of the crested tit into Deeside there seems no reason why the birds should not ultimately spread to the great pine wood of the Dee Valley from Moray and west Aberdeen. Although in the past reported from the Great Glen and Glen Garry, Argyll, Dunbartonshire, Perthshire and Angus, the crested tit has not so far appeared in the old forest at Kinlochewe and Coulin, Wester Ross. The bird nests in the Cairngorms National Nature Reserve.

Some now regard the Scottish crossbill as a race of the parrot crossbill (*Loxia pytyopsittacus*). Nethersole-Thompson considers that the Scottish crossbills have a diagnostic call, similar to that of the parrot crossbill, which helps to keep them separate from immigrant *Loxia curvirostra*. The range is considerably wider than that of the crested tit, but it is as yet uncertain what subspecies (or race) breeds in Aberdeenshire, Angus and Kincardine, but the Scottish crossbill breeds in Perthshire, Banff, Nairn, Moray, Inverness, Ross and sporadically in south-east Sutherland. They probably also nest in Aberdeenshire, and have been seen with young in Mar on two occasions. The main strength is concentrated in Strathspey and after the young are fledged large flocks are seen there working through the ripening crop of cones. The crossbill nests high in the pines in the National Nature Reserves of Cairngorms and Beinn Eighe, where, in recent years, it has been seen feeding young in the old pines of Coille na Glas Leitire.

Such attractive birds as the goldfinch (*Carduelis c. britanica*), the redpoll (*C. flammea*) and the siskin (*C. spinus*) must have been greatly affected by deforestation. Industrialization followed deforestation and the dwellers in the new Lowland towns began to lose a daily contact with the countryside. Many found solace in the song of a cage-bird and these species were badly hit by bird-catchers. The goldfinch was common in the Moray Basin until the third quarter of the 19th century when the bird-catcher brought it to extinction. A decrease in the siskin had taken place by 1900 in Kincardineshire and Aberdeenshire due to bird-catching. Now all three species are protected by law and are returning as breeding birds to many of their former haunts.

The great spotted woodpecker (*Dryobates major anglicus*) became extinct as a breeding species in Scotland about 1850 (Harvie-Brown, 1892). Evidence of its former abundance in the Highlands was seen in the remains of old timber well worked by the birds. After the turn of the century the birds came back to the Highlands and have spread consistently up to the present day. It is now breeding in most of the Highland counties and an account of the truly remarkable return of this lost species is given by Baxter and Rintoul (1953). The extensive wartime felling of woods recolonized by the birds caused local setbacks, but these have not been enough to halt the spread. The young forests of block-planted conifers are not likely to prove as attractive to the woodpecker as those with an element of the old forest still present with its variety of tree holes and rotten insect-infested trees.

The new forests are providing breeding habitats for many birds which formerly inhabited the old forest and which have since only bred in local pockets. The monocultures of conifers are, however, a poor substitute for the diverse woods of the old days which harboured an equally diverse community of song birds. The chaffinch, redstart, tits, treecreeper and willow warbler are now much more numerous and the woodpigeon is extending its range into the reafforested glens. The plantations give cover to the migrating flocks of storm-bound redwings and fieldfares. The hen harrier (*Circus cyaneus*) is staging a comeback in the young plantations. In Glen Garry, between 1837 and 1840, sixty-three harriers were trapped, and the vermin lists of other estates tell the same story of the abundance of the bird and its widespread slaughter by gamekeepers. Such birds as harriers, buzzards, short-eared owls, kestrels and sparrow-hawks were the countless victims of the pole-trap, one of the most cruel of man's inventions which is now happily outlawed, but which is still secretly used.

The wood warbler (*Phylloscopus sibilatrix*) has extended its range in the last seventy-five years in the Highlands. Harvie-Brown frequently mentions the enlarging range of the species, but in 1904 he was careful to say that it had not yet reached Wester Ross. He puts 1895 as a year of great expansion. Sixty years before that time, very few wood warblers had been seen in Scotland. It now breeds in woodlands of the West Highlands as far north as Ross and south-east Sutherland, but is much more common in the wooded glens of Perthshire and Inverness-shire. Breeding has been reported from Mull and Rhum.

The chough (*Pyrrhocorax pyrrhocorax*) is possibly on a different stage of the same journey as the woodpecker. The woodpecker went as its habitat was cut from under it and has come back to a more tree-clothed Scotland again. The chough faced the blast of 19th-century game preservation and went under. Whether it was truly inimical to game is highly questionable, but being one of the crow tribe it therefore must be bad. By the middle of the 19th century choughs had gone from inland districts of Scotland and were to be found only on the wilder coastal districts. Thus reduced in numbers by game preservers, the impact of predators such as the peregrine and the egg collector probably contributed to a further shrinkage of their range.

Graham mentions three pairs breeding in Iona in 1852, where they were left alone, but there are none there today. Skye lost its last chough by 1900 and no one knows how few were left on the cliffs of Islay when at last their rarity made them birds to be protected. However, they are still present on extreme south-west coasts, and have bred recently in Jura, Mull of Kintyre and in Islay where they are fairly common. If we can achieve effective protection in future the chough is in the way of coming back. There is probably no time more propitious than the present for the chough to increase with a new sympathy from land-owners along the coast and the establishment of the Rhum National Nature Reserve. More than protection may be needed to bring back the chough, however ; its habitat of mixed farmland on the rocky coast has changed and is now in possession of jackdaws in some areas.

The spread of the starling (*Sturnus vulgaris*) in the Highlands must be recorded as a remarkable change in status. This spread of a bird unknown for many years over most of the area does not apply to the Outer Hebrides where there has been for a long time a population of the sub-species *Sturnus vulgaris zetlandicus*. Even on North Rona there is a flock of a hundred or more of this form and there is another very old-established flock on St. Kilda. But on the mainland the country-wide colonization began about 1840 and was complete by 1890. Harvie-Brown (1895) gives a detailed account of the spread suggesting that the West Highlands were colonized by two thrusts, one coming down from the north, and the other pushing north by the Callander-Argyll route. Harvie-Brown could not know then what Bullough discovered only in 1942, that the British and Continental starlings behave differently, have different thresholds of gonadic development, and are for this reason physiologically isolated, and Harvie-Brown's fears of what might happen as a result of the vast increase and spread of the native starling have proved unfounded. The enormous incoming flocks to the east coast in September may be of no consequence in the spread of starlings as a breeding species in Scotland.

Human habitation and cultivation are not a necessity for the starling's spread, though such factors are clearly helpful. It is attracted to good green grass to feed on insects, harvestmen and spiders which abound and it is an industrious worker of the cow-pats, where it feeds on flies, beetles and earthworms. It does not frequent the spacious heather moorlands in anything like the same numbers as for example the *machair*.

The fulmar (*Fulmarus glacialis*) (plate XIII) has undergone an extraordinary expansion in range in the last two hundred years. The spread seemed to come from Iceland and to affect the Faroes, Shetland, Orkney and the Western Isles and mainland coasts in turn. James Fisher (1952) and Wynne-Edwards (1962) believe that in the 17th century there were two stocks in the north-east Atlantic, one at Grimsey off the north of Iceland and one at St. Kilda, and that colonization came to Britain from the north and not the west. This spread Fisher attributes to the birds being attracted into lower latitudes by the offal of whaling and trawling; Wynne-Edwards does not agree that this hypothesis fits the facts and considers that the spread may have been caused by the descendants of a genetical mutant in Iceland.

St. Kilda has a very ancient colony of fulmars thought to be about 30,000 strong today. These birds were the staple food of the St. Kildans, and have increased since the fowling ceased in 1930. The Rev. Neil Mackenzie (1829-43) put the annual take of fulmars at about 12,000 and Eagle Clarke (1911) put the number at 9,600. The diminishing population took progressively fewer and in 1929 about 4,000 were taken to sustain the thirty-two islanders. The ecology and distribution of the fulmar is more fully described in chapter 10.

Man's behaviour, aimed directly at a species, or in interfering with a habitat, is one of the primary factors affecting the welfare of wild life, as much in the remote Highlands as the fields of England. These hills and glens and rivers and lochs are not wild, even if unpeopled. The ground is covered by observant men engaged in their own particular use of the land. Many who fell, burn, drain and kill in the name of good land management are usually short-sighted. The waters are fished, sieved and polluted. The story could be continued at great length with fish and invertebrates, and some of these are mentioned in chapter 12.

Young herring (*Clupea harengus*) used to come close inshore in summer on the West Highland seaboard. They still do to a lesser extent than in the past and in a less predictable way. The migration as a whole from spawning grounds to the feeding areas has remarkable regularity but the detailed inshore routes undergo modification in the course of the years. This is discussed more fully in chapter 9. These changes of detail used to be of great significance to the townships on the way. For example, throughout the 18th century the herrings came up the length of several of the sea lochs of the west, but before 1850 the fish had changed their route slightly and passed the lochs. During the 1700's and for centuries before, the herring had been so regular as to pass up one shore of Loch Broom and down the other, with the result that fishing took on an almost clockwork precision. Fishing stations such as that on Isle Tanera were built at great expense, and paid handsomely for fifty years or more as the herring passed so near and so regularly. The fish were cured and exported to the West Indies and to southern Ireland. All the local people filled their barrels for the winter. There seemed no reason to believe this seasonal manna would ever cease. Unfortunately it did. The fishing passed almost entirely to the East Coast boats, Zulus and Fifies and later the steam drifter, except for the fleet of smallish boats at Stornoway and the Loch Fyne smacks.

The seasonal herring fishery in the Minch is as follows: from November to February the Minch fishery is at its maximum; there is a lull in April and May and then the summer fishery from June to September. The latter is always on a smaller scale than the former. There is little fishing from September to December except in the Clyde area, famous for its high quality Loch Fyne herring, and a small fishery of spents and recovering spents in the Minch usually in the sea lochs with ring-nets.

A decline has taken place in the West Coast herring fisheries as it has in the East, since the first world war. The main factor was the collapse of the trade in cured herring with the Continent after that war. This and changing tastes in Britain destroyed the demand for herring. Herring fishermen turned to other fish, and there grew a vigorous seine net fishery for white-fish which is still prosperous. The main reasons for the general decline were therefore economic rather than biological. West coast herring have never been heavily exploited. The main shoals of herring may not be inexhaustible, though they have been fished for hundreds of years and have shown great resilience. Modern methods of fishing, not only with improved boats and gear from home ports but fleets of catchers and mother ships from Continental ports, may serve dangerously to deplete stocks.

The fishing ports for the Minch are at Kinlochbervie, Ullapool, Gairloch, Kyle, Mallaig and Stornoway; from the Sea of the Hebrides boats go to Oban. No landings are now made at Castlebay where the foundations of the old herring-curing station serve as a reminder to the Barra men of the nights when their fathers could walk across to Vatersay on the decks of the drifters; from the North Channel and Firth of Clyde boats go to Ayr, Campbeltown and Tarbert (Loch Fyne). The Herring Industry Board have a processing plant at Stornoway, and at Mallaig there is a large kippering station.

Where there is a tradition of fishing on the West Highland coast, there is seasonal activity which affects the inshore natural history. Lobsters (*Homarus vulgaris*), for example are fished at all times of the year on the sheltered coasts. In summer the fishery becomes more intense as the weather improves and boats work at exposed stations as far out as the Monach and Flannan Islands. No British boats at present fish lobsters at St. Kilda and North Rona but smacks from Brittany go there to fish illegally when weather permits. H. J. Thomas (1958) has provided a detailed account of lobster and crab fisheries of Scotland, in which the intensity of fishing, methods used and markets are given for each Fishery District.

Of the 966,000 lobsters landed on average each year in Scotland in 1950-54, 402,000 came from the West Coast. The Oban District, which includes Mallaig and the coasts between, the Small Isles, Mull, Tiree and Coll, had about 30 per cent; the Barra District, which includes the Uists and Benbecula, had about 26 per cent; the Stornoway District, which includes Lewis and Harris, had about 15 per cent, and the other four Districts, Loch Broom, Kyle of Lochalsh, Campbeltown and Ayr had less than eight per cent each. In all Districts except Loch Broom, which extends from Cape Wrath to Loch Torridon, the fishery is most intense between May and October due partly to increased fishing effort and partly to increased activity of the lobster itself. On the Sutherland and Wester Ross coast more lobsters are landed in November, December and January than in other months of the year, and in Lewis and Harris there are higher landings from August to December than from January to July.

Dr. Thomas states that harbour facilities, anchorages and beaches dictate the type of craft used for lobster fishing, and this in turn determines the methods of fishing. The commonest trap is the double, open-eyed, braided creel, and parlour traps are used locally. Mechanical creel-haulers are uncommon in the West, where the crofter-fishermen are practised at the accurate siting of individual creels. Tagging experiments show that lobsters do not move far and this is important in the movement of creels within an area and the seasonal fishing of the grounds. Lobsters are usually stored in boxes floated a short distance off-shore in a sheltered creek or sunk if there is a risk of freshwater overlying saltwater near estuaries. After a suitable number have been collected they are despatched by the fishermen.

The storage and marketing problems are severely limiting in many localities. The winter fishery in the Loch Broom District, for example, is due to heavy transit losses which make a summer fishery unprofitable. Marketing organisations in Kyle and Oban have stimulated the fishery with profits for the fishermen. Those in the Outer Hebrides had summer transit problems equal to those of Wester Ross and Sutherland until a buying organisation established depots at Bernera (Lewis), Rodel, Lochboisdale and Lochmaddy. Experimental marketing by air has been done by fishermen from Benbecula, South Uist and Eriskay.

Where transport difficulties are most acute, fishermen store the lobsters in small ponds (dammed-off sea inlets) to await favourable market and transport conditions. There are large ponds at Bernera (Lewis), Stockinish (Harris) and Cullipool (Luing). Ponding takes place in summer and early autumn when prices are low and transit from the Outer Isles difficult ; the ponds are refished during the late autumn with baited scoops.

In all areas of the west coast and particularly in the crofting districts, localities adjacent to the anchorage are more intensively fished than distant reaches of the coast ; most inshore grounds are exploited while off-shore ones remain almost unfished. On the ocean end of the Sound of Harris, for example, the North Uist and South Harris men fish the coasts of the Monachs, Pabbay, Shillay and Coppay until winter sets in, but seldom go to Haskeir, the Taransay Glorigs and Gasker where there are good grounds but no shelter. Without a considerably increased demand, most of the exposed coasts of the west are unlikely to attract sufficient exploitation to give an optimum level of fishing in relation to stocks.

The edible crab (*Cancer pagarus*) is fished off the rocky coasts on pockets of shingle and sand between the rocks, at depths of more than ten fathoms. Due to poor communications in the West Highlands and Islands, the Scottish crab fisheries are mostly confined to the east coast. In recent years, however, a crab fishery has developed in the Skye and the Small Isles district based on the ports of Kyle and Mallaig.

R. H. Millar (1961) in his report of the Scottish Oyster Investigation 1946-58 summarizes the historic evidence of the distribution of the oyster (*Ostrea edulis*) in Scottish waters. Oysters were present in the Outer Hebrides from Lewis to South Uist. In the Sound of Scalpay, Skye, the spring tide sometimes uncovered "fifteen, sometimes twenty, horse load of them on the sand". They were found in Mull in Lochs Scridain, na Keal, Spelvie and at Ulva. There is a record from Kilchoman in Islay with other sites in Argyll in Lochs Creran, Feochan, Melfort, Caolisport,

Sween, Craignish, West Loch Tarbet and in Morven. Sutherland records are from Durness, Eddrachillis, Lochs Eriboll, Inchard, Laxford, Glendhu and Glencoul. There were also beds in Gairloch, Lochewe, Little Loch Broom and Loch Torridon.

Oysters are now much rarer but comparisons with the past are difficult due mainly to the numerous scattered and unrecorded introductions of oysters from England, France and Holland. They are known to exist in Lewis, North Uist, Skye, Mull and on various parts of the mainland from Sutherland to Argyll. In some of these places oysters have in living memory, become scarce or have disappeared altogether. Fisheries were present in West Loch Tarbert, Skye, Loch Sween and in the Cromarty and Moray Firths.

Dr. Millar has carried out experiments in growth and fattening of oysters in West Loch Tarbert, Lochs Craignish, Gair, Creran, Melfort, Feochan, Don, Striven, in Balvicar Bay, Clachan Seil, Lewis (five places), Soay, Lochs Torridon and Ewe and in the Inverness Firth. Breeding experiments were carried out at Loch Sween, Easdale and West Loch Tarbert.

Several localities proved satisfactory for growing relaid French oysters and further survey would reveal more, but fewer places have conditions suitable for breeding, such as Loch Ryan, Linne Mhuirich in Loch Sween and probably Little Loch Roag in Lewis. Loch Ryan provided the best site with a combination of good growth and breeding conditions. It is concluded that with careful selection of sites, proper management and assistance from fishery scientists it should be possible to establish a sound commercial oyster fishery in Scotland. The commercial fishery was, in fact, revived in Loch Sween in 1963.

The salmon fishing industry is important in Scotland. In recent years between 1,500 and 2,000 tons of salmon, grilse and sea trout at a market value of usually over £1,000,000 have been caught. Between 15 and 20 per cent of the catch was usually taken by rod and line, and the remainder by sweep nets in estuaries and fixed nets on the coasts. The influence of the industry on the salmon (*Salmo salar*) is considerable, but had it not been a game fish—a human classification in itself—the effects of the sea fishery would probably have been much more devastating on the species than it has been up till now. Primarily the salmon was the king's fish and salmon rights were in the gift of the crown. The rights of salmon fishing are owned and jealously guarded. The right to net salmon in fixed nets off-shore is let to fishermen by the proprietor on condition that the nets are made unfishable from 12 noon on Saturday to 6 a.m. on Monday. These 42 hours out of the 168 in the week mean that for over 25 per cent of the time the salmon are in the shallow water seeking the river they are free from danger as far as human beings are concerned.

The numbers of nets are also closely limited so that there should remain an ample number of salmon having the chance to run up the rivers. Now once the salmon, grilse and sea trout are beyond the tidal waters, they are not fished by methods which would give the maximum catch. In fresh water they are angled by rod and line and lure, a method which allows the vast majority of fish that get into the river to run up it. Even then, no salmon are fished on Sunday, so for one-seventh of each week and for the hours of darkness the salmon is free from human danger. Salmon poaching is unfortunately common and one's feeling of repulsion at the practice need not be bound up with class distinction and notions of what are private rights, but with the necessity for an easily-caught species to be given its free time. The men who poach salmon with a splash net, or explosives or cyanide are breaking a law for their own ends which society as a whole has made for the benefit of the salmon. Each year well over 100 persons (209 in 1960) are proceeded against for offences under the Salmon and Freshwater Fisheries (Protection) (Scotland) Act, 1951. Population studies on salmon and trout are discussed in chapter 12.

The greater part of the sea fishery is located on the east coast, but stations are scattered along the entire west coast, usually with bag nets fished with a coble. There are stations in the track of salmon migrating to the sea-lochs and river mouths usually on the prominent headlands such as Rhu Stoer, Rhu Coigach, Cailleach Point, Greenstone Point and Ardnamurchan. There are also salmon fisheries on the coasts of Mull near Tobermory and Lochbuie, Skye near Staffin and the Firth of Lorne. The fishing is strictly a summer one. Since 1960 drift-netting for salmon in the sea has occurred mainly on the east coast. Under the recent Sea Fish Industry Act, 1962, it is now illegal to take salmon with drift nets between 15th September and 15th February.

A common factor to both east and west coast fisheries is the damage to nets and fish by both common and grey seals. Much research is now in progress with a view to cutting down damage of this sort by modifying the design of nets, permitting the netsmen to shoot both grey and common seals in their locality all year round and by learning more about both seals and salmon with a view to the conservation of both stocks.

One of the greatest disturbing factors in the freshwater fisheries in recent years has been the development of hydro-electricity. The North of Scotland Hydro-Electric Board have been alive to this from the outset, however, and have installed the best types of fish passes together with instruments for counting the salmon and grilse in seasonal passage in the rivers. The effects of hydro-electricity together with all the others which the salmon must endure on its long and arduous migrations from the minnow reach to the distant ocean feeding grounds, and back, may in the long-term be enough to destroy for ever the present stock of Scottish salmon. Another hazard is the pollution of fresh waters by toxic chemicals from new sheep-dips, herbicides and industrial effluents. If plans for the rehabilitation of the Highlands with light industry come to be, there will be an increasing danger to the already much meddled-with lochs and rivers.

The ethics of sport is a subject on which we may argue without end, but it is not our main concern here. We have to consider the influence of sport as part of the human factor operating on the natural history of the Highlands. It has been a depressing factor on many species such as most birds of prey, herons and saw-billed ducks, foxes, otters and the smaller carnivorous animals, but we must not neglect the fact that the red deer owes its survival to sport, the salmon probably as well, and with the assignment of large areas of wild country to sport, certain animals have had a better chance of survival. The golden eagle is an example: it is accepted and even protected in the deer forests, though it is often shot out on sheep farms and grouse moors. Sport in the past has not taken the form of exploitation of game stocks, as for example has depleted the stocks of white fish in the North Sea. One cannot be so certain of the future, however, because sport in the Highlands has become syndicated and commercialised with a consequent loss of personal responsibility.

The effect of reafforestation, by the method of close stands of conifers with occasional rides kept clear of undergrowth, has yet to be seen in the long term. We believe that with a little give and take, planting of fringes with berry-bearing bushes and keeping the unplantable ground free of sheep and deer, retaining as much scrub woodland as possible within the forest enclosures, seeding wet ground with alder and allowing for a long rotation of hardwoods and conifers, our forestry policy could have a buffering and even stimulating effect on much of our wildlife, as distinct from the certain depressing effect of sheep farming. Unfortunately, forestry policy in Great Britain in the past does not appear to have been influenced by the ecological outlook, and by some skilled observers is thought to have been a hundred years behind that of France. This policy may have been geared to the present needs of the country as a whole, but it is to the needs of the Highlands that we must look if they are not progressively to become more of a desert than they now are.

Good ecological research puts us in possession of knowledge which enables us to control the country's wild life, preventing extinctions and undue increases in numbers, and conserving natural resources of minerals, soil, vegetation and animals. It will soon be for the British public to decide whether the knowledge gained shall be applied and further knowledge gained, or whether some other poor natural historian a century hence will have to write another chapter more generally dismal than this one.

Chapter Five

The Deer Forest, Grouse Moor and Sheep Farm

The fact of two and a half million acres of deer forest in the Highlands has been a matter for congratulation or for execration, dependent on whether the comments come from a sportsman, a naturalist, a sheep farmer or crofter or a member of a political party opposed to the existence of what is considered an anomaly in modern times. Our approach is that of the biologist who wants to know truth and how far the deer forest fits into the natural history of the Highlands and Islands. The rise of the deer forests in the middle years of last century has been debited wrongly with the Clearances which were mostly conducted half a century earlier when the sheep farmers came. It is equally wrong to work up a personal animus against a very beautiful and useful animal because its hunting has been confined to a certain class of society.

It may be said that 90 per cent of the 2,500,000 acres of deer forest (now reduced by refforestation) is among the poorest and roughest grazing in Scotland. For the remaining 10 per cent there is little excuse for its remaining as deer forest and nothing else. But in fact, and especially since the outbreak of the second world war, much of the total ground styled deer forest has also helped to graze some sheep and cattle. The remainder is of such a nature that it will not carry domestic stock economically. Much of the ground is over 2,000 feet high, with a lot of bare rock and scree, difficult to travel through and without much wintering. On such ground there is no animal, wild or domesticated, better adapted than the red deer to use it and to render a crop to humanity.

The red deer is a thriving, colonizing species and needs

to be kept in check or it would soon be all over Scotland again, making the pursuit of arable farming almost impossible ; but having said that, let us value our Highland red deer. There has been a rise of shepherds' wages to about £700 a year. Nobody should be more thankful than the shepherds themselves for this change of financial status. But it should be remembered that even before the second world war when shepherds were getting the poor wage of £70-80 a year cash, with certain perquisites, some of the rougher ground was going out of sheep and reverting to deer. There are three main reasons why ground may go out of hill sheep farming: (a) it may be too high and therefore short of winter grazing ; (b) it may be so rough and broken that shepherding is altogether too difficult for one man to manage a full hirsel, i.e. about 600 ewes ; (c) the ground may be so poor that it will carry no more than a sheep to 10 acres, when shepherding a full hirsel again becomes almost impossible for one man. Unless one man can manage 500-600 ewes with a little extra help at lambing time, shearing and dipping, hill sheep farming is uneconomic. It is almost certain that the progressive rise in shepherds' wages will put the poorest, highest and roughest of Highland ground out of sheep farming. The only other animal able to use such ground is the red deer.

By far the largest stretches of deer forest are north and west of the Great Glen, where it is possible to walk over a length of a hundred miles of continuous deer forest with the only fences along the railways. The north-western stretch of deer forest has for the most part a different vegetational complex from the Central Highland region where the deer forests are on ground so high and snow-bound that it is little use to sheep in winter. As has been pointed out in an earlier chapter, the deer forest country of the north-west is grassy and sedgy rather than heathery. Large expanses of thriving heather or ling (*Calluna vulgaris*) are not common although poor heather is widespread ; bell heather (*Erica cinerea*) grows in tufts on the drier spots and among cairns and rocks ; but most of the low ground is covered with purple moor grass or flying bent (*Molinia caerulea*), bog myrtle (*Myrica gale*), a variety of sedges and rushes, and cross-leaved heath (*Erica tetralix*). Throughout the lower ground there are patches of birch, alder, and willow scrub, and many plantations of conifers in various states of careful fostering or neglect, as well as solitary rowans.

The flowers of this lower zone are predominantly milkwort (*Polygala serpyllifolia*), tormentil (*Potentilla erecta*), bedstraw (*Galium saxatile*), wood rush (*Luzula* spp.), with yarrow (*Achillea millefolium*) and bird's-foot trefoil (*Lotus corniculatus*) on dry places and old crofting lands. Wild white clover (*Trifolium repens*) also occurs freely among sweet vernal grass (*Anthoxanthum odoratum*), wavy hair grass (*Deschampsia flexuosa*) and bents (*Agrostis* spp.) in favoured places and on roadsides. All these serve in greater or lesser degree as food plants. Plants occasionally eaten by the deer and generally present are soft rush (*Juncus effusus*), furze or gorse (*Ulex europaeus*) on the drier places, bramble (*Rubus* spp.) on edges of woods, withered nettle (*Urtica dioica*) at past human habitations, ragwort (*Senecio jacobaea*) on river flats, coltsfoot (*Tussilago farfara*) on gravels near rivers, dog and marsh violets (*Viola* spp.) and wood sorrel (*Oxalis acetosella*).

The plants obviously present but not normally eaten include primrose (*Primula vulgaris*) in sheltered situations and woods, devil's-bit scabious (*Succisa pratensis*) on the open bog, silverweed (*Potentilla anserina*) on river flats and shieling sites, spotted, butterfly and fragrant orchis (*Orchis ericetorum, Platanthera bifolia* and *Gymnadenia conopsea* respectively) generally in the herbage, marsh thistle (*Cirsium palustre*) in woods, bracken (*Pteridium aquilinum*) generally spreading over drier places, bog myrtle and bog asphodel (*Narthecium ossifragum*) in the wetter bog. Species of eyebright (*Euphrasia officinalis* agg.) are widely distributed in dry and wet places alike. Ferns of various species occur in birch woods and some such as polypody (*Polypodium vulgare*) are epiphytic on the trunks, along with many mosses and lichens. These are mentioned later in the discussion of woodland communities.

The herbage floor which provides the main winter food of the deer lies between 250 to 1,750 feet. Many plants noted already as occurring on the low ground extend upwards as long as the conditions for them remain. But there is a big increase in the sedges such as stiff sedge (*Carex bigelowii*), ribbed sedge (*C. binervis*), carnation sedge (*C. panicea*), flea sedge (*C. pulicaris*) star sedge (*C. echinata*), heath rush (*Juncus squarrosus*), deer's hair grass (*Trichophorum caespitosum*) and bog cotton (*Eriophorum vaginatum* and *E. angustifolium*). The cotton grass is thinly distributed through much of the bog, but it occurs as the main constituent of certain flats of deep peat which are wet but not too wet. The cotton grass or drawmoss comes earlier than anything else in spring so that it is usual to see both sheep and deer grazing on it in February. Sedges, soft rush, club moss (*Lycopodium selago*) and such lichens as *Cladonia* spp. and *Cetraria* spp. provide good grazing in a country where good grass is not to be had. The better flats have a variety of sedges including bottle sedge (*Carex rostrata*), mud sedge (*C. limosa*), common sedge (*C. nigra*) and carnation grass (*C. flacca*).

This main mass of the hill also grows a quantity of wavy hair grass, tufted hair grass (*Deschampsia caespitosa*) and heath grass (*Sieglingia decumbens*). Milkwort and tormentil continue to be common and wild thyme (*Thymus drucei*) is locally abundant. Blaeberry (*Vaccinium myrtillus*) and crowberry (*Empetrum* spp.) are fairly generally distributed throughout this zone but only become prominent in dry places.

The high ground above 1,750 feet varies in its vegetational composition depending on the geological formation and the land form. As has been mentioned when describing the gneiss areas, the characteristics of the lower slopes may be carried well up to 2,000 feet and more. But where the peat ceases at about 1,750 feet, the herbage becomes richer in proteins and minerals and is grazed whenever there is sufficient growth and when the weather allows.

The grasses are sheep's fescues (*Festuca ovina* agg.) bent grasses, sweet vernal grass, mat grass (*Nardus stricta*) and wavy hair grass. They occur with sedges and woodrushes (*Luzula* spp.) in a thin carpet of dwarf shrubs such as heather, blaeberry, crowberry and dwarf azalea, on wind-swept shoulders and ridges. Wild thyme goes to well over 3,000 feet. The mossy cyphel (*Cherleria sedoides*) is to be found among the herbage at higher levels and is well eaten down by the deer. The cushion moss *Grimmia* is common on the boulders. It is after a sudden melting, which can be quite spectacular in the maritime west, that the deer go high to graze the newly-washed herbage. This habit of the deer to go high in winter whenever weather allows and to get down again in plenty of time before it changes, is but one example of this animal's efficiency in its total environment.

In *Plant Communities of the Scottish Highlands,* McVean and Ratcliffe (1962), following the pioneer work by Watt and Jones (1948) in the Cairngorms and Poore (1954) in Breadalbane, describe the vegetation of the deer forests and sheep runs as grass heaths, grasslands and bogs, extending from the forest zone to the summits of the mountains. Those which occupy the ground above 2,000 feet are termed montane and are described in chapter 8 of our book. This chapter deals with the sub-alpine grass-lands upon which the economy of Highland sheep and cattle farming is based, and upon which the red deer depends for most of its food. These vary greatly in their floristic make-up and in their nutritional value.

The variation in water content and acidity of the soil is reflected in the presence or absence of particular grasses, sedges and rushes, a few of which assume a dominant role. Within the mosaic, dry ground is dominated by bent grasses and fescues, the wet by mat grass and the very wet by the soft rush. The tufted hair grass, which is affected to a greater extent than others by snow-lie, is also some-times dominant, but its distribution does not fit so neatly the pattern of soil conditions as do the others.

The fact that the tufted hair grass does not fit into the general scheme of things may point to its origins being different from those of the others. The expanses of bents, fescues, mat grass and rushes are man made, but the tufted hair grass may owe more to winter snow than to man's activities. The whole complex has, however, been fashioned by man in deforestation, intense grazing with continued heavy extraction of nutrients and by muirburn over the last two hundred years or more. There are species-poor and species-rich types of the various associations; those on the acid ground are usually poor and those on the basic soils usually rich.

The species-poor bent-fescue grassland, which extends from about 100 to 2,700 feet, has brown bent grass (*Agrostis canina*), common bent grass (*A. tenuis*), sweet vernal grass and sheep's fescue (*Festuca ovina*) as dominants with the heath bedstraw, tormentil, violet, mat grass, pill-headed sedge (*Carex pilulifera*) field woodrush (*Luzula campestris*), many-headed woodrush (*L. multiflora*), and the mosses *Rhytidiadelphus squarrosus, Hylocomium splendens, Pleurozium schreberi* and *Thuidium tamariscinum* also present. This close-cropped grassland is widespread in the Highlands, but is found mostly in the drier south-east extending in tracts covering many hundreds of acres on the steeper hillsides. In the north-west it is usually restricted to patches of a few hundred square yards on well drained alluvial flats. As the soils become wetter on the more gentle slopes this bent-fescue community grades into one dominated by mat grass.

In widely separated parts of the Highlands in Breadalbane, Clova and Drumochter, to mention the more important, there is a bent-fescue grassland between 1,100 and 3,300 feet which is rich in alpine lady's mantle (*Alchemilla alpina*). This community is found on the calcareous schists and is one of three types of species-rich bent-fescue communities. In addition to the species already mentioned these communities contain self heal (*Prunella vulgaris*), meadow buttercup (*Ranunculus acris*), thyme, red fescue (*Festuca rubra*), lesser club moss (*Selaginella selaginoides*), carnation sedge (*Carex panicea*), flea sedge

(*C. pulicaris*), purging flax (*Linum catharticum*), *Polygonum viviparum* and yellow mountain saxifrage (*Saxifraga aizoides*).

In the sub-alpine mat grassland the community is similar to the bent-fescue grassland except that mat grass is dominant over the others. The mat grasslands are well developed on moderate slopes in the Central Highlands. The altitudinal range is from 1,000 to 2,300 feet. On wet slopes in the Eastern Highlands heather moor is replaced by mat grassland. The richest examples occur on calcareous mountains such as Ben Lawers. Where the ground becomes really boggy the dominance changes with the soft rush replacing the mat grass. As might be expected, most of the soft rush bogs are found in the wetter northwest, usually on gentle slopes between 1,500 and 2,600 feet. Extensive mixtures of mat grassland and soft rush bog occur on Ben Wyvis and at the head of Glen Fiadh (Affric).

The tufted hair grass dominates many acres between 1,600 and 3,000 feet. This grassland is more montane than man-made, associated with a wet climate and late snowlie, and may contain the rare grass *Phleum commutatum*. Sometimes the tufted hair grass becomes overwhelmingly dominant at the expense of the other members of the association and relatively pure grasslands of this species occur around perennial springs of icy water in the Cairngorms and on the high plateaux of Ben Alder, Creag Meagaidh and Ben Nevis. On Beinn Bhan in Applecross there is a large ledge inaccessible to grazing animals holding a luxuriant growth of ferns, blaeberry and greater woodrush (*Luzula sylvatica*) on unstable scree. This is one of the numerous ledge sites in the Highlands which are instructive in the effects of grazing and burning. The community on this particular ledge is also characteristic of extensive areas on Ben Loyal.

To get a sharp impression of the difference between sheep farm and deer forest the naturalist may compare the hills of Morven, which are green to the summits and are sheep farms with deer on the ground in moderate

density, with hills like Arkle or Suilven in Sutherland, or
the upper reaches of the Cairngorms. The deer forests of
the Grampian or Central Highland zone are much more
heathery in character because of lower rainfall and better
drainage. The main reason for their unsuitability as
sheep farms is their lack of winter grazing. Nevertheless,
many of these forests do carry a summer sheep stock. In
the Central and Eastern Highlands there is more wintering for sheep on heather moorlands. In the East the ground
is managed for sheep, deer and grouse with emphasis on
grouse, and the distribution and numbers of grouse depend
on the coverage and quality of heather.

The heather moor is essentially a sub-alpine dwarf
shrub heath ; it may extend upwards from near sea level
to almost 2,500 feet, but is chiefly situated between 500 and
1,500 feet. The wine-red moors have become emblem-
atical of the wild romance of Scotland, so much so that
they are almost accepted as native to the Highlands and
Islands. The plant sociologist, however, declares that the
heather moor is almost a complete artifact, brought about
by the clear felling of the native forest which occupied
the ground before the heather and by subsequent grazing
and burning. None of the Scottish bards knew the native
forest ; it was before their time. This, however, does not
detract from the serenity of the Central and Eastern Highlands during the heather bloom in August. It is sad that
man's hand cannot be withheld and that the wonderful
tracts of beautiful and fruitful heather cannot be preserved
for ever, but the heather coverage is shrinking in most districts and being replaced by less fruitful grasslands.

The heather moor occurs in mosaic with the bent-fescue,
mat grass and soft-rush grasslands and the frontier between
heather and grass is usually sharp and man-made. Some
of these show the limits of burning, draining, cultivation,
grassland regeneration schemes, tree felling or peat cutting. The dry heather moor is substantially a monoculture
of ling (*Calluna vulgaris*) with an association of mosses
including *Dicranum scoparium, Hylocomium splendens,
Hypnum cupressiforme* and *Pleurozium schreberi,* and the

lichen *Cladonia impexa*. Bell heather (*Erica cinerea*) and cowberry (*Vaccinium vitis-idaea*) share dominance locally with the ling (fig. 7). In the Eastern Highlands, mostly in Speyside and Deeside, a heather moor rich in bearberry (*Arctostaphylos uva-ursi*) occurs, and in it are found also wavy hair grass, bitter vetch (*Lathyrus montanus*), intermediate wintergreen (*Pyrola media*), petty whin (*Genista anglica*), bird's-foot trefoil (*Lotus corniculatus*) and violet (*Viola riviniana*). In localities of poor drainage and deep peat, the heather often has a heavy admixture of deer grass or cotton grass.

In the damp heather moors the bearberry among the ling is replaced by crowberry and blaeberry. In the northwest these moors are rich in oceanic liverworts, and in the east have cloudberry (*Rubus chamaemorus*), dwarf cornel (*Chamaepericlymenum suecicum*) and the cotton grass (*Eriophorum vaginatum*), all of which are absent or rare in such moors in the west. *Sphagnum nemoreum* may dominate the moss layer. This community is somewhat similar to the heather-blaeberry community of the pine forest which is mentioned later.

The long-term effects of moor burning are usually completely ignored by landowners which is surprising, since it is in them more than in any other section of society that the inheritance of the land is vested. It is a traditional practice, and tradition dies hard in the conservative mind. Many do not take enough personal interest in the longterm consequences of muirburn as it will affect the future economy of the farms and estates. Ronald Elliot (1959) has shown that the amount of minerals in the drainage waters from moors is doubled immediately after burning, and amounts to more than one per cent of the total amount available. This is not significant at one burning but over the space of 200 years it might be serious. Dr. Elliot suggests that in such situations there should be a re-establishment of woodland as a gradual means of restoring fertility. This is simply large-scale rotational land use.

Charles Gimingham (1959) describes the growth of heather in five stages over a period of thirty years: three

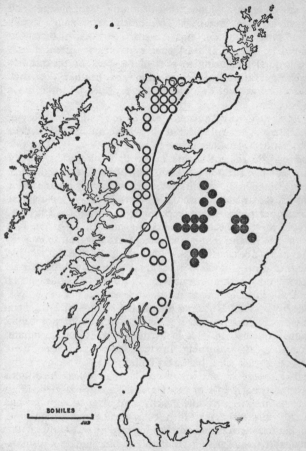

FIG. 7. Heather moors: the black dots are the Callunetum vulgaris (mainly ling) and the circles are the Arctostaphyleto-Callunetum (bearberry and ling) showing the distribution of the two types of moor to the east and west of a line AB. Many more sites could be added to those shown

From McVean and Ratcliffe, 1962 by permission of the Controller of H.M. Stationery Office

to six years as a pioneer, ten in building, ten in maturity and five in degeneration. This timetable probably varies from moor to moor but it seems essential that moors should be burned at least once every ten to fifteen years, the general aim being to prevent the bulk of the heather reaching maturity. If regeneration of heather does not take place without delay it may be replaced by moor grass or bracken.

The practice of muirburn can be a blessing or a curse, and there is no rule of thumb which can be followed in relation to it. Heather is one thing in the Southern Uplands and on the eastern moors of Scotland, but it is a very different plant through a large part of the West Highlands. It grows so well, and is such a close carpet on the drier moors, that regular burning is necessary if a good crop of young heather is to be kept going for sheep and grouse. Great care is practised in burning and all leases contain clauses to ensure that sufficient men are engaged to control the fire; notice must also be given to the laird. The heather crop is so important that no one, whether sheep farmer or sporting tenant, wishes to burn out of season or badly or in such manner that the fire will get out of control. But in the West Highland hills where sedge and flying bent are much denser on the ground, a hill ought not to be burnt oftener than once in twenty years. It is unfortunate, therefore, that burning in the West and North-West Highlands is such a haphazard affair.

The theory is that where burning has taken place there comes an early bite of moor grass. Those early stems may appear in extremely low density, and after the ground has been punished by the sheep it lies derelict for a long time, giving neither food nor shelter for bird or beast. By far the biggest trouble arising from this indiscriminate overburning in the west is the subsequent spread of bracken, which grows strongest where it is free from competition; and in the west its rhizomes are but little affected by frost in the winter. The deep rhizomes are also unaffected by the burning and send up the first shoots thereafter. Tormentil is the second plant to appear, or first if bracken is not present.

The careful burning of good heather moors needs a knowledge of what is going to happen. If heather gets too old before it is burnt there is no regeneration except occasionally from seed. If the ground is too dry (which does not happen often) the heat may be so fierce as to kill the seeds and again there is no regeneration. Heather on a sheep farm has to be burnt in sufficiently large pieces otherwise the animals congregate on the newly burnt ground densely enough to injure and retard new growth. Further, such heavy densities of sheep bring about undue worm infestation in the flock and there is serious trouble. Much of a shepherd's time is taken up with keeping his animals well dispersed, and it is one of the fortunate natural tendencies of the Blackface breed to disperse in grazing instead of working the ground as a flock. Burning not only prevents regeneration, but aids the establishment of deer's hair grass. In cases of neglected moors it is often not a question of what is the best method of burning, but of whether the ground should remain as moor or be returned as quickly as possible to forest.

Heather burning done properly preserves a *status quo* which would otherwise be a temporary phase in the natural succession. Persistence of a dense growth of *Calluna* is assured and colonization by, or regeneration of, trees is prevented. Ground-nesting birds are deprived of cover, but such as do nest on burnt ground are not seriously put out, because moor burning finishes on April 15, or on special occasions with written consent of the landowner it may be continued until April 30. Here again burning in the west falls short of good traditions, for fires may be seen well after the closing date and into May; burning then becomes plain vandalism.

The life history of the red grouse (*Lagopus s. scoticus*) is that of a bird excelling in hardiness. Surely the ptarmigan (*Lagopus mutus*) is the only other bird which can exist in more unsheltered and inclement conditions. When the snow falls heavily and without wind, the grouse will move elsewhere, but if there is sufficient wind to keep patches of heather and grit clear of snow, the grouse will stay where they are and endure the weather. The birds move upwards immediately the snow begins to melt.

Breeding time on a grouse moor is full of chances of failure. The birds and their eggs have certainly evolved characteristics which minimize loss. For example, a snow-storm is a common thing during April and there is even the "cuckoo" snow of May with some very cold nights; the hen birds are laying at this time and they do not sit until the clutch of 5-15 eggs is complete. This means that the eggs may be quite severely frosted, yet experience seems to show that they rarely fail to hatch from this cause. Some keepers say that the eggs of grouse and ptarmigan will not split with frost as soon as those of other game birds.

The hen grouse at all times sits close and may herself be buried for a day or two under snow, but in winter grouse as a species do not get snowed up as sheep do under a peat hag. The birds tread with their feet all the time and rise with the snow. Hoodie crows are the greatest menace to sitting grouse in spring. These sharp-eyed birds quarter the moor ready to suck eggs. Usually, the cock and hen grouse between them are pugnacious enough to repulse the crows during incubation but when the chicks are hatched it is difficult for the hen to keep the brood close together.

The red grouse of the Outer Hebrides are assigned to the Irish sub-species *Lagopus scoticus hibernicus*. The plumage differences are extremely slight, and then only in winter, but Campbell (Witherby *et al.* 1940) notes several differences of behaviour. They are more silent than the typical form, they are very unobtrusive and do not flush and swing away for a long distance in the characteristic fashion of *L. s. scoticus*.

The food of the grouse is varied, but the shoots of young heather are the mainstay of its diet. As sheep favour the same food, the management of heather moors is the most profitable type of husbandry of non-arable ground in the Highlands. *Calluna* is grazed throughout the year—in winter, the green shoots growing in the shelter of the canopy; after May, the new green shoots of the canopy; in August and September, the flowers in bud, in bloom and in new seed head; and throughout the autumn the seed capsules are picked from the plant. After January, when the seeds have fallen, they appear to have no significance for the birds. The shoots, stalks and leaves of blaeberry are also eaten, and the berries which appear in season on the herbage floor, such as crowberry in June, blaeberry a little later, then the bearberry and in July-September the cloudberry of the high bogs. The young flowers of cotton grass and the seeds of woodrush are also taken. Thus we see that the grouse must be grouped with the sheep and the deer as grazing animals.

The grouse population of the north-western deer forest country is not nearly as large as on the true heathery hills of the east. The sport of deer-stalking makes no effort to preserve grouse, because the sudden uprush of one or more of these birds in the course of a stalk can move the deer well beyond reach; as a result, the eagle, peregrine falcon and wild cat breed in fair security. The fox is as much hunted in the deer forest as on the sheep farm because deer calves are preyed upon as much as are lambs.

In Wester Ross the preferred wild prey of eagle and fox is scarce but there may be as many as one pair of eagles and two pairs of foxes breeding in every 16,000 acres of deer forest (Lockie, 1963). Both take similar food. In winter foxes eat 63 per cent carrion (deer and sheep), 26 per cent vole and the remainder birds, hares and shrews; in summer foxes eat 32 per cent deer calf, 31 per cent vole and 15 per cent lamb. In winter eagles eat 38 per cent carrion, 32 per cent hare and 30 per cent bird; in summer eagles eat 60 per cent birds and 30 per cent lamb. The deer calf and lamb figures include carrion. The percentages are calculated from contents of castings and droppings. Watson (1955) found foxes in the east Cairngorms eating ptarmigan, red deer, blue hare and rabbit in winter.

The deer forest, then, tends to be a greater repository of wild life than the sheep farm and grouse moor, not only through the toleration already mentioned, but because it can allow and even encourage a greater variety of habitats. Birch woods remain, and the ground is not burnt anything like so frequently as it is on sheep farm and grouse moor. Deer come into long heather when the snow is new and deep. They shake the stalks so that the snow falls off and the green top is then browsed. It is no uncommon thing to see a head and two long ears silhouetted against the snow in a winter's dusk. A hind is lying there protected from the wind and very snug.

In the past six years the grouse has been the subject of special study, and now stands at the centre of a large-scale investigation of moorland ecology in the Highlands. The development of this study has been a process of outward enquiry from the grouse itself to the living web of the moorland, of which the grouse population is but a part. The factors which control the numbers and behaviour of this bird have involved its food, cover, territorial system, predators and reactions to climate and management by man.

The enquiry was started in 1956 and ran for three years with funds provided by the Scottish Landowners' Federation. In 1960 it became the Nature Conservancy's Unit of Grouse and Moorland Ecology, directed from the Department of Natural History, Aberdeen University. Summing up the work of the first three years, Professor V. C. Wynne-Edwards, the Director of the enquiry, states that the important factor which emerged from the trial period was the predominant part the grouse themselves play in regulating their own population-density. Thus, attention focused on the changing productivity of heather as being the primary condition on which grouse appear to adjust their numbers. The factors affecting and controlling the productivity of heather extend considerably beyond the interests of grouse management and make contact with the

work of the Hill Farming Research Organisation and the
Nature Conservancy's wider interests.

The work has been done by David Jenkins, Adam
Watson, Gordon Miller and others, lately in Glen Esk and
now in Deeside. They have overcome many of the problems of theory and practice, not the least of which has
been how best to count grouse and subsequently interpret the
figures. They have been compelled to come to terms with
the natural history of the whole moorland. In their understanding of the grouse they have reached up into the
ptarmigan country and down into the arable land to find
parts of the puzzle.

The results of the enquiry so far, as interpreted by Dr.
Jenkins, show that numbers and breeding success of grouse
depend on the quantity and quality of edible heather available. The cocks take up territories in August-October when
some of the old cocks are ejected and some of the new young
cocks take their place. More aggressive birds occupy larger
territories and barring accidents all territory holders are
still in possession in spring. Those which do not find territories in autumn disperse or die and thus an autumn
decrease in numbers is apparent.

In autumn, while the cocks are staking their claims to
territory, the hens are wandering about in small flocks and
some birds can survive in poorly held territories. As the
winter progresses, however, they gradually pair with the
territorial cocks; all the territories are fiercely defended
and the surplus birds are expelled. In the late winter breeding pairs are formed in the territories. If heather has
survived the winter well and food is abundant, the adults
lay more eggs subsequently and also rear more young. If
there is little food because of poor growth in the previous
summer or because of severe die-back due to winter
browning (Watson, Miller and Green, 1966), the health of
the adults is affected and they breed poorly. These data
on food and breeding were recently published by Miller,
Jenkins and Watson (1967).

Dr. Watson has made a close study of territorial
behaviour of grouse and notes three grades in a hierarchy

of cocks: paired owners of territory, unmated owners and wanderers on communal or stolen ground. Birds intergrade from one step of the hierarchy to the other in the constant jockeying for position in the society. Only healthy birds are able to hold territory and consequently to breed; this is the mechanism of natural selection. Since this edition was prepared several papers have been published by Jenkins, Watson and Miller between 1963 and 1967 and the references are in the selected bibliography.

We now see the gradual piecing together of the life of the grouse and realise that the theory that numbers are regulated by the amount of edible heather is not a simple one. Numbers appear to be regulated by the aggressive behaviour of the birds and the size of territories they defend. This aggressive behaviour is related to the numbers present, and hence to the numbers of young reared in the previous summer. This in turn is related to the food supply available to the young birds' parents in the previous spring.

We should not pass this point without mentioning the black grouse (*Lyrurus tetrix britannicus*) for it is a grazer as well as a browser. It is not so much a bird of the open moor as of thin scrub. It eats the buds and young catkins of birch and the buds of conifers and such other trees as it may find in its habitat. Its influence on limiting the growth of birch scrub may be negligible, but the Forestry Commission finds the bird sufficiently damaging in young plantations of conifers to make some efforts to reduce it. The rest of its food is similar to that of the red grouse. Both species, especially when young, consume many insects and are certainly beneficial to the sheep farmer. The crop of black grouse may be packed at certain seasons with *Bibio lepidus,* a fly, the maggots of which live in the roots of herbage and are destructive to it. The heather beetle (*Lochmaea suturalis*) is also consumed in quantity when present. This latter insect is of great importance in the heather districts of Scotland because it destroys patches of ling, giving it a frosted appearance. Cameron (1943) conducted a study of the life history of the beetle and shows the importance to the species of damp mossy patches in moors, in which one stage of the beetle's life is spent.

The black grouse is probably on the increase in the Highlands generally; yet in many localities it has completely disappeared. It is at present common in the Eastern Highlands but has declined in the west. Coccidiosis has been blamed in the past for the decline of blackcock; but at least three other factors are involved. These are man's use of the land in timber, sheep and deer, the social factor in bird behaviour and the wrong idea of the sportsman in shooting cock birds. The black grouse is a species in which the males gather at traditional places in spring and join in a ceremonial posturing. This dancing ground is called the "lek". It is after these meetings and stimulating evolutions of the male birds in concert that mating takes place with the hens, which are not far away. The formalized postures of attack indulged in between the males at the lek are nevertheless distinct from the actual sexual display which precedes coition. Meetings at the lek occur in March before mating is to be expected, as well as later. Morley (1943) drew attention to the phenomenon of autumnal display in many species of British birds, in which the blackcock is a prominent participant. On the theory of social stimulation to breeding condition outlined by Darling (1938), it may be that the blackcock mates much less successfully if he is unable to join with sufficient of his fellows at the lek in spring. The display of the blackcock is described in detail by Lack (1939) and the lek as a control mechanism of breeding density by Wynne-Edwards (1962).

The complex social system of our British grouse has developed alongside but independently from that of human society in the Highlands and Islands. Thousands of years have passed with the black and red grouse fulfilling their formalized social meetings, certain manifestations of which are taking place in the half light following the dawn, with man observing in part, but not comprehending. The beauty and complexity of this social life is the challenge to research which ecologists have now taken up in the Highlands.

One of the main vegetation types in the deer forests and sheep runs, and one of the most uniform and extensive in the Highlands (with the exception of man-made forests), is blanket bog. The Moor of Rannoch, Claish Moss in Sunart and Strathy Bog in the flow country of north Sutherland are outstanding examples. The vegetation of Highland bogs and mires is described in detail by McVean and Ratcliffe.

The main constituents of the Western blanket bogs are ling, bog myrtle, cotton grass (*Eriophorum vaginatum* and *E. angustifolium*), moor grass (*Molinia caerulea*), deer's hair grass, bog asphodel (*Narthecium ossifragum*), sundew (*Drosera rotundifolia*) and the mosses *Sphagnum papillosum, S. rubellum* and *Hypnum cupressiforme*. The Sphagna are usually dominant and the few-flowered sedge (*Carex pauciflora*), the great sundew (*Drosera anglica*) and some mosses are selective for the blanket bog. Depending on the level of the water table, the vegetation varies from almost pure *Sphagnum* in the wettest places with pool and hummock systems, to drier areas possessing the grasses and ling. The pools may be overgrown with bogbean (*Menyanthes trifoliata*), cotton grass and mud sedge. Rannoch Moor is the only Scottish station of *Scheuchzeria palustris,* which grows in the wet depression in the bog. Just over 3,700 acres of Rannoch Moor are now a National Nature Reserve.

This type of bog is present in the Western Highlands on gentle slopes below 1,500 feet and also extends south-east of the Great Glen into the Monadliadhs, Badenoch and Atholl. In the east, however, the characters of the blanket bogs are generally different, possessing much more ling, crowberry, blaeberry and cloudberry with Sphagnum and cotton grass. Bog myrtle is usually absent. These eastern bogs occur mainly above 1,000 feet and cover square miles of ground on the broad watersheds in the Monadliadhs, Atholl and the hills of Angus. The small cranberry (*Oxycoccus microcarpus*), the dwarf birch (*Betula nana*) and the black bearberry also occur locally and more lichens are present than in western blanket bogs. Above 2,500 feet on the high plateaux of the Cairngorms and on Lochnagar there is high-level blanket bog in which crowberry replaces ling as a co-dominant with cotton grass.

These blanket bogs occur in mosaic with the grasslands and the dwarf shrub heaths both in the alpine and subalpine zones, and the dominance of the various species is strong or weak depending on ecological conditions. Widespread throughout the Highlands there are bogs dominated by deer's hair grass and ling. In the west this association occurs up to 2,000 feet and in the Cairngorms to about 3,000 feet. Also widespread, especially in the west where it covers many square miles of gently sloping lower hillsides, is the heather-moor grass bog, with the change in dominance from cotton grass or deer's hair grass to moor grass, with the ling.

The vegetation of the blanket bog receives its vital substances from the atmosphere; it is cut from the mineral soil by a deep layer of peat and it receives no drainage water from higher ground. The input of nutrients comes from the rain, dust and animal debris. The vegetation of other valley-bogs does have access to the soil. In these mires the following species are locally dominant: deer grass, cotton grass, moor grass, bog myrtle, rushes (*Juncus effusus, J. acutiflorus*), sedges (*Carex echinata, C. panicea, C. nigra, C. aquatilis, C. curta, C. rariflora, C. pulicaris, C. rostrata,* and *C. saxatilis*), bog rush (*Schoenus nigricans*) and the mosses *Sphagnum* spp., *Arcocladium cuspidatum* and *Campylium stellatum*.

The birds of the deer forest, hill sheep farm and grouse moor are distinctive if small in numbers of species. Commonest bird of all is the meadow-pipit (*Anthus pratensis*) whose small, thin song and parachuting display flight are linked inevitably in one's mind with this country in April and May. Throughout the west the twite (*Carduelis flavirostris*) also an inhabitant of the long heather, is not averse to coming into the settlements and lonely gardens of stalkers, keepers and farmers, where it is much to be preferred to the absent sparrow (*Passer domesticus*). Twite and sparrow may meet and mix, however, on the border of their habitats, e.g. in the streets of Ullapool.

The wren (*Troglodytes troglodytes*) is a surprisingly common bird in a particular type of deer forest country in the west ; that which is steep and rocky, with long heather and occasional birches and rowans. In such ground the wren may be found up to 1,250 feet. In the Cairngorms its habitat is less rocky and it is found up to 1,700 feet. On hillsides with bog myrtle, at moor-edges and roadsides the whinchat (*Saxicola rubetra*) is locally common nesting often in gorse or long heather. A little bird sitting on a telegraph pole or wire in a Highland glen is likely to be a whinchat or a stonechat (*Saxicola torquata*). The wheatear (*Oenanthe oenanthe*), skylark (*Alauda arvensis*), pied wagtail (*Motacilla alba*), reed bunting (*Emberiza schoeniclus*) and in winter the snow bunting (*Plectrophenax nivalis*) are all characteristic birds of the moors.

The meadow-pipit is important food of the merlin (*Falco columbarius*) (plate v), while that falcon is present in summer. The merlin is, however, not a common bird, and certainly not as numerous in the Highlands as the buzzard (*Buteo buteo*). The main food of the short-eared owl (*Asio flammeus*), which inhabits this terrain, is small rodents, and this is discussed more fully in connection with vole plagues in chapter 7.

The snipe (*Capella gallinago*) is fairly common on the river flats and low bogs, but is not a bird of the heather or deer grass flow. The common sandpiper (*Tringa hypoleucos*) and other waders, ducks and gulls are so closely associated with streams and lochs that they are dealt with in chapter 12 on freshwaters. It is here, however, that we must mention the greenshank (*Tringa nebularia*), which is really a bird of the sub-arctic forest marshes, but in Scotland mostly nests in the treeless deer forests of the North-West Highlands.

Desmond Nethersole-Thompson states (pers. comm.) that there are more breeding pairs in Sutherland than in any other county. In north and west Sutherland green-shanks are birds of the gneiss hills and wet flows with their innumerable burns and lochs. There and in Ross most pairs nest beside rocks and stones, but in north-east Sutherland some lay their eggs close to snow-fences beside the rail-way. In the upper Spey Valley greenshanks are largely birds of the forest marsh, nesting beside stumps or branches of dead or burnt pine or even occasionally beside living trees. In other parts of Inverness-shire they sometimes nest in peat-bogs or beside stones on hillocks or hillside. The altitudinal extremes in the greenshanks' range are from under 50 feet above sea-level in Suther-land and Ross to about 1,700 feet and exceptionally 2,000 feet in the Spey Valley.

In *The Greenshank* Nethersole-Thompson (1951 and pers. comm.) estimated that 300 to 500 pairs might breed in Scotland. This may have been an underestimate, but in the last few years greenshanks are certainly scarcer on their more southerly nesting grounds. In the five years from 1935 to 1939 the population of two Speyside deer forests fluctuated between 5 and 13 pairs; in the five post-war seasons from 1945 to 1949 numbers fell to 5 to 8 pairs; and in the five years from 1959 to 1963 numbers further dropped to 3 to 5 pairs. Comparable data for two adjacent forests is equally depressing.

It is difficult to give precise causes. Greenshanks certainly lost ground in the war from which the Speyside ones never seemed to recover, perhaps because their natural predators also increased in the war years. There are other factors: a possible contraction of range; the dry-ing up of favourite food tarns; and an overgrowth of rank heather on many nesting braes—greenshanks and their chicks dislike long heather and sometimes desert good nesting grounds if the heather grows too long. Disturb-ance of nesting and feeding grounds by humans and crows Nethersole-Thompson believes to be less significant in the decrease; in a Sutherland deer forest in 1964-65, he found 15 to 17 breeding pairs and a non-breeding surplus of over 6 birds, in 1966 there were six pairs and two birds

respectively. Fraser Darling's (1947) estimate of 1 pair
to 3,000 acres on the Lewisian gneiss and 1 pair to 6,000
acres on Torridonian sandstone probably still stands.

Greenshanks, redshanks (*Tringa totanus*), green sandpipers (*T. ocrophus*) and common sandpipers all apparently have different ecological niches and differing feeding behaviour, although they all share some feeding
grounds. Many greenshanks complete their clutches between 3 and 10 May; but some pairs are habitually later.
There are greenshanks which regularly feed several miles
from the nest and only probably change duties with each
other twice each day, usually in the early morning and
late in the evening. Some cocks brood only at night. Some
north-west greenshanks feed on rivers and burns rather
than beside large lochs, even when these are available. The
off-duty bird of one pair regularly fed about 1,000 yards
from the nest; two others generally fed 1.7 miles away.
The favourite feeding ground of eight pairs averaged 1.1
miles from the nest. In unknown country searching for
greenshanks' nests is a considerable test of observational
power. In over forty years the Nethersole-Thompson
family have now found 194 nests.

The golden plover (*Charadrius apricarius*) is fairly
widely distributed about the moors rather than among the
high hills but breeds up to 4,000 feet in the Grampians.
There are records from St. Kilda, Sutherland, Ross-shire,
the Cairngorms and the Central Highlands of the so-called
northern or black-faced race (*C.a.altifrons*). On this point,
Wynne-Edwards (1957) says that breeding populations
of golden plovers in Britain contains varying proportions
of *altifrons* and *apricarius* types with intermediates as
normal varieties; trinomial designations are therefore
inappropriate.

In personal notes to us Nethersole-Thompson states that
in mild winters golden plovers return to the nesting moors
in the first half of February. From mid-February onwards
you hear the mournful *O-dee-ah* of the cock as he slowly
flaps high over the moor. On the ground, cocks threaten
and leap-frog over one another, and single out and pursue

hens on the ground and in the air, and have display centres from which they chase rivals. Early spring snowstorms, however, often force the pairs back to the flocks. Courting and fighting cocks swish through the air or half roll almost like displaying ravens. Once firmly paired the cock periodically tilts forward and runs away from the hen to show off his white undertail coverts.

On south-east Sutherland and Speyside moors, early hens sit from the third week of April, but on the Cairngorm and Grampian tops laying is considerably later. Golden plovers lack an elaborate egg-laying ritual, but some hens take long to produce their four eggs ; one took from 2nd to 12th May. Brooding patterns vary ; some pairs share equally while with others after the first few days, the cock alone broods. Golden plovers have long brooding periods ; 11 clutches averaged 30.7 days. The beautiful golden chicks sometimes stay in the nest for as long as 24 to 36 hours, and usually fly in their fifth week.

The ring ouzel or mountain blackbird (*Turdus torquatus*) is not common in the Highlands but is fairly evenly distributed on high ground. We have heard it when in the forest with deer, in the very early mornings of June and on one memorable occasion in Coire Mhic Fearchair of Beinn Eighe we had around us fledged families of ring ouzels, ptarmigan and a golden eagle soaring above.

The survival of the golden eagle (*Aquila chrysaëtos*) (plate v) depends to a great extent on the goodwill of the owners of Highland estates and their servants, and on sheep farmers. This goodwill is conditioned by a knowledge, however sketchy, of the food of the bird.

Since 1945 the population of golden eagles in the Highlands and Islands has been assessed, in some parts more carefully than others. Over a period of years the breeding population in many areas has remained fairly static with some local increases and sites continue to be occupied for a long time by a succession of different birds. The principal food includes mountain hare, red grouse, ptarmigan, rabbit and carrion, mostly of red deer and sheep. Fox, stoat, weasel, crow, pigeon, caper, blackcock, pheasant, wild ducks, gulls, fulmar, shearwater, waders, red deer and roe deer calves and lambs (mostly as carrion) also form a proportion of the eagle's diet.

The most important hunting ground is on the sub-alpine grasslands and dwarf shrub heaths—the range of the mountain hare and red grouse in the east, and the sheep and red deer in the west. The Scottish eagles take more mammals than birds, and a high proportion of the birds are game species. On the western seaboard the tables are reversed, and the eagles of the Hebrides and possibly the mainland coast probably take more sea-birds than mammals. There is more live food for the eagles in the Eastern Highlands than in Argyll. In the west about 18 per cent of the lamb crop may die before marketing in September, but in poorer areas more than 30 per cent of lambs may die in a late spring. Lockie and Stephen (1959) found 28 dead sheep in a walk of two miles in Lewis. Brown and Watson (1964) state that there would be roughly one dead adult deer to 600 acres of average forest per annum, and one dead deer calf to 300 acres per annum. Freshly dead deer are unlikely after 10th May. The hunting area of a pair of eagles varies from about 11,000 acres in the Eastern Highlands to about 18,000 in the west, but they do not regularly hunt all the ground available. Birds invade the territory of others and there are some areas such as the plateaux of the Cairngorms which are hunted by several pairs. There appears to be mutual tolerance which has biological advantages in winter.

Golden eagles in captivity eat about 8 or 9 oz. of meat per day. Brown and Watson suggest that it is unlikely that a wild Scottish eagle will require more than 12 oz. of meat per day and that after a large meal of over two pounds a bird can probably go for several days without food. Apparent overprovision at the eyrie is used up sooner or later ; there is not much waste, probably about twenty per cent in small kills and thirty per cent in large ones. The weight lifted by an eagle can be as much as 11 lbs. Such a piece of carrion was seen in a Ross-shire eyrie by one of us in 1963 and there is the classical account of a golden eagle lifting a struggling wild cat to a height of 1,500 feet over Mam Rattagan, Kintail, before dropping the cat to its death (Anon. *Scot. Nat.*, 1949).

Throughout the Highlands and Islands there is more than sufficient food for the adult golden eagle; the lack of favourable living prey in the west is made up with carrion and other species not important elsewhere. Eagle territories maintained not by fighting but by advertisement probably ensure that a surplus of food is available. In some areas, however, persecution by man is perhaps the most important factor in determining breeding density.

Eyries are usually on precipices or pine trees. A pair may have as many as nine and as few as three eyries, which it uses over the years. Watson (1957) and Sandeman (1957) showed the breeding success of eagles in the North-East Highlands and the southern Grampians to be 0.8 and 0.4 young per pair per annum respectively, due mainly to human interference. Two eggs are laid and on many occasions two eaglets are hatched; two survive only occasionally. One is usually older and stronger than the other and eventually ousts it in scrambles for food in times of food shortage. The rearing of two eaglets is a rare occurrence in the North-West Highlands.

Toxic chemicals are now suspected as the main cause of a rather dramatic fall in the breeding success of the golden eagle in the Scottish Highlands (Lockie and Ratcliffe, 1964). A sample of the population in a wide area in the West Highlands showed that the numbers of pairs rearing young was 72 per cent during 1937-1960 and 29 per cent during 1961-1963. Disturbance by man, climatic change and change in food supply all seem unlikely as possible factors causing the decline. Ten eggs from seven eyries were contaminated with dieldrin, gamma BHC and DDE (a metabolite of DDT). These substances are contained in sheep-dips, but a young eagle found dead in a remote area without sheep or arable farms, where it had lived for six months, contained insecticide residues (Watson and Morgan, 1964).

The cuckoo (*Cuculus canorus*) is extremely common in the Highlands, the meadow-pipit once more being the basis of another bird's existence, this time as foster-parent and not as food. The rowan tree so commonly found by the byre may hold three or four cuckoos in the sunny weather. As many as seven together have been seen in moorland country with birches.

If there is one bird which we should wish to see reduced in numbers it is the hoodie crow (*Corvus corone cornix*); but there is little likelihood that the species will become conveniently rare. The hoodie is abundant throughout the Highlands and Islands, is full of guile and most difficult to kill. It is extremely destructive to other bird life, being much given to emptying nests of eggs and taking young birds. It will even take young rooks able to fly. Yet it is also a fine scavenger. The hoodie interbreeds with the carrion crow (*C. c. corone*); the former inhabits the country north of the Highland line and the latter that to the south of it. There is, however, a zone of hybridisation, and carrion crows have bred recently in Kincardine, Aberdeen, Banff, Moray, Inverness, Easter Ross and southeast Sutherland.

The raven (*Corvus corax*) is neither so destructive in its habits nor so numerous. It is our opinion that the corbie does more good than harm by his assiduous clearing of carrion. His tremendous beak is the first to puncture the hide of the dead stag in spring and clear the entrails before the blowflies get busy. Some species of these blowflies are the scourge of the sheep farmer throughout the summer. The hoodie crow too must be given its little bit of credit for feeding on maggoty carrion.

Derek Ratcliffe (1962) states that the breeding populations of the raven have not fallen below fifteen per cent of the maxima since 1955. The same was the case with the peregrine (*Falco peregrinus*) which, however, has recently shown signs of decline. As in the case of the golden eagle, food is unlikely to be a direct limiting factor and, though availability of suitable nest sites controls the nesting density at lower levels, it does not impose an upper limit. Mutual tolerance between adjacent nesting pairs is regarded as the factor controlling nesting density. When space is limited, ravens and peregrines may compete for nest sites ; the golden eagle is a dominant competitor for nesting cliffs and perhaps for food in the Highlands and Islands. Deliberate persecution by man can reduce breeding success, especially in the peregrine, but does not limit breeding density over a period of years, except in unsuitable country. In the study regions used by Dr. Ratcliffe, current breeding success together with migration were sufficient to maintain numbers in both species. High breeding success was not followed by an increase in the nesting population.

Grazing as an ecological factor is the meeting ground of natural history and pastoral husbandry. We cannot dissociate the two ; indeed modern natural history has ceased to concern itself only with the collection of wild species and the tabulation of habits ; it is now a finite study of animals in relation to their environments ; and environments have a habit in this country of being created, shaped, modified or destroyed by man and his domesticated animals, so we must include him and his beasts in a broad study of natural history. We have seen the forest disappear and the herbage complex of the hill sheep farm arise. For well over a hundred years we have seen positive attempts to maintain that complex. The sheep farmer knows in his way, as the plant ecologist in his, that the resulting vegetational complex is not a climax. If the density of the sheep grows less or the farmer ceases to burn regularly, the grazing floor will disappear in its present state. The seedling trees which even now appear plentifully if we look for them, would raise their little heads above the herbage. Without the firestick and the selective muzzles of sheep we should soon be back to scrub over much of the ground below 1,000 feet and later to a new recuperative forest age of the Highlands.

Deer and cattle will browse and kill young trees by rubbing, but deer and cattle between them would not prevent a fair amount of regeneration. The reason rests not on appetite but on numbers of mouths. One cattle beast represents five sheep and five sheep in place of one cow means five grazing tracks and five times the destructiveness of a cattle stock. The cattle beast tears several square inches of herbage at a time, and the hill plants being well rooted lose only their tops, but the sheep pokes its muzzle farther down and pulls. The grazing action of deer approximate to those of young cattle.

The decline in cattle grazing in the past 150 years in the Highlands and the intensification in sheep grazing and the associated muirburn has brought about a fairly general increase in the area under bracken. Bracken is poisonous to cattle and horses, giving rise to "bracken staggers", but it has less effect on sheep and deer. An enquiry by the Department of Agriculture and Fisheries for Scotland showed that there are nearly half a million acres of bracken in Scotland, most of which are in the Highlands. The increase between 1943 and 1948 is thought to have been about 10,000 acres. Cattle and deer tend to depress bracken by mechanical action *when it is not unduly common*, but they cannot limit it when the bracken covers large areas as it now does. The spread of bracken is a direct result of hill sheep-farming practice, and much of it is on slopes inaccessible to machines and uneconomic to cut frequently by hand. Many attempts at eradication have been made. Planting with conifers has been an answer to the problem with expensive fencing of inconvenient places. Flooding and burning have also been tried without success, and now work is proceeding on control by herbicides.

In chemical treatments the plant is favoured by the wet climate, the nature of the terrain, and the fact that it grows quicker in some localities than in others. Such substances as sulphuric acid, sodium chlorate and sodium borate are quickly removed by rain on steep slopes. A chemical specially developed for bracken control, 4-chlorophenoxyacetic (4-CPA), must be sprayed at a certain stage in the growth of the plant to be effective; since the rate of growth varies from place to place locally, this necessitates repeated treatments, best done from the air. The cost is prohibitive, and we must look to putting cattle and trees back on the sheep grazings as the only satisfactory method of control of bracken in the long term.

Sheep grazing in the southern Highlands has resulted in the spread of mat grass and heath rush, as on the lower slopes of Ben Lawers. The sheep eat little of these plants and the result is the formation of a dense mat of decaying vegetation between the true soil base and the young shoots. An acid herbage of comparatively little food value results. Fenton (1937) has shown that the areas where mat-grass is spreading graze the heaviest densities of sheep in Scotland, in the Southern Uplands and the South-Central Highlands. The only hope in such places is to reduce the sheep density and stock heavily with cattle. Smith (*cit.* Fenton 1937) showed that mat-grass could be depressed and replaced by a bent-fescue association by the cessation of grazing.

The basic social unit of the sheep flock is the family group, a number of which are bound together by behaviour and preference among different hill pastures. When Scottish Blackface lambs were separated from ewes in groups at different ages it resulted in differences in both the social behaviour and the ground grazed by each group. The ewe-lamb bond appears to be critical in determining social structure. Grazing intensity of Cheviot sheep is more clearly related to CaO and crude fibre content of the pasture than to any other factor (Hunter 1962, 1963; Hunter and Davies, 1963).

There is no doubt that if we are to preserve some of our natural woodlands in the Highlands we must be prepared to fence for a few years until the leaders of the young trees are above grazing or browsing height. It would be a good measure of rehabilitation if private ownership of a deer forest (or public ownership for that matter) was contingent on a certain amount of fencing being done to allow regeneration of natural woodland. And certainly any comprehensive land-use policy in the Highlands should include a scheme of rotational fencing for this purpose. Darling (1937) suggested that a density of more than one deer to 25 acres is overstocking, and that on most of the poorer forests one to 40 acres is probably sufficient. Unfortunately, some forest owners favoured much heavier stocking and boosted their numbers by winter feeding of stags and by allowing the hinds on to good low ground. Wars have put a stop to that and it is unlikely that economic conditions will allow such extravagance again for many years to come.

The blue hare (*Lepus timidus*) is an important herbivore of both the alpine and sub-alpine zones frequenting the best pasture. Its food is mentioned later in chapter 8.

The rabbit is an important local grazing factor but has become less so since myxomatosis in the mid-1950's. The effect of the rabbit was disastrous where it was allowed to increase unchecked for a long period. Just as we found the sheep more destructive than the cattle because there were more mouths quartering the ground, so can rabbits denude an area completely of regenerating timber because their mouths are legion. Rabbits also destroy heather and encourage moss in the pasture. The rabbit often does not go much above 1,000-2,000 feet in the Highlands. Though in the wet hills of the west it quite disappears from the peat covered ground, it occurred and may still do on the green, porous-soiled basaltic cap of Ben Iadain (1,873 feet) in Morven. In the rocky glens and talus slopes below cliffs the rabbit finds perfect harbourage.

Finally, as a grazing factor we consider those tiny feeders on the bases of the stalks of grass, the voles. The field vole (*Microtus agrestis*) is much the most important in the Highlands as elsewhere in Scotland. The several works of Elton and Middleton (1929-1942) have shown that a 3-4 year cycle of increase and decrease is common in many areas, but that local advantageous conditions may alter this rhythm by a year if the voles are approaching a peak year at the time of onset of favourable conditions. Vole plagues have at times devastated large areas of sheep-farming ground in the Southern Uplands of Scotland, but the Highlands seem to have escaped such extreme increases and their results. Elton (1942) considers it remarkable that the great Southern Uplands plagues of 1875-6 and 1892-3 should have occurred in the area where sheep farming had gathered such a heavy stock. Moor grass and tufted hair grass, both tussocky in character, are able to create sufficient mat to protect and provide succulent stalks for the vole, despite a heavy grazing density, while the common deer's hair grass does not. In the Southern Highlands, where the ground cover is altogether more luxuriant, the vole years such as 1929 are fairly common. Summerhayes (1941) has shown that the grazing of voles tends to preserve a varied herbage of moor grass, tufted hair grass, creeping soft grass and mosses.

The frog (*Rana temporaria*) is common on mainland deer forest moors up to 2,000 feet, though absent from many islands. They have been seen at 3,400 feet, in the Cairngorms. Toads (*Bufo bufo*) and newts (*Triturus* spp.) are widespread on the lower moors. The adder (*Vipera berus*) is widespread and occurs also in the Caledonian pine forest in the Eastern Highlands, with the common lizard (*Lacerta vivipara*) occurring up to about 2,000 feet though rarely above 1,500 feet in the east.

It is appropriate in this chapter to discuss the life history of an organism which is common to most of the hill land of Scotland, which has significance for all warm-blooded animals. This is the sheep tick (*Ixodes ricinus*), sometimes called the grass tick or castor-bean tick. This animal swarms in the damp herbage of some moors throughout the Highlands and Islands. Apart from the discomfort it causes to other animals, it is the cause of mortality, particularly in sheep, because it is the vector of at least three diseases.

The engorged female drops from the host and lies in the grass and sedge for three weeks or so (Macleod, 1932). She then lays 500-2,000 eggs which hatch after four weeks in summer or possibly eight months if winter intervenes. One to three weeks after hatching the larvae climb the herbage stalks and attach themselves to any passing mammal or bird. The six-legged larva clings to the tip of the leaf or stalk with its two hind pairs of legs and uses the front pair for making contact with the host. Macleod says: 'Several larvae may be found in this position on the tip of each blade. It was found that immediately an air current was set up, e.g. by breathing on the grass, or if the grass were otherwise disturbed, the larva became excitedly alert and, relaxing its hold with its middle pair of legs, rose on its hind legs, ready to transfer itself to the disturbing object."

After they have punctured their host the larvae become engorged in about four days. They then relax their hold and fall into the herbage, where they secrete themselves and await ecdysis into the nymphal stage. This may take six weeks or nine months and once again the creature climbs the stalks of fortune and hopes for the best. Five days are needed for engorgement, then another falling off and a three or four months' interval for metamorphosis and ecdysis to the adult stage. There is now the last climb, attachment to the host, seven to eleven days of engorgement for the female and the last fall back into the herbage. The male walks actively about the skin of the host seeking an engorging female. He remains minute because he does not feed ; the female swells with blood to the size of a small gooseberry, dull grey in colour and shiny.

Sheep and cattle are very heavily infested in spring and autumn ; horses are also attacked but to a lesser extent. The ticks congregate at those places where the skin is thinnest, such as between the fore legs and behind the elbow, between the hind legs and on udder and scrotum. The larval ticks have a habit of aligning themselves along the eyelids where they cause much irritation. The larvae and nymphs infest and engorge on all wild mammals and birds with which they come in contact. It is therefore impossible to clear a piece of ground from ticks by removing domesticated stock for a long period. Rabbits, blue hares, hedgehogs, field mice, voles and grouse—all do quite well for the tick.

As far as can be ascertained (Macleod, 1933), the tick has no natural enemies in this country ; it is realized now that land cleared of sheep can still carry a heavy larval population of ticks and the species could keep going with but a small number of adults. The three diseases already identified as being carried by ticks are red-water fever in cattle, caused by a piroplasm; louping-ill or trembling, caused by a virus; and tick-borne fever, also caused by a virus. At one time the two last diseases were thought to be more or less one; Macleod (1932) showed the two to be quite distinct but that the fever was often an aggravating factor towards louping-ill. Cattle reared in the Highlands show a fair degree of immunity to red-water fever or may escape with a light attack. Highland cattle which are out on the hill all the year round are sometimes attacked by louping-ill, but as yet have not been shown to take the tick-borne fever. A good many young grazing cattle turned out in the deer forests for the summer contract louping-ill in June. Deer may also have louping-ill, and here again the wild and the domestic animals cannot be dissociated.

It is fortunate that many parasites are specific to their hosts. For example, the warble fly of cattle is *Hypoderma bovis*; that of deer is *H. diana*. Red deer cannot be blamed as a reservoir of this pest for cattle. Equally, the sheep nostril fly (*Oestrus ovis*), is not the same as that of deer, *Cephenomya auribarbis*. The sheep ked (*Melophagus ovinus*) is wingless; the deer ked (*Lipoptena cervi*) is a winged insect, commonly found in the vicinity of deer wallows in September, until it gets on to a deer, when the brittle wings break off.

Throughout the deer forest and sheep farm, man and his animals and the wild animals are plagued with Tabanid flies—the big ones such as *Tabanus sudeticus* (rarely), *T. distinguendus*, *T. montanus* and *Chrysops relictus,* and the smaller clegs *Haematopota pluvialis* and *H. crassicornis*. Once more it is the female that does the blood-sucking; the male is a harmless creature feeding on plant juices. If the female like the female tick does not get her blood meal she does not ovulate. The Tabanid flies are of interest in relation to the ecology of grazing in the deer forest, for these more than anything else may drive the deer to the high grazings—in the daytime only during the middle or third week of June when the flies first appear, but in July when the Tabanids are their most active on the hot, sunny days, the deer stay high altogether.

Insects are extremely numerous in species, but few can receive attention here. There are, however, several butter-flies and moths characteristic of the deer forest and grouse moor and too obvious to be neglected. The butterflies are rather local in occurrence and the high grounds have only one. The mountain ringlet (*Erebia epiphron*), a glacial relic, is found only in the western half of the Grampians, as far as Ben Nevis and near the head of Loch Fyne where it may reach as low as 1,500 feet. The Scottish form is larger and brighter-coloured than the English (Ford, 1945). Its food is the mat grass frequently mentioned earlier in this chapter, but only the tips are eaten. The Scots argus (*E. aethiops*) is of wider distribution in the Highlands, reaching north to Sutherland, but it does not appear in the Outer Hebrides. Its food is moor grass. The meadow brown (*Maniola jurtina*) is as common in the Highlands as elsewhere in Britain; the Highlands and Islands hold the sub-species *splendida*. The small heath (*Coenonympha pamphilus*) is found in the Southern and Eastern Highlands and though absent in the north is yet quite common on the Outer Hebridean moors. The large heath (*C. tullia*) is also local in distribution. It is the sub-species *scotica* which is found, a type almost lacking eye spots on the wings. Ford mentions this butterfly par-ticularly in that its several types through Britain provide a good example of a *cline*, the term devised by Huxley (1939) to describe a regular geographical gradation in type within the range of a species (see also chapter 10, p. 270). The fritillaries are of fairly general distribution, the dark green fritillary (*Argynnis aglais*) showing a very dark form in the Islands. The marsh fritillary (*Euphydryas aurinia*) occurs in the Inner Hebrides and to some extent on the western mainland. The small pearl-bordered fritillary (*Argynnis selene*) shows a sub-species *insularum* in the Outer Hebrides. The sub-speciation of butterflies in the

Highlands and Islands is of great interest; reference should be made to the papers of Heslop Harrison and to Ford's volume in this series.

The small blue (*Cupido minimus*) occurs here and there south of the Great Glen. The common blue (*Polyommatus icarus*) is found throughout the Highlands and Islands except on the high ground. The green hairstreak (*Callophrys rubi*) is also fairly well distributed in the deer forest and moorland country where there are brambles or furze. Rarities occur in the Highlands, e.g. the chequered skipper (*Carterocephalus palaemon*) near Fort William.

Some of the moths of these high moors such as the rare northern dart (*Agrotis hyperborea*) need two of the cold brief summers to complete their development. The most conspicuous moths of the moor are the emperor (*Saturnia pavonia*), northern eggar (*Lasiocampa quercus* var. *callunae*) the fox moth (*Macrothylacia rubi*) and the drinker (*Cosmotriche potatoria*). The toll on the larvae and pupae of these moths by chalcid and ichneumon flies is enormous. Most emperor cocoons have chalcid infestation. The fox moth, and to a lesser extent the eggar, are subject to violent fluctuations from year to year.

There are many geometrid moths in this general habitat of such genera as the carpets (*Xanthorhoe* spp. and *Entephria* spp.), and there is the allied argent and sable (*Eulype hastata*) whose larvae feed on birch and bog myrtle. The purple bar (*Lyncometra ocellata*) is commoner in the Highlands than elsewhere in Britain, bedstraw (*Galium* spp.) being the food plant. The rivulets and the pretty-pinion (*Perizoma* spp.) feed on the rattles and eyebright (*Bartsia* and *Euphrasia*). Throughout the West Highlands and Islands, the magpie moth (*Abraxas grossulariata*) is a common insect wherever *Calluna* is found in quantity. The caterpillar in this habitat has taken entirely to a diet of *Calluna,* but if man plants gooseberry or blackcurrant bushes—or particularly worcesterberries—in such surroundings, the moth comes to them and is a serious pest. Many moorland Lepidoptera of the Inner and Outer Isles are recorded in Heslop Harrison 1937 and in Campbell (1938).

The house fly (*Musca domestica*) is common throughout and in the Central and Eastern Highlands is found up to 2,000 feet. Almost everywhere in the Highlands below 2,000 feet there are vast hordes of midges (Chironomidae) which affect the movements of mammalian life, including man, to a considerable extent. The stags are terribly irritated by midges while their antlers are in velvet. Midges are doubtless preyed on by the several species of dragon-flies, by the frogs which are numerous on the wetter moors and by the palmated newts. But nothing keeps pace with the ubiquitous midge which has been the subject of enquiry by a special committee and a research team. The place of the midge in human ecology is such that a greatly increased tourist industry to the West Highlands could be encouraged if the midge could be controlled. One of the midgiest places in the West Highlands is the narrow line of decaying seaweed left high by the March spring tide and untouched again by the sea until October. This line, found on every bay and beach and inlet, is hundreds of miles long between Cape Wrath and Mull of Kintyre. But every square yard of Highland and Island moors has its midges. Little if anything can be done by way of control which would not cause extensive damage of agriculture, forestry, game and freshwater fisheries. The Committee's Reports (Cameron *et al*, 1946, 1948) mention the several species found, chief of which is *Culicoides impunctatus*. Mosquitoes (*Aedes* spp.) are fairly common in the old pine forests of Deeside and Speyside and locally common in the Eastern and Central Highlands up to 3,000 feet.

Chapter Six

The Life of the Red Deer

The red deer (*Cervus elephus*) is Britain's largest wild mammal and it is only in the Highlands that it has remained truly wild without a break from earlier times. The large areas of rough country enabled the species to survive the period of the medieval chase and later the age of gunpowder, until, when it was in a much reduced state in the 19th century, a certain William Scrope, Esquire, enjoyed himself thoroughly for ten years in the Forest of Atholl and wrote a book describing his stalking days after the red stag. It was published in 1845. "Literature?— Heaven help us!" he exclaimed, "far from it ; I have no such presumption ; I have merely attempted to describe a very interesting pursuit as nearly as possible in the style and spirit in which I have always seen it carried on ... The beautiful motions of the deer, his picturesque and noble appearance, his sagacity and the skilful generalship which can alone ensure success in the pursuit of him, keep the mind in a constant state of pleasurable excitement." The book is rather crude ; the illustrations by Edwin and Charles Landseer are in our opinion bad, but William Scrope, Esquire, and his book, had an influence on the Highlands almost as great as the Rising of a hundred years before. *Days of Deerstalking* was widely read at a time when a lot of money was being made out of the heavy industries, distilling, brewing and what not. Here was the very thing: a few of the sporting aristocracy set the pace by acquiring deer forests, the manufacturers and brewers and shipbuilders were not far behind and the impoverished Highland lairds found themselves in clover at last. There were nothing like enough deer or deer forests, and the sheep farmers who had come some time in the previous sixty or eighty years after the eviction of the crofters from the glens now found themselves heaved out from some parts. The ground cleared of sheep was left quiet for a while to let the deer increase and thoroughly heft themselves. Large rents were forthcoming and there began a period of lodge-building in some of the most fantastic styles of architecture Britain can ever have seen. Queen Victoria built Balmoral in the fictitious Scots baronial style more reminiscent of the German *schloss*. It was widely copied, but other original spirits got almost as far as the French *château*, and this in extremely remote places.

Before the advent of the express rifle the deer were hunted with deerhounds, and this amounted in many cases to a military operation. While the gentry slept, several hundred of their retainers were out in the darkness taking up positions over a wide area of country to drive as many of the deer as possible to a central point in the bottom of a glen. The drive would commence at first light and by the time the gentry had breakfasted and taken up their positions in hiding in the glen bottom the herdsmen moving the deer downwind were closing in. With most of the deer on the flats the ambush was loosed and they were run to ground and killed by the powerful hounds. Guns, arrows and daggers were used by those who could get close enough, and in the high-spirited free-for-all the carnage was wrought. Probably ten times more escaped than were killed, but there was no refinement in the choice of those which were killed. The herd was savagely mauled with stags, hinds and calves being killed indiscriminately. The precision of modern stalking was unknown.

A strict code of etiquette grew round the new deer-stalking, and it was a good one. Deer were driven no more. Even the use of deerhounds in the way of slipping them cold at a stag in the middle of a forest was given up, and these whimpering creatures (that was ever their fault) were crossed with working collie dogs to produce a truly beautiful and intelligent dog that would track a wounded deer and keep quiet until it was needed. The increasing precision and killing power of the express rifle did away with the need for these dogs and by 1890 hardly any were used.

The method was now of the quiet forest, the stalker who knew his ground and who could handle his master's guests, and a straight battle of skill in getting up to the old stag. Frock-coats were left at home: Lord Tomnoddy donned the tweeds of the country and was happy to crawl in a burn or through the glaur of peat hags in the acute discomfort which the Highland weather of September and October can impose. He shot his stag clean, at a time of year when it was strongest and on the high ground. He picked his beast and kept to it and did not pot at anything which came along. When he returned to the lodge and had a hot bath in his room (water all carried up and carried down, for interior plumbing was not equal to the façade) Lord Tomnoddy had an individual experience to talk about for the rest of the evening in the smoke-room. It made him feel pretty good. O'Dell and Walton (1962) give interesting accounts of the history and economy of deer forests.

The influx of such large sums of money created a prosperity of a sort. Many men found jobs as stalkers, ghillies and pony boys. Others did all manner of estate work. County Councils in the Highlands found their incomes rising, as all sporting properties carried a high assessment. The Highland deer forest, nevertheless, was a grand place for those who lived there and the deer themselves entered upon a good century. They were allowed to increase and take back lands from which they had been banished, and in bad winters were given food on the low ground, with the intention of improving the size and quality of the antlers, which were the criteria of a good beast.

It is rather surprising that in these two generations of deer forest prosperity so little good natural history of the deer was published. There were rows of volumes of anecdote and concern with "heads", but the daily life of the animal was largely left to the stalker to watch, and he, being a fine intelligent sort of man, knew much but he did not put it in writing. J. G. Millais obviously loved his deer and Cameron (1923) wrote what was probably the

first scientific book on them, based on Henry Evans's carefully kept notes of many years on Jura Forest. The animals
pose us many fine problems and much remains to be done
by an inquiring mind that can give the time to their study.

Following the pioneer work done by the senior author
(Darling, 1937) in Ross-shire, 1933-36, a great deal of time
is now being given to the study of the ecology and control of deer populations in the Highlands and Islands by
scientists and field staff of the Nature Conservancy, and
the Red Deer Commission. On Rhum, which is owned by
the Nature Conservancy, an intensive study of the red deer
population has been made by Pat Lowe. When the island
became a National Nature Reserve in 1957 the sheep were
taken off. Before 1957 about 1,700 sheep were sharing the
island with about 1,600 deer, which was a rate of stocking
far beyond what the grazings were able to sustain. The
fall of grazing intensity which resulted from the removal
of the sheep is aimed at slowing down the drain on soil
nutrients, soil fertility and the feeding value of the vegetation. The harvesting of one sixth of the deer population
annually—about 100 stags and 140 hinds—has resulted in
a fairly stable spring population of between 1,450 and 1,500
in the last few years. Parallel to the population study which
also includes investigations on the age structure and social
structure, behaviour and parasites of the deer, there is also
a study of the food. The vegetational productivity of the
grazings is being measured and evaluated as nutrition for
deer.

The Nature Conservancy has also investigated on two
private deer forests, Invermark and Glen Fiddich, whether
the findings on Rhum apply equally to the Scottish mainland. An attempt is being made by Brian Mitchell to assess
the consequences of traditional management practices on
those two forests, which contrast with Rhum in climate
and vegetation. On the mainland the Conservancy is co-
operating also with the Red Deer Commission in supple-
menting the general census work throughout the Highlands
and Islands which was started by the Red Deer Survey in
1953 under the direction of the senior author with the help

of three stalkers of long experience: Archie Macdonald, Philip Macrae and Louis Stewart. By the autumn of 1960 the survey had covered some 1,600,000 acres of deer ground and counted almost 53,000 deer. The provisional findings of the Red Deer Survey, as detailed in the Conservancy's Annual Report of 1960, state: "If the existing sample data are representative of the whole of Scotland and if the population changes in recent years have been comparatively small, it is reasonable to think that the total spring population is now of the order of 125,000 (excluding the previous year's calves) or 155,000 (including the previous year's calves)."

Red deer are social animals; their herds are more than mere aggregations. For ten and a half months out of the twelve the sexes remain in separate herds, the hinds in large, close family groups, and the stags in loose companies that do not show the same quality of cohesion as the hind group. Nearly all the one- and two-year-olds, and a few three-year-old stags, usually remain with the hind herd through the year. The hind herds are very well hefted to their own ground and only exceptionally severe winter conditions cause them to move far from it. Stags dislike wintering on ground which carries a heavy stock of hinds and this may be one cause of their breaking out of deer forests and invading arable land. Both sexes in their herds observe home ranges which cover an area of one, three or four, or even more square miles. They know the topography of their own ground intimately.

Deer make seasonal movements within their home ranges—to high ground in summer as soon as the flies trouble them and there is grass on the tops, and down again in the fall for the winter. Although these two movements are broadly true, there is both variation and definite movement at other times. For example, a wild storm in July—which is quite common—will bring the deer down as low as they come in winter; if there is calm frosty weather after a time of snow and drifting east wind, the deer will go high to the places drifted clear of snow, even if it is January; and in June, when days are hot and nights cold, the deer will

make quite long treks to high ground in the day and
down to their winter quarters at night. In summer, heat,
flies and disturbances such as sheep gathering and parties
of climbers, cause deer to move out to comparatively
high and undisturbed ground.

The nature of the ground influences the distances and
frequency of these treks through 3,000 feet of altitude. In
the steep hills of the west coast where the climate is mild,
and the tops are often clear of snow, movement is freer,
and the distance from 2,000 feet to sea level may entail
no more than three miles of a trek. The distance travelled
in the east is probably greater on the whole, due to the
conformation of the forests and length of the glens. In the
winter of 1950-51, which was the worst in the Cairngorms
since 1895, deer did not move more than a few miles out
of their normal range.

The social life of the red deer is founded on a matriarchy and on the fact that the sexes keep apart for so
much of the year. Each hind group usually possesses one
or more old hinds usually with calf at foot. The old hinds
are more wary than the young beasts and wariness could
be said to increase with age. If you are stalking a herd to
watch them for a while, you will make a reconnaissance
with your glass first to find out from their wary behaviour
which are the elders. Thereafter you will give your most
careful attention to them. Any small staggies which will still
be running with the hinds you can deceive by remaining
motionless. Any one of the old hinds may give the show
away and it usually assumes leadership. If a young beast
gives the alarm, one of the old hinds will come forward to
lead. Hinds are more wary than stags and once you have
drawn the attention of the older hinds there is little chance
of the herd settling.

The first sign that you are seen is one loud bark, very
sharp and far-carrying, made by the hind which has seen
you. Every beast in the herd stands with its fore feet
together, head high and perfectly still. You also lie still
and face down. There will be another bark, and another,
and suspicion may be allayed. If so, the young beasts,

8

especially the staggies, will start grazing first and the self-appointed leader will be the last to drop her head. Perhaps you are still considered a source of danger, but the leader does not know what you are. She may do the bold thing and come to find out while the main bulk of the herd walks slowly with stilted, disconcerted steps in the opposite direction. Or if the ground is right she may climb farther uphill so that she has a better angle of vision on this object in the herbage. Failing in this she will circle in an attempt to get downwind. Should she get a good view or find your scent she will bark and move off quickly. The rest will follow her, and if she stops to look around they will stop also and not go in front of her. She will be the last to settle to grazing, looking back again and again in the direction of the danger.

That hind group will be made up of hinds and their off-spring up to coming three-year-olds and occasionally over three years. Even then it is only the stags which leave to join the stag companies on different ground. The three-year-old hinds stay in the original group. The size of the group could increase to the limit of sustenance of the range, but budding-off two or three hinds, often an old one and her daughters, usually occurs before the limit is reached. There are indications that boundaries of home ranges change somewhat as years pass, but on the whole the hind ground remains hind ground for a long period of time. The colonization of new country, such as sheep ground, is usually done by stags, which are forced to break out by the over-stocking of hinds. If a heavy stock of hinds have not actually spread out themselves from the fore they will certainly displace the stags in winter and cause them to move to new ground.

The pattern of a departing company of stags after human intrusion is quite different from that of hinds. Often if you are spotted by a single stag he seems unable to warn his companions as a hind would, and unless they see him behaving restlessly they will remain until most see or smell the danger. At some times of the year a stag may be loath to leave his companions, but in summer and

F

autumn a single stag when locating danger will spring to his feet and depart, leaving the others to follow if they will. When a stag company is disturbed as a whole it is each stag for himself and the devil take the hindmost.

The deer forest owner has always regarded the stag as much better sport than the hind (plate VIII). He and his tenant are usually concerned with the shooting of a large number of stags in autumn, but the killing of hinds in winter is left to the stalker. Throughout the last half-century the number of stalkers which estates employ has fallen greatly and the job is no longer so attractive to young men. Estates have become grossly understaffed and consequently hinds are generally much more numerous than they should be. This is perhaps the main cause of deer marauding on agricultural land. To shoot an insufficient number of hinds is bad forest management; if an owner neglects his responsibilities the resultant damage may not occur in his own locality, but far from the area from which the trouble stemmed. Also, large tracts of country have been fenced for forestry without any attempt being made to dispose of the deer which formerly occupied the ground.

The calves are born from late May till July with the bulk in the first three weeks of June (plate VII). The hind leaves her calf hidden in the heather or bracken for the first few days, feeding it two or three times a day and not allowing the calf to follow. She does not give the calf much mothering in these first days, but as soon as she allows it to follow her, she is extremely attentive and feeding takes place every hour or less. The hind will spend almost an hour, in the slack part of the day, licking her calf's head and ears with her rough tongue. This appears to be sheer joy to the calf. The hind suckles her calf until February at least, and as the death rate is heavy among calves anyway—about 40 per cent in the first winter and heavier if the winter is severe—this drop of winter milk is probably of great benefit to those that survive. The dappled calves of June lose their spots as the weeks go by and at the end of August are the same colour as their mothers. A dappled calf in October means there has been a late birth.

Rutting begins in the third week in September. During the summer the stags have been grazing by themselves in the high corries, free from the irritation of flies on their growing antlers. Their antlers of last year dropped off in April and May while they were on their winter ground, but almost immediately the new ones began to grow, first as velvety knobs and later bifurcating and further ramifying until the complete "head" was to be seen, still covered by the nutrient skin known as "velvet". The antlers are bone, extending from two cores on the frontal bones of the skull, and not horn and permanent structures as in the antelope. In August, increasing growth of the "coronet" at the base of the antler strangles the blood-vessels to the velvet, with the result that this nutrient skin dies, putrefies and begins to peel off. The stag feels rather tender at this time and likes to rub his antlers gently against some high-growing tree. All through the growing period he has been careful not to involve his antlers in possible damage. Quarrels are settled by rising on the hind legs like hinds and boxing a few strokes with the fore feet.

The growth of the antlers, the stoppage of the blood flow to the velvet and the development of the gonads are physiological problems. Wislocki (1943) working in America with Virginia deer (*Odocoileus virginianus borealis*) suggested that the increasing daylight in the spring caused the pituitary body lying at the base of the brain to release a hormone which stimulates antler growth and activity of the testes in the late summer. Activation of the testes in late summer produces another hormone, testosterone, which halts growth of the antlers and causes the velvet to die and be shed. As the activity of the testes diminishes after the autumn rut the amount of testosterone also falls off and the antlers are eventually cast, possibly when testosterone production ceases. Deer have been castrated soon after shedding the velvet, and the antlers dropped off shortly afterwards. Wislocki, Aub and Waldo (1947) have demonstrated experimentally that the

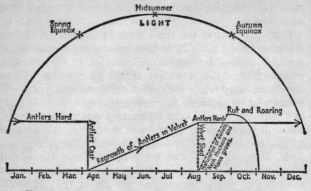

FIG. 8. Growth of antlers and maturation of the gonads
*From Darling: 'A Herd of Red Deer', by permission of
Oxford University Press*

shedding of the velvet was precipitated by the level of
testosterone in the blood. It is supposed that the hormonal
products in the ovaries of the hind prevent antler growth.
The reindeer is antlered in both sexes but the rhythm of
shedding is quite different between buck and doe. Bucks
shed in October and does in April. If a doe fawns out of
season, her antlers are also shed out of season. Experimental work with other species has shown that antler
growth in a doe can be achieved by injecting testosterone.
Hummels (antlerless stags) and female deer which have
antlers and do not normally possess such, probably
express hormonal derangement. The secretion of the
testicular hormone brings about a thickening of the stag's
neck, development of the larynx and activity of the infra-
orbital glands. Fig 8 from *A Herd of Red Deer* (Darling,
1937) summarises the sequence of events.

Considering the annual drain that antlers are on the
system of the animal, it is difficult at first sight to see why
they should have developed at all. A heavy and numerously
pointed pair of antlers is a handicap to the animal rather
than a help in fighting and it is doubtful whether such a
head is much of an intimidation, for a "switch-horned"

stag is a powerful antagonist and a hummel is very often a master-stag. Huxley (1930) has said that in evolution antlers may have increased in size and branching with increase in body size of the deer, so the elaborate antlers may be merely a by-product of size.

The red stags in their companies on the high ground remain together and in amity until some time after the velvet is peeled completely from the antlers. Then one stag will suddenly leave and then another until all the mature ones are gone. With the onset of the rut the stag adopts a running gait and often travels distances of several miles to hind ground in another forest. The majority make their way to hind grounds nearer home. It is principally at this time that the stags wallow in dubs of peat, from which they emerge quite black and awesome.

When the stag, newly come into rut, breaks into the hind grounds he proclaims his presence by a loud roaring. If he is a big fellow he may not roar on arrival and may be with the hinds for a week or more, ousting the smaller stags and herding his hinds. This rutting activity is accompanied by much roaring and this is the only time of year that his voice is used ; the stag seems incapable of emitting the sharp bark so characteristic of the hind, though on rare occasions a stag has been heard to bark. His roaring is intended as a warning for other stags to keep away from this ground which is his territory for the time being. He is always much more noisy and active when other rutting groups are close by, and he is attempting to isolate his hinds. There is a definite attempt on his part to keep the hinds on ground over which he can run easily, and he is fairly successful in this, for he herds them as a collie dog would. Otherwise the hinds are indifferent to his presence ; the traditional matriarchy of the hind group is not broken.

Suppose there is a large herd of hinds and followers in a corrie ; at the beginning of the rut one stag may have the corrie and the herd to himself, but this state of affairs will not last. Stags arrive insistent on claiming hinds, and the first stag will have too much to do to keep all the hinds to himself. Here are some actual figures from observation in a Ross-shire forest in 1934 (Darling *op. cit*.). A big, dark stag with wide-spreading antlers had 77 hinds and followers in a corrie at 1,700-2,000 feet on September 28. These deer were grazing on an area of about twenty acres and the stag was running round them continuously, roaring every minute or less, scraping the ground with his fore feet, lying down for thirty seconds, up again and running round the group with his muzzle outstretched and roaring as he ran. Two or three hundred yards away on each side of the group were some youngsters and a few rutting stags. The young stags took no notice of the central group of hinds and either grazed or lay quiet. The others were in no way equal to the big stag who was with the hinds, and though roaring occasionally, they made no positive challenge. If they came a little too near, within two hundred yards of the hinds, the master stag would run towards them at a swift trot, muzzle far extended. They did not stay to meet him and he never ran so far as to lose touch with the hinds. Two days later this stag had 46 hinds and followers and three other stags had the rest in much smaller territories well below him. They were closer to each other than to him. On October 4 the big stag had only 23, and 11 on October 7. Then he disappeared.

It is common to hear of great fights between stags at the rutting season. Nature fictionists and artists like Landseer find such fights a source of perennial interest. Sometimes there are such fights, pursued with terrific vigour, and occasionally to the death, but they are not common. Broadly, it may be said that fighting occurs more frequently later in the rutting season than at the beginning. The vast majority of encounters are no more than challenges and a sparring of antlers. After all, it may be to the interest of the species that the stronger male should be dominant and therefore the sire of calves, but the species does not benefit by heavy fighting among the males in which one might be killed and the other exhausted. The whole subject of fighting has received far too much attention in popular natural history, probably because man is himself a bellicose species.

The encounters between stags are occasions of formality and punctilio, or, in biological terms, threat display. There are definite forms: two stags will roar at each other across what they conceive to be a boundary of a rutting territory. They will walk up and down each side of that imaginary line, occasionally trying to take a dig at each other's ribs or flank. But both of them know the game quite well; the one thus attacked whips round head on, with antlers lowered. The attacker does not waste time driving in, so the watchful march up and down begins again. Perhaps they will face up, stretch their hind leg—which depresses the rump—lower the head and tuck in the muzzle so far that the antlers are extended forwards, and then fence with the tips of the antlers as lightly and cleverly as two men with foils.

The activity shown by the stags in the rutting territories cannot be maintained by the same animals for the whole six weeks of the rut. As it is, stags eat very little at this time and show the same nipped-up appearance of the abdomen as a race-horse in training. Rests are necessary and a stag which is found resting in a peat hag will probably have been driven out by a better beast and will rise in the late afternoon in an effort to reclaim a few hinds. Often the high corries become progressively deserted during the rut with only the feeble beasts remaining at the end. The stags are quiet in the high corries feeding on club moss and lichens, and though they do not feed in the close group seen earlier in the summer, it is noticeable that there is no display of animosity or of challenge. They rarely even roar. These high corries are neutral ground, movement into and out of which usually occurs at night, but occasionally in daylight a stag trots away towards the hind grounds.

The rut ends soon after the end of October and the stags resume life in their companies; when bad weather sets in they gather in herds, moving on to the low ground of their home ranges for the winter (plate VIII). The interrupted but not disrupted matriarchy of the hind groups continues normally again. Had we wished to see how apparent rather than real was the temporary dominance of the stag in the rutting territories, we could have shown ourselves instead of keeping in hiding, and then we should have seen the leading hind go off with her group in good style; normally the stag bundles along with them or he might strike a course of his own and sheer off not to return to the hinds until dark or perhaps not at all. Such a stag has probably just joined the hinds and is not fully established.

The senses of the red deer are acute, particularly that of scent, upon which it relies for much that a human being might ascertain by sight and touch. The tactile organ of the deer is the muzzle. Hearing is acute and may be looked upon as the inquisitive sense. The observer may also correctly deduce some of the emotions of the deer by studying the movements of the ears and the angles at which they may be held. You may squeak and make whistling sounds which may draw hinds nearer to you, for they like to know, but if you try scraping the metal of your shoe against the rock, the deer will move away. The sight or sound of flushed grouse will make deer move quickly without waiting to find the reason, and if the deer hear the explosive hiss of a disturbed sheep they will move immediately. Wild goats also hiss with an explosive snort through their nostrils, but the sound carries farther than that of the sheep. The deer take note from as far away as half a mile if it happens to be a still day in a high corrie where the sound is carried and amplified.

The sense of scent is linked with a great deal of the behaviour of a herd of red deer; weather much affects the acuteness of this sense and therefore is one of the primary stimulating influences to action of one kind or another. The state of the weather may make the deer either hypersensitive or apparently tame. The temperature and relative humidity of the air are of particular importance. If you run a thermograph and hygrograph, you notice that as the temperature rises, the needle of the hygrograph falls.

Suppose our hygrograph shows a relative humidity varying from 20 to 40 per cent at intervals of from 2 to 3 hours ; this is a fairly dry atmosphere which might occur during a light east wind in February. We should expect to find the deer fairly tame on such a day, and they would stand still to satisfy themselves of the nature of a possible disturbance before moving. But suppose there is a warmish south-westerly breeze with a hygrograph showing minute to minute variation in relative humidity of 60-100 per cent, this would mean that the deer would be in a highly nervous state and very irritable. The wind may veer to the north, with a fall in temperature and steadying of the hygrograph ; then the deer become less irritable. If there is a state of constant saturation, as in frost, or supersaturation, as in mist, the stalker can make a close approach.

A warm moist atmosphere conducts scent better than a dry cold one. The olfactory system of animals needs a moist outer receptive surface if it is to function properly. It is known that the thing called "scent", that which is smelt, consists of minute droplets of volatile oil exuded by the animal smelt. These droplets carry much easier in a damp atmosphere and become gaseous sooner if it is warm. Scent will reach an animal from a greater distance in a moist warm air than in dry. The same scent reaching an animal's nostrils from the same distance under different degrees of humidity will give different strengths of stimulus to the olfactory senses, and will therefore result in different degrees of the type of behaviour induced.

Super-saturation, resulting in mist and dead-steady hygrograph record, offers a physical obstacle to the rapid passage of scent through the air. Furthermore, in mist the temperature nearer the ground is lower than that above the mist, and results in restriction of air movement. A clear night of frost, which gives easy approach to the deer, means a still air and a steady hygrograph record at about 95 per cent relative humidity. A thaw sets the needle of the machine oscillating and the deer will immediately become irritable.

Lastly there is the influence of coming snow on the behaviour of the deer. They move downhill to the glens well before the coming of the snow, ignoring territorial boundaries. Long strings may be seen steadily making their way down the hillsides. The straths fill with hundreds of them, and when it is remembered that the average density of red deer on the forests of Scotland is about one to thirty or more acres, the deer of large areas of country are to be seen together at these times. There is real enjoyment to be had—in which the animals join— when the drifting by the east wind ceases and frosty weather follows. Then the snow will carry us, the sun shines, and despite the cold the still air gives the illusion of warmth. The deer go up to the patches drifted clear on the hillsides. When a thaw is imminent they will come down again beforehand lest they be isolated on those bare patches by a sea of soft snow, but if they have been kept low by soft snow the deer go up again in the rapid thaw knowing that the snow will soon disappear and they will have newly washed grazing.

The deforestation of the Highlands, the Clearances, the coming of the sheep and the rise of the deer forests all played their part in moulding the population of deer as we know it today. Red deer feature as greatly now in the Highland scene as ever they have done in a long history of controversy and deep cleavage of feeling. Between 1872 and 1952 no fewer than six committees or commissions were appointed by Parliament to make recommendations on the deer problem. Following the demand for more agricultural productivity during and immediately after the first world war, the Stirling Maxwell Committee, 1919, reported that the sport of deer stalking was on the decline, and that most forests were incapable of carrying sufficient numbers of sheep and cattle to make them profitable. In 1921 the Buccleuch Committee, or the Game and Heather Burning (Scotland) Committee as it was officially known, reported that deer forest owners should be obliged to fence their forests where they were contiguous with agricultural land and that occupiers should be permitted to kill deer damaging their crops. These two committees achieved little or nothing but the introduction of the Deer and Ground Game (Scotland) Bill in 1939 showed promise of reaching the heart of the problem. The Department of Agriculture was to have powers in the control of deer, but the war intervened and the Bill was shelved.

In the second world war the stocking of the Highlands with sheep and cattle increased and has remained higher than before 1939; the sporting value of deer had been on the decline since the end of last century and the wartime conditions gave it an extra push on the down grade. Forest owners have increasingly brought back sheep to the deer ground to reduce rating assessments and to supplement their income. In 1948, the Agriculture (Scotland) Act took up the deer problem where it had been left in 1939. It provided powers in the control of deer to be exercised through Agricultural Executive Committees. Though the provisions were agreed in principle by the various interests they proved ineffective in practice, and the Act was eventually revoked by the Deer (Scotland) Act, 1959.

While the mainstream of legislation had been concerned with control of the deer, there was another concerned with cruelty. During the second world war large numbers of deer were slaughtered indiscriminately to supply meat for the lawful and "black" markets. The means of killing by spotlighting the herds by the roadside in the winter nights and shooting them down with all manner of shotguns and rifles was a repetition of the cruelty of the deerhound ambushes previously described. After the war this practice continued unabated and in 1949 the Scott Henderson Committee on Cruelty to Wild Animals was appointed. Its findings (1951) brought into official light what had been common knowledge in the Highlands for years: the shooting of red deer at night by gangs of poachers. These operations resulted in deer escaping with ghastly wounds and dying in what must have been the greatest agony. We have seen this with our own eyes. Bloody trails in the snow lead up the hillside for several hundred yards to the poor maimed beast with shattered pelvis or protruding bowel. We have also been present at 2 o'clock in the morning during the apprehension of a "gang"; three "unemployed" youths in a hired car with an old shotgun, home-made shot, a home-made spotlight and the carcases of three half starved stags in the car.

The Scott Henderson Committee recommended that an increase be made in poaching penalties and the introduction of statutory close seasons. Following this the Poaching of Deer (Scotland) Bill was introduced in 1952, but agreement was not reached between the various interests, especially on the vexed question of close seasons. The Secretary of State for Scotland appointed yet another committee to enquire into the desirability of close seasons for deer; in 1954 it reported that there was difficulty in reconciling the various interests and the Poaching of Deer (Scotland) Bill did not go forward.

Following this break-down the Nature Conservancy took a lead in marshalling the residual goodwill among the various interests. Discussions culminated in an agreed representation to the Secretary of State for Scotland, which was the basis of the Deer (Scotland) Bill which was laid before Parliament in 1958. Controversy still raged but the protagonists of conservation and control coupled with prevention of cruelty had the wind of change blowing with them stronger than ever and the Red Deer (Scotland) Act received the Royal Assent on 14th May, 1959. The Act provided close seasons for stags from 21st October to 30th June and for hinds from 16th February to 20th October. By it the Red Deer Commission was founded.

Thus ended the long tale of bitterness. The first Report of the newly constituted Red Deer Commission began with these words and we think them appropriate to bring this chapter to a close:

"... Whatever success may attend the efforts of this new permanent Commission, its setting up, as a matter of history, marks the first comprehensive attempt to deal with a controversial problem which has vexed the Highlands for the best part of two centuries."

Chapter Seven

The Pine Forest, Birch Wood and Oak Wood

THE PINE FOREST

The destruction of the old Wood of Caledon has been bewailed at sufficient length already in earlier chapters of this book. Let us now study those small portions of it which are left and which have recently been listed and described by Steven and Carlisle (1959). The largest piece is the basin between Aviemore and the Cairgorms comprising Rothie-murchus (plate VI), Abernethy and Glenmore Forests, from about 700 feet altitude at the level of the Spey to the tree line at 1,500 feet. The stretches of undisturbed pine forest are, however, very much smaller than the few thousand acres of the whole. The old trees appear also in the Forest of Mar, occurring in several patches down the valley of the Dee but particularly in its tributary glens. There are some fine examples at Ballochbuie on the Royal Estate, and Lord Glentanar conserves a few hundred acres on the slopes south of Aboyne (fig. 9).

There is the Black Wood of Rannoch on the south side of Loch Rannoch. The term black wood means the same as dark wood—which is pine as opposed to oak and birch. Some stretches of pine exist in Forestry Commission ground on the south side of Glen Carron in Ross-shire, where the old trees' greatest danger is from the Forestry Commission itself which in the past has been obsessed with monocultural methods and non-native conifers. Farther north still, at Rhidorroch near the Sutherland border, there is a wood of native pine above the Old Lodge, but with heavy grazing by cattle, sheep and deer the wood is heading for extinction. Many old pines occur in the Inverness-shire glens. There are the remnants of the Loch Eil forest

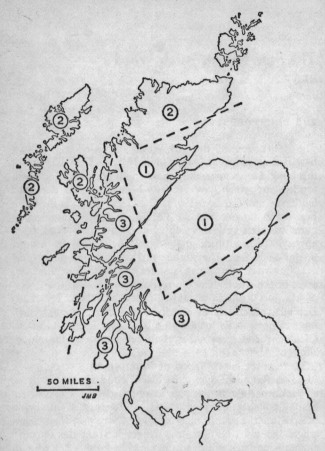

FIG. 9. The main tendencies of woodland distribution in the Highlands and Islands:
1. Predominant pine forest with birch and oak
2. Predominant birch forest
3. Predominant oak forest
*From McVean and Ratcliffe, 1962 by permission of the
Controller of H.M. Stationery Office*

on the shores of Loch Arkaig and within the Forestry
Commission enclosure in Glen Loy, but the best surviving
stretches of the old western forest are the Coille na Glas
Leitire on Loch Maree-side, and the woods at Coulin in
Glen Torridon and at Coille Creag loch in Glen Shieldaig.
In remote Knoydart there are surviving trees along the
shore of Loch Hourn and in Glen Barrisdale. Extensive
even-aged stands were to be found until recently in Glen
Affric (plate i), Glen Cannich and Glen Strathfarrar. There
is little or no regeneration in the western remnants except
locally at Guisachan, Coille na Glas Leitire, and Coille
Creag loch. But in the eastern forests there is in places
good widespread regeneration examples of which can
now be seen along the road from Aviemore to Glenmore
Lodge and around Loch an Eilean in Rothiemurchus.

In Argyll, areas of old pine still precariously exist in the
Black Mount on the south side of Loch Tulla and at
Crannach Wood on the east side of the railway as it
reaches the Moor of Rannoch. In west Perthshire there is,
besides the extensive Black Wood of Rannoch, a patch of
black wood at the foot of Cononish Glen and there are
solitary trees in Glen Falloch.

Tansley (1939) gave special attention to the fragment
of old forest at Coille na Glas Leitire on the south-western
shore of Loch Maree (plate xi), where regeneration of
old pine is now occurring. This wood is now part of the
Beinn Eighe Nature Reserve, and the Nature Conservancy
aim to rehabilitate it. About 200 acres of woodland have
been fenced against deer and sheep and a boost given to
the natural regeneration of pine, birch, oak, rowan and
alder. On one bit of ground pines are absent and
Sphagnum becomes dominant with some *Carex* spp.,
Drosera spp., *Potentilla erecta, Rhynchospora alba* and a
little *Calluna* and *Erica tetralix*. Dwarf pines appear on
rocky islands in this terrace (Gimingham *et al*, 1961).

One factor which is of great help to regeneration of
native pine forest is a breaking of the ground surface which
tears up the moss and lays bare the soil. In the natural
forest this occurs through the uprooting of old trees in

storms or by avalanches. Seedlings grow on the upended root platform of the blown tree, out of reach of grazing animals and with plenty of sunlight from the gap in the canopy. Every year small numbers of trees are blown down which provide a continuous supply of such sites, while at long intervals in the Highlands there are hurricanes which devastate wide areas of forest but which at the same time prepare the ground for the next generation of trees. In many areas where ground has been fenced and ploughed for reafforestation in the neighbourhood of old mother pines, regeneration on the mineral till has been spontaneous. The disturbance of the soil by the haulage of timber and by the driving of extraction roads through the old pine woods has provided opportunities for regeneration, but in the western forests and also in the Black Wood of Rannoch seedlings will survive only in areas which have been fenced against deer, sheep and rabbits. The visitor to Loch Tulla can see how freely the pines regenerate on the roadside plots some of which were planted and others left entirely to natural seeding from the old pines in the 1930's. Today those plots are healthy young pine with birch and rowan scrub, and are an example of what would happen in the West Highlands if grazing pressure on sparse woodland was relaxed.

But the fencing out of grazing animals is often not enough to produce good saplings. In most of the western forests where sheep and deer have grazed for two centuries and which have not been burned for twenty-five years or more, the mat of heather, blaeberry and bracken may be almost waist deep. In the soft places there is a thick mat of flying bent grass and *Sphagnum*. This is a most inhospitable place for pine seeds; those that send up a shoot have a wet, sunless and poverty-stricken life for the first five years of growth. There is competition for nutrients and mechanical damage to the seedling by the brushing action of the heather. During this time the young tree has been out of sight, but immediately its delicate green leader appears above the general level of the woodland floor it is seen by deer, especially after heavy snow. Only then does the natural seedling require the protection of the fence.

In the natural forest the wind prepared the forest floor for regeneration by uprooting old trees. The other natural agency of regeneration is even more devastating—fire. The deep mats of heather and blaeberry require to be cleared periodically to allow the sun to reach the soil and the improvement of surface conditions for the reception and germination of seed. Spontaneous forest fires are caused by lightning, but these are few compared with man-made fires in the Highlands. Controlled burning of the heath which followed the felling of the old forest has caused the gradual replacement of heather and herbaceous floor with flying bent grass and bog myrtle in which no pines will grow. In Coille na Glas Leitire the Nature Conservancy has succeeded in obtaining good regeneration of alders on this type of floor by direct seeding with phosphate and an inoculum of alder rootnodule added.

Typical untouched and regenerating pine forest does not show a great variety of plant species. Systematic lists from Highland forests can be found in the monograph *Plant Communities of the Scottish Highlands* by McVean and Ratcliffe (1962) and also in Stevens and Carlisle (1959). The complex is one of large, mature pines which are many-branched and flat or round-topped; young trees, whose crowns are still conical, with needles longer than those of the large trees; juniper (*Juniperus communis*), some of which is tall and conical (10-20 feet) and some spreading and low (4-8 feet), and which forms an understorey to the pine in the eastern forests but occurs only occasionally in the west; there are birches and rowans in plenty and alders become thicker along the sides of burns and lochs. The ground vegetation is dense ling (*Calluna vulgaris*) and bell heather (*Erica cinerea*), where the shade is not too great and juniper not dominant, but under the trees the ling and heather thin out and their places are taken by blaeberry (*Vaccinium myrtillus*) and cowberry (*V. vitis-idaea*). Where the trees (even of native strain) have been planted and stand in close formation they lose their round-topped form, grow long trunks and provide a general shade which precludes regeneration. The undergrowth becomes almost a pure blaeberry heath with some *Hylocomium* moss.

Other plants characteristic of the pine woods but thinly distributed therein are *Linnaea borealis* (in Eastern Highlands only); creeping lady's tresses (*Goodyera repens*); some of the wintergreens (*Pyrola* spp.); that arctic-boreal relic allied to the loosestrifes, chickweed wintergreen (*Trientalis europaea*), coralroot orchid (*Corallorrhiza trifida*) in the Black Wood of Rannoch and in the Eastern Highlands; and lesser twayblade (*Listera cordata*), found much higher on the hills and reaching into the alpine zone. Those which are shade lovers, such as the common wintergreen (*Pyrola minor*) and the wood anemone (*Anemone nemorosa*), may occur in treeless gullies seemingly far outside the habitats in which we are accustomed to see them. Chickweed wintergreen also occurs locally in the northern birch woods.

The plant communities in the pine forest have been described by McVean and Ratcliffe (1962). There are two distinct associations. The first is a Pinewood *Vaccinium*-moss association which consists mainly of Scots pine (*Pinus sylvestris*), ling (*Calluna vulgaris*), blaeberry (*Vaccinium myrtillus*), cowberry (*V. vitis-idaea*) and the mosses of *Hylocomium splendens* and *Rhytidiadelphus triquetrus* and is characteristic of moderately dense pine forests in the Central and Northern Highlands and in Scandinavia. The second is a Pinewood *Vaccinium*-heather association consisting mainly of those plants already mentioned except *R. triquetrus,* together with wavy hair-grass (*Deschampsia flexuosa*), the mosses *Plagiothecium undulatum* and *Ptilium crista-castrensis* and *Sphagna.* Creeping lady's tresses and twayblade are often found within this association which is characteristic of the more open pine forest in the Western Highlands and at higher altitudes in the east, and of pine-birch mixtures and even pure birch wood which has colonised former pine ground. It is somewhat similar to communities found in the spruce forests of

central Europe. In many localities intermediate stages
between the two associations are found.

The birds of the pine forest may be bound to it completely, or they may be species which have a greater range
of habitat or powers of adaptation. The Scottish crested tit
(plate 11b), the sub-species found only in Scotland, has its
headquarters in the Spey Valley, though it is now common
in plantations on low ground near the sea as well as on the
hills in Nairn, Moray and Banff, almost to the Aberdeenshire border near Huntly and Keith. The crested tit is not
nearly as scarce as is sometimes suggested. In good years
before the second world war Nethersole-Thompson
(personal communication) calculated that there were
roughly 120-125 pairs in the 208 square miles of deer
forests east of the Rivers Feshie and Spey. On 3,400 acres
of old pine forest which he specially worked there were in
good seasons 11 to 15 pairs of crested tits per 1,000 acres.
It has been more difficult to estimate numbers since the
war, but he believes that there are at least 300-400 pairs
breeding in Scotland. After particularly severe winters like
that of 1946-47 crested tits certainly suffer population-
crashes, but numbers always seem to recover quickly; by
1949 crested tit populations in Rothiemurchus were almost
back to normal.

These little birds seem to prefer tracts of old forest, but
are not averse to planted woods after they have been
thinned out and in which there are plenty of old stumps.
Crested tits not only nest on the edges and interior of old
pine woods, but also often choose forest clearings in which
there are old stumps and trees; they also nest in old pines
and pine stumps on the greenshank ground in forest-bogs.
Nethersole-Thompson has made notes on the position of
341 nests, of which 246 were in pine stumps; 45 in dead
pines; 8 in forks of living pines; 7 in birch stumps; 6 in
alder stumps; 5 in strainer posts; 4 in great-spotted woodpecker's holes or borings in pines or birches; 4 in dead
alder trees; 4 in dead birch trees; 2 in nesting boxes; 1 in a
gate post; and 1 in a squirrel's drey 15 to 20 feet up a pine

tree. The height of the nests above ground level varied
from under 1 foot to 45 feet.

Altitudinal range varies from braes or ridges under 50
feet above sea-level in some Findhorn forests to roughly
1,500 to 1,600 feet—or exceptionally higher—in woods in
the upper Spey Valley. Clutch dates and clutch size vary
considerably from year to year. First clutches have been
as early as 8th April and at least as late as 12th May. One
pair was double-brooded.

A second bird bound to the pine forests is the Scottish
crossbill. Its range is considerable on the eastern side of
the Highlands from Dunkeld to Sutherland, and were
observers more numerous in the west we think a good many
occurrences would be recorded.

Nethersole-Thompson, who is writing a book on the
Scottish parrot crossbill, states in personal notes to us that
Scottish crossbill populations vary greatly from year to
year in any particular forest. Although they sometimes do
feed on spruce and larch seeds their principal food is seed
of Scots pine. The Scots pine cone crop varies from year
to year, roughly running in a three- to four-year cycle.
Scottish crossbills, therefore, wander from forest to forest
seeking a good crop ; large numbers seldom, if ever, nest in
any particular forest in consecutive years. In 1936, 1952
and 1958 hundreds of pairs nested in the parts of the Spey
Valley in which he was working, but there were very few
pairs on the same ground in each of the following years.
In 1953, however, he found many Scottish crossbills
nesting in some of the Moray forests ; in good years the
breeding density of Scottish crossbills was up to 16 pairs
to 1,000 acres of suitable forest and in "bumper" years—
1936—six pairs nested in a wood of 26 acres.

Scottish parrot crossbills (see page 94), continues
Nethersole-Thompson, start to pair in late winter flocks.
Later small mating groups of cocks and hens break away,
leaving a widely wandering "rump" consisting of cocks and
one or more unmated hens. Meanwhile, the mating groups
move from wood to wood and pairs "hive off" and
ultimately settle on territories. In years of low numbers a

cock isolates himself and sings loudly sometimes in different woods until he attracts a hen.

In the Highlands, crossbills occasionally build in larch or spruce, but normally in Scots pines. Nests are usually placed high, generally well out on lateral branches. The cock usually escorts the hen while she is choosing the nest-site, and, although he seldom builds, he often accompanies the hen while she is doing so. In early seasons a few hens complete their clutches in February, but in many years the last fortnight in March is the normal peak for full clutches. The usual clutch contains 3 to 4 eggs, occasionally 5, and 6 has been twice recorded. 148 clutches from Ross-shire and Inverness-shire averaged 3.77 eggs. Normally the hen alone broods, but each day the cock feeds her about 8 to 10 times. The hen also leaves her nest three or four times daily, but only stays off for up to 6 minutes. While the hen is brooding, some cocks roost close to the nest, others in their feeding territories. Eggs take $12\frac{1}{2}$ to $14\frac{1}{4}$ days to hatch, after which the hen generally eats the eggshells.

For the first four to five days the hen broods her chicks for much of the day and night. While she is covering the chicks, the cock feeds her and she regurgitates food to the brood, or allows the cock to feed them. Later both parents feed the young, but, for at least ten days, the hen briefly broods them after each meal. Some hens brood chicks at night throughout their stay in the nest. Hens, and sometimes cocks, eat faeces of small chicks, but towards the end of the fledging period the nest becomes a solid mass of faeces. Young crossbills spend 17 to 25 days in the nest, after which the parents continue to feed them for several more weeks. Scottish parrot crossbills are normally single-brooded.

"Invasions" of continental crossbills take place in the summer and autumn of some years. The foreign birds pass through the Highland pine forests, but do not seem to affect the native race in any significant way (see page 94). The natural history of the crossbill is described by Campbell (1955).

The capercaillie (plate III) is not confined to pine forests but eats the shoots of spruce or larch, the buds and shoots of Scots pine and the fruits of blaeberry, cowberry, crowberry, rowan, hips and juniper berries. The cocks meet in April and May in small numbers for aggressive display with head held high, wings drooped, neck ruffled and tail fanned ; display before the female takes on different forms. The nest is usually a scrape at the foot of a pine and five to eight eggs are usually laid and hatched between mid-April and mid-May. The fledglings are tended by the hen only. The natural history of the caper is described by Palmar (1956).

The siskin (*Carduelis spinus*) (plate III) is a fairly common bird in conifers on the eastern side of the railway from Perth to Inverness and farther north as well, but there is no doubt that it is breeding in the West Highlands also where conditions allow, and certainly breeds in several places in north and mid-Argyll.

Siskins love sunlight. On warm, bright days, several cocks together take part in pretty song-flights. With slowly beating wings and fanned tails, they fly erratically over the trees and then suddenly, with those sad little cries, they go lilting through the forest. Although siskins are not strongly territorial, cocks fight fiercely with rivals and engage them in claw-to-claw flight duels. The cocks chase hens high in the air and in corkscrew flights in and out of the trees ; they also feed them. The hen chooses the nest site high in a conifer and builds the nest, but the cock escorts her and often sings nearby. In years with bumper cone crops, some Strathspey hens clutch in the second week in April, but the second week in May is often the peak for laying.

Siskins lay three to five, exceptionally six, eggs. 46 Spey Valley clutches averaged 4.04 eggs. The hen alone broods, but the cock feeds her at $\frac{1}{2}$ to 2 hour intervals. A cock is very quick in feeding his hen, diving through the trees at high speed, calling to the hen, who answers him from the nest. He is at the nest and away like a flash. Each egg takes $10\frac{1}{2}$-$11\frac{1}{2}$ days to hatch. Young siskins stay in the nest for 15 to 16 days and the parents regularly rear two broods (all from notes by D. Nethersole-Thompson).

The goldcrest (*Regulus regulus*) is another bird of the coniferous woods joining the great tit (*Parus major*), coal tit (*P. ater*), blue tit (*P. caeruleus*), long-tailed tit (*Aegithalos caudatus*), and the tree-creeper (*Certhia familiaris*). The willow-tit (*Parus atricapillus*) is rare in pine woods.

Tree-pipit (*Anthus trivialis*), robin (*Erithacus rubecula*), wren (*Troglodytes troglodytes*), chaffinch (*Fringilla coelebs*), wood-pigeon (*Columba palumbus*) and bullfinch (*Pyrrhula pyrrhula*) are regular inhabitants of the pine forest. The mistle thrush (*Turdus viscivorus*) far out-numbers the song thrush and blackbird which are scarce in the high forests. In winter the only passerine birds are the coal tit, goldcrest, tree-creeper, wren and crows. Long-tailed and other tits are scarcer then and chaffinches, wood-pigeons and many other species leave the pine forests for the winter. The nightjar (*Caprimulgus euopaeus*) used to occur throughout the mainland Highland region and in the Inner Hebrides but is decreasing and is in fact rare in the area.

The commonest bird of prey of the old pine forest is probably the tawny owl (*Strix aluco*). The long-eared owl (*Asio otus*) is uncommon. The sparrow-hawk (*Accipiter nisus*) has recently become rare or absent as a breeding species in many districts. It has been greatly persecuted but is now protected by law. The hawk lives largely on small birds and wood-pigeons and the owls take many bank voles and woodmice as well as birds.

The old pinewoods lack shrubbery beloved by warblers and finches. The wood-warbler (*Phylloscopus sibilatrix*), however, prefers a wood of high mature trees with light low ground cover where it can build its domed nest hidden among leaf debris and moss, and where the male high above can sing his two contrasting songs. The willow-warbler arrives at the end of April or a day or two later in the far north, and suddenly one morning one wakes to the song of many of these birds. The willow-warbler is more common than the chaffinch in these northern woodlands (Yapp,

1962). The whitethroat (*Sylvia communis*) occurs in the pinewood but is not common. The spotted flycatcher (*Muscicapa striata*) is a characteristic bird of the pine forests particularly in Speyside and Deeside. The redstart (*Phoenicurus phoenicurus*) is more plentiful in the central and eastern pine forests than in the west. Like the siskin it tends to sing in the tops of the trees so that the beauty of its plumage is lost to the observer on the ground. The cock redstart catching insects in a sunlit break in the pine forest is outstandingly beautiful: orange and chestnut on a background of deep green foliage, russet bark and purple shadow.

The grey wagtail (*Motacilla cinerea*) is not usually considered with trees, but in the Highlands certainly occurs more often along woodland than moorland burns. The ring ouzel (*Turdus torquatus*) inhabits the open moorland and watercourses above the woods and in Rothiemurchus it disappears from the banks of the stream entering the woodland and is replaced by the grey wagtail.

The jay (*Garrulus glandarius*), that noticeable predator of small birds in deciduous woodland, is now fairly common in the Southern and Eastern Highlands, and in recent years has increased in numbers in the Black Wood of Rannoch and in Forestry Commission plantations in Stirlingshire, Argyll and Angus.

We cannot say there are any distinctive mammals of the pine forest, for even the pine marten has taken to the open hill and is now a rare inhabitant of the old pine woods. Squirrels are certainly commoner in coniferous woods than in deciduous ones, and are entirely absent from the birch woods. Mice, voles and shrews are present but seldom in abundance; there are probably larger populations of these in the drier eastern forests than in the west where the field layers of the pine woods are thick with moss and are comparatively poor in herbs, grasses and invertebrates. Mountain and brown hares occur in open heathery parts of the forest all year round, but are especially numerous when there is heavy snow on the hills. They breed in open parts of the forest near the edge.

The woodlands of the Highlands whether they be coniferous or deciduous are frequented by both red deer (*Cervus elaphus*) and roe deer (*Capreolus capreolus*). The story of the red deer is told in chapter 6, and it would be appropriate at this point to make special mention of the roe which has been much less studied than its larger cousin, but about which much is now known (Fooks, 1960). A major contribution to our knowledge of roe deer in Scotland has been made by David Stephen who has kindly given us the following summary of his unpublished information.

The roe deer (plate VII) *is* widely distributed on the mainland and is plentiful in parts of the Highlands. In the islands it is reported so far only from Bute, Islay and Seil. Although usually looked upon as a woodland species, the roe is also a deer of birch thickets and scrubland; in addition, it is frequently found living on the open hill for months at a time, often far from tree cover. Roe may be completely independent of woodland as in the Grampians near Dalwhinnie. Fawns have been recorded, newly born, at 1,500 feet, two miles from the nearest trees. Sometimes they may be seen moving between forest and hill in regular rhythm.

In this species twin births are the rule. Although there are no useful statistics for Scotland, there is no reason to suppose that Highland roe differ markedly from those in Denmark in this respect. In Denmark, during the complete kill-out of a population for purposes of restocking, it was found that, in 46 mature does, 39 had had twins, four had a single fawn and three had triplets. The survival rate in the herd was 1.8 fawns per mature doe (Andersen 1953).

Roe fawns are born from late April into June. May is the main month. But there are records of births in July and August. In this species there is delayed implantation of the embryo in late December; but obviously there must be some variation in this, to the extent of several weeks. Like the calves of the red deer, the fawns of the roe have white spots at birth, but these disappear in the autumn, when the fawns are about 3 months old. Spotted fawns after the end of August mean late births.

The buck fawns begin to show their pedicles or antler rudiments after they lose their spots, and these will be fully developed, depending on birth date, in November or December. An April fawn has a head start over one born in June. Buck fawns usually produce a pair of "buttons"; these are horny tips to the pedicles, without coronets, which pierce the skin of the skull. They are shed between January and March, again depending on the age of the fawn, and its condition.

The young buck is left with two raw sores when he throws his buttons, but these are quickly covered with skin and the first proper antlers begin to grow. Since fawns may throw buttons from mid-January until the end of February, the completion of antler growth varies as widely. The young buck may clean his first antlers any time between April and June. Good living, heredity and other factors, probably influence antler growth in roe much as in red deer.

Roe rut is usually in the last third of July and the first third of August; but it can begin earlier and go on later. At this time the bucks chase the does a great deal, and usually in rings. These chases appear to be initiated by the does, which are often mated on the rings. Rings are usually cast about a birch tree, a mound, a rock, a stand of tall vegetation. They may be in standing oats. They vary in diameter from 8 to 20 feet.

Roebucks often fight fiercely during the rut. After a head-on collision between two bucks, followed by hook, stab and thrust, the weaker animal retreats at once and gives up. But, more often than in red deer, the fight may be prolonged, and bloody; yet fatalities are rare. Yearling bucks are usually prevented from mating by older bucks. A mature roebuck stands up to 29 inches at the withers. Many never reach within 2 inches of that height. Some bucks are fully grown at a little over a year old. Does are smaller and lighter.

Unlike the red deer the roe has no close season, and

may be killed at any time by almost any means, including snaring. The Forestry Commission has officially given up snaring, and recommends the rifle for killing roe on its ground. The Commission has also issued a booklet on the humane control of roe deer and is in this respect ahead of Parliament and many private foresters. Roe damage trees when bucks are cleaning their antlers of velvet, scent-marking their territory and thrashing.

The roe's most usual vocal display is a gruff bark. Does are much more talkative. Bucks grunt like pigs during the rut. The doe's call to fawns is a clear *whee-yoo*. The fawn's answering call is a low-pitched *eep-eep-nee-eep*.

Vole (*Microtis agrestis*) plagues are not common in Britain; nearly all recent plagues have been in young plantations where the tall ungrazed ground cover forms a favourable vole habitat. There is usually a four-year cycle in the populations, but the peaks may not always reach plague proportions causing noticeable damage to trees and crops.

One such plague has been studied in detail in the Carron Valley Forest, just south of the Highland line in Stirling-shire, by James Lockie (1955) and Nigel Charles (1956) of the Nature Conservancy. Voles were abundant in the area during 1952 and 1953, numbering probably over 1,000 per acre, and the ground vegetation suffered wholesale destruction. Most of the tussock-forming species such as bent grasses, hair grasses and fescues were undermined and killed, leaving bare areas which were later colonised by creeping soft grass and show up as a bright green patchwork on the hillsides. It was noted that the soft rush (*Juncus effusus*) was eaten only at times of high density and extreme winter poverty. Young trees suffered greatly and many planted in 1939 were ring-barked. Severe damage was done to Norway spruce (*Picea abies*), Sitka spruce (*P. sitchensis*), Scots pine and beech (*Fagus sylvatica*), but lodge-pole pine (*Pinus contorta*) and ash (*Fraxinus excelsior*) were almost untouched. The setback in the forest was short-lived, and by 1955 most of the effects of the plague had disappeared.

Short-eared owls were living in territories of between thirty and forty acres during the 1952-3 vole plague. As the numbers of voles fell, many owls dispersed, and those that remained bred in territories of between 300 and 400 acres. The influence of the voles on the owls was great, but the influence of the owls on the course of the plague was slight. Subsequently the effects of the whole predator force of owls, weasels and stoats was studied. The predators were responsible for all the measured losses of voles between January and May, when the voles were reduced from ninety per acre to forty-six per acre. When the voles began to breed, after May, however, the predators could not cope with the increase and other checks to population growth came into play.

If these coniferous plantations are devastated by fire in their fairly young state, the regeneration which takes place is of birch. Fire rarely burns the trees thoroughly. It runs through them as a scorching flame exploding the tops and more slowly removing any ground cover there may be. The bark splits and peels part away and the exposed living tissue of the cambium scorches, turns orange and the trees are dead. The cost of a fire in a young forest is not merely the value of the trees which as saplings may be of little worth, but in the years lost and in the amount of work in clearing the ground afresh for replanting. The cost of fire in a mature forest is less since the flames destroy the scrub and canopy of the pine wood but leave the big trees charred and undamaged within. Removing these fire-stunted trees to the sawmills is dirty but remunerative work.

The community ecology of the fauna in Highland woodlands is almost unknown. The natural history of most of the birds and to a lesser extent the mammals is known in part, but the intricate relationships which bind them with the physical and other biotic factors in the comprehensive ecosystem, are undescribed. Charles Elton and his colleagues, working in Wytham Wood near Oxford, have shown the complexities of the system and the sustained effort which is required over a period of many years to elucidate it. Systematic work on the animal ecology in

Highland woods has yet to be started. The systems in the Highlands are possibly simpler than at Wytham since there are fewer species present and this is particularly the case in the North-West Highlands, but the traceable connections between the communities of the present day and those of native forests are of great scientific interest. Many excellent sites for such studies now lie within National Nature Reserves and many opportunities await the animal ecologist in the forests of the Cairngorms and the woodlands in the Loch Maree basin.

The study of the invertebrate fauna of the Highlands has not got beyond the inventory stage and in most woodlands has not been begun. Most papers dealing with insects and other invertebrates are simply systematic lists with taxonomic and ecological notes. The account of the spiders in the Black Wood of Rannoch is typical—in 1914 A. R. Jackson wrote: 'this (the Black Wood) seems a very rich locality'; yet it was not until 1954 that A. B. Roy (1955) found the richness of the area "amply confirmed". Especially widespread species were the ubiquitous *Meta segmentata mengei*, *Philodromus aureolus* on pine branches, *Dictyna arundinacea* on heather and the three species of *Lycosa* in sawmill debris. Roy (1962) also describes the spiders of the Cairngorms. Forty years can see a great change in the ecosystems of Highland woodlands, in which, due to lack of wildlife conservation, diversification is lost. Roy was in fact unable to find the rarities of forty years earlier; his criterion of richness was in total numbers of spiders collected, that of Jackson embraced both variety of species and total numbers. In forty years the taxonomy of many groups of invertebrates may also have changed so much as to cast doubt upon determinations based on outdated keys.

Writing in 1948 on the distribution of the larger moths and butterflies in Scotland, F. W. Smith pointed out that the great majority of records were then between thirty and ninety years old, and large areas of the mainland were a complete blank. That the chequered skipper (*Carterocephalus palaemon*) should have remained so long undiscovered in the Fort William area, he says, is a measure of what may still wait discovery. The situation is the same today, and when one considers that the macrolepidoptera are among the best known groups of invertebrates, our knowledge of others is indeed small. In the same volume of the *Scottish Naturalist,* there are notes on the extension of range of the peacock butterfly (*Vanessa io*) with new records from the Small Isles, Barra, Lorne, Mull and Knapdale, but it is against the outdated and incomplete record for the country as a whole that one must judge the "spread".

The problems attendant upon the piecemeal study of the Highland fauna may in the long run only be solved by the establishment of permanently staffed field stations. In the management plans and the Reserve Records of the National Nature Reserves, the Nature Conservancy have taken a positive step in the direction of a continuous study of the woodland, moorland and mountain habitats. Most of the work on the fauna is still in the primary survey stage of recording exactly what is there. This has resulted in the Reserve Records for Cairngorms and Beinn Eighe, to quote only two examples, containing a number of valuable unpublished papers on the woodland invertebrates by G. W. Harper, A. G. F. Dixon, J. E. Knight, R. M. Mere, E. C. Pelham-Clinton, O. W. Richards and others.

The vast, diverse concourse of the Highland invertebrate fauna is beyond description in this book, even in brief summary. A great stimulus would be given to Highland entomology by a monograph of similar stature to that on the plant communities by McVean and Ratcliffe. More simply, the vast amount of information which is at present widely scattered through the literature and in museum collections might be gathered together in one volume. All this needs sifting and arranging into a concise foundation work which could be revised periodically. It is a suitable *magnum opus* for an up-and-coming group of entomologists.

There are more species of ant in the Rothiemurchus-Abernethy-Glenmore pine country than elsewhere in the

Highlands, and it is interesting to compare the communities there and in the Coille na Glas Leitire at Loch Maree. The dry, deep, highly porous ground layer, with the right kind of detritus and other factors which characterise the Speyside forests provide a suitable habitat for at least ten species (Donisthorpe, 1927). In the wet north-facing pine wood in Wester Ross only three species have so far been discovered *Myrmica ruginodis, Leptothorax acervorum* and *Formica lemani.* There are no mounds of the wood ant *Formica* spp., though it is present in other western woodlands not only of pine near Tyndrum, but of oak in Lorne and alder in Cowal. Ants play a part in maintaining the porous organic floor of the forest and give a rapid turn-over of energy in the soil. C. A. Collingwood (1951 and 1961) has produced papers on the distribution in the Highlands of *Formicoxenus nitidilus, Myrmica laevinodis, M. ruginodis, M. scabrinodis, M. sulcinodis, M. sabuleti, M. lobicornis, Tetramorium caespitum, Leptothorax acarvorum, Lasius niger, L. flavus, L. umbratus, Formica aquilonia, F. sanguinea, F. lugubris, F. exsecta, F. fusca, F. lemani.*

Damage to pines is caused by the weevil *Hylobius abietis* which breeds in stumps and roots and when adult feeds on the bark of young trees. The bark beetle (*Myelophilus piniperda*) breeds under the bark of dying or recently dead pines and the adults make tunnels in the leading shoots of growing trees. *M. minor* is common in the old pine forests of the Eastern Highlands. *Hylastes ater,* another beetle, injures pine by feeding at the root collar. In natural woodlands of mixed age and uneven spacing these insects are not usually a threat to natural regeneration. The Forestry Commission in their fine series of leaflets has described the natural history of these and other forest pests.

The Rannoch looper moth (*Thamnonoma brunneata*) has its classic site in the Black Wood of Rannoch. Its caterpillar feeds on blaeberry on the forest floor and the moth is small, bright chestnut in colour, and may be seen on the wing in June and July. Juniper which makes up the underbrush in parts of the old pine woods is the food plant of the caterpillar of the juniper pug moth (*Eupithecia sobrinata*). This moth shows much variation and an extensive series of varients can be seen in Rothiemurchus.

THE BIRCH WOODS

The birch woods of the Highlands are far more extensive than the pine (fig. 9). Strange as it may seem, none of the latter shows regeneration within the limits of the existing wood, while birch regeneration is rife throughout the Highlands in felled and burned pine forest and on the open moorland such as between Tullochgrue and Achnagoichan in Rothiemurchus. Most of the existing birch wood will disappear quickly since it is a comparatively short-lived tree, but the Nature Conservancy is studying the possibility of perpetuating the fine birch wood of Craigellachie, much of which has been declared a Reserve. The absence of internal regeneration is partly caused by over-grazing by sheep, red deer and rabbits and partly by other factors not yet investigated, since cessation of grazing is not usually followed by spontaneous regeneration. This is discussed at greater length both by Yapp (1953) and McVean and Ratcliffe (1962). Fortunately, as the kaleidoscope of the Highland vegetation turns into the future and the old birch woods disappear, the birch wood habitat will evolve elsewhere in localities like the Moor of Dinnet which are now thick with saplings. It is perhaps in these woods more than in any other in the Highlands that the ever-changing nature of the woodland habitat mosaic can be seen.

The uncontrolled burning of heather and grass on sheep runs and grouse moors takes its toll of young woodlands of all types. Birch perhaps suffers most of all, both as young regeneration and in the middle-aged groups. The soils under birch woods are suitable for the more exacting coniferous species such as Douglas fir (*Pseudotsuga taxifolia*) and Sitka spruce (*Picea sitchensis*), and the Forestry Commission have a hand in clearing the birch woods from the lower slopes of the glens. Birch was also

exploited in Glen Urquhart and Glen Affric for the manufacture of bobbins for the Indian jute mills. The bobbin
mill was at Drumnadrochit, but the market closed when
India became independent in 1947. Most continental
foresters are surprised at the indifference shown to birch
by their colleagues in the Scottish Highlands, where it has
long been considered a weed species. In Scandinavia, for
example, it is used freely for pulping and the development
of the pulp mill at Fort William may be followed by yet
greater reduction of Highland birch woodland.

McVean and Ratcliffe have assigned the vegetation of
the Highland birch woods to one or other of two types.
The first is the *Vaccinium*-rich birch wood consisting of
birch (*Betula pubescens*), rowan (*Sorbus aucuparia*), hard
fern (*Blechnum spicant*), blaeberry, wavy hair-grass, heath
bedstraw (*Galium saxatile*), wood sorrel (*Oxalis acetosella*), tormentil (*Potentilla erecta*) and the mosses *Hylocomium splendens, Pleurozium schreberi and Thuidium
tamariscinum*. The woodland floor may be dominated by
blaeberry, greater woodrush (*Luzula sylvatica*) or the
mosses. There is usually no scrub layer but occasionally
juniper, hazel and honeysuckle may be abundant.

The second type, the herb-rich birchwood, includes most
of the species already mentioned except blaeberry, together
with violet (*Viola riviniana*), mountain fern (*Thelypteris
oreopteris*), bent grass (*Agrostis tenuis*) and sweet vernal
grass (*Anthoxanthum odoratum*). Tall shrubs are even less
common than in the first type and although there are no
species which occur exclusively in the herb-rich birch
woods, such species as the wood-anemone (*Anemone
nemorosa*) and earthnut (*Conopodium majus*) are
characteristic. Many birch woods may be intermediate
between the two types, but further study would probably
lead to an even more distinct separation.

The birchwoods are found up to 2,000 feet. In the
Southern and Eastern Highlands oak often replaces birch
up to nearly 1,000 feet and 500 feet in the west, with little
change in the floor of the woodland. In many birch-oak
woods which grow on steep slopes with block scree, the

tumultuous floor is moss-covered. The trees also have moss and lichen jackets and on the wet scarps there are luxuriant growths of liverworts. Excellent examples of these woods can be seen at Meall nan Gobhar at the head of Loch Etive, Ariundle Wood in Sunart (a Forest Nature Reserve) and Stack Wood, Sutherland. Birch woods in the West Highlands are distinctly different from those of the Spey Valley where conditions are drier. The birches themselves are different for the most part, the birch (*Betula pubescens*) being the common species in the west and north and the silver birch (*B. pendula*) in the Central and Eastern Highlands. Contrasting examples are the birchwood in Coille na Glas Leitire in Wester Ross with heather and blaeberry and lack of scrub, and Morrone Wood near Braemar with its thick juniper scrub. On Eilean Mor of Loch Sionascaig in the Inverpolly Forest there is a small rowan wood surrounded by a birch-rowan mixture and this is common throughout the birch woods of the West Highlands.

The birch woods are golden on the mountainsides in the autumn sunshine and in winter they are a purple-brown mist merging into the grey-green tops of the alders in the bottoms of the glens. In sparkling February sunshine the shadows of the exquisite tracery of branches are cast upon the mountain snows, showing the intertwined tracks of rabbit, hare, mouse, vole, roe, red deer, fox, badger, wild cat, otter, weasel and ermine. In the dripping thaw the woods look faded and impatient for the spring, which comes suddenly in April with returning tits, warblers, wrens, thrushes, robins, pipits, finches and others. There is an extraordinary feeling of joy in being among the birches when the leaves are opening from the bud. The colour of the trees is good; and the young green of the leaves is pleasant in conjunction with the sound of willow-warblers. Redstarts are fairly common and long-tailed tits also nest in the birches, the lichen-laden boughs of which provide well-camouflaged nesting sites. The redwing (*Turdus musicus*) bred in the birchwoods of north Sutherland in 1941 and has bred on many occasions since in Inverness-shire, Sutherland, Wester Ross and Morayshire.

Craigellachie is one of the most extensive, mature and pure birchwoods in the Highlands. It contains the associations of plants and animals characteristic of the habitat and many moths of special interest. The angle-striped sallow (*Enargia paleacea*), great brocade (*Eurois occulta*), Rannach sprawler (*Brachiomycha nubeculosa*), scarce prominent (*Odontosia carmelita*), orange tip (*Euchloe cardamines*) and Kentish glory (*Endromis versicolora*) are all found in this woodland Nature Reserve. The Kentish glory has disappeared from its English localities and from all Scottish ones except those on Speyside and near Forres. The birchwoods contain many conspicuous fungi, especially on dead or dying trees. The bracket fungi *Polyporus betulinus* and *Fomes fomentarius* are common.

Yapp (1962) discusses the birds of birch woods in Ross-shire, Sutherland and Caithness on information obtained from surveys in twenty-six of them lasting in all about eighteen hours in the summers of 1951 and 1952. The relative abundance of species was similar to that obtained in the pinewood of Collie na Glas Leitire in the same district (plate IX). The willow-warbler was more abundant than other main species. The coal-tit was more abundant and more widely distributed in northern birch woods than it was in the south, where blue and great tits predominated.

THE OAK WOODS

On south-facing slopes, in open rolling country and on islands in lochs where brown earths survive, oak woods of varying purity are found ; the oaks are usually mixed with birch, ash (*Fraxinus excelsior*), wych elm (*Ulmus glabra*) and alder (*Alnus glutinosa*) in the wet places. The loch-sides and glen floors have patches of these woodlands which are grazed through by farm stock, deer and rabbits and which seldom possess a shrub layer or show much regeneration, either inside or on the fringes, except on inaccessible ledges. When present, the shrub layer consists mostly of hazel, rowan and bird-cherry (*Prunus padus*). Many, such as those on Loch Sunart, have been felled and the ground planted with exotic conifers, and others, like those in Glen Garry, have suffered by loch-raising for hydro-electricity. Most of them have timber of little commercial value, but occasionally one comes across a wood like Coille Mheadhonach in Glen Creran with marketable oak. This is a type of Highland woodland habitat which may be on the way out and the Nature Conservancy has declared a number of the best areas as Nature Reserves (plate x) and has notified many others to the owners and local planning authority as Sites of Special Scientific Interest.

The pristine Highland oak forest was concentrated in Argyll (fig. 9, p. 164), with strips extending up the Great Glen to Easter Ross and along the coast of West Inverness-shire to Wester Ross. As would be expected, the scattered relicts reflect this ancient coverage of the country. A few small areas are present in Islay (Kildalton), Jura, Colonsay, Knapdale (Loch Sween) and Cowal (Glendaruel), but it is on the Lorne plateau that they are most extensive. Coille Leitire and Coille Driseig on the steep slopes at the north end of Loch Awe add great charm to the scenery in the Pass of Brander, and as one travels westward on the main road to Oban the same charm of oakwood, mountain and loch is repeated in the Taynuilt district. In Glen Nant the woods have a heavy admixture of birch, but at Clais Dhearg there is perhaps the finest parcel of semi-natural oak wood in the Highlands. Other fine woods occur in Benderloch between Ardchattan and Lochawe in which there is a well developed hazel scrub, and on the coastal flats from the Lynn of Lorne to upper Glen Creran. The finest of the oak woods in Mull is at Laggan opposite Ulva, and in this locality there are good stretches of hazel scrub extending far beyond the limits of the existing oak wood.

Long strips of oak wood on the slopes of the Great Glen in the past are represented today by small patches on the

shores of Loch Lochy, Loch Arkaig, Loch Garry, Loch
Ness and in Glen Moriston. What applied to the Great
Glen also applies to the Sunart, Ardnamurchan and west
Inverness-shire coastal strip. The best surviving fragments
between Ardnamurchan Point and the Ross-shire
boundary are at Glen Beasdale in Morar and Coille
Mhialairidh high on the slopes above the north shore of
Loch Hourn. The Glen Beasdale wood shows well the
effects of wind exposure and shelter.

The oak woods of Letterewe in Wester Ross are the
most northerly of this series apart from very small stands
in Strathoykell. Mention has already been made in
chapter 4 of their connection with the iron-smelting era in
the Highlands. Many parts of Letterewe are of pure oak
but in others it is mixed with birch and ash and there is
little or no understorey of hazel scrub. The oaks are even-
aged and there is a symmetry and spacing of trees which
suggests that they may have been planted in the early 19th
century. There is very limited regeneration and the owner
is co-operating with the Nature Conservancy with a view
to improving the growth of young trees.

The floor of this wood is mossy with a sparse covering
of bent grass, sheep's fescue (*Festuca ovina*), sweet vernal
grass, blaeberry, heath bedstraw, wood-sorrel and
tormentil. Slender false-brome (*Brachypodium sylvati-
cum*), primrose (*Primula vulgaris*), wood-anemone, hazel
and wild hyacinth (*Endymion non-scriptus*) are common,
and rowan, pine, juniper, hawthorn (*Crataegus monogyna*)
and dogrose (*Rosa* spp.) are also present.

There are red deer and roe and if one goes carefully
they can be seen in a wonderful setting of chequered mossy
woodland, loch with pine-clad islets and the distant mountains of Torridon. The Loch Maree basin is the classical
north-west: wild cat and marten country, with fox and
badger, buzzard and eagle, peregrine and merlin, ring ouzel
and dipper, black-throated diver and grey lag goose.

There are some good oak woods in the Southern and
Eastern Highlands also. For example those in the
Trossachs, in the Tay Valley between Dunkeld and Aber-

feldy, in the Pass of Killiecrankie, along the shores of Loch Earn, Loch Tay, Loch Rannoch, at Ballater and Dinnet in Deeside and at Spinningdale. Nowhere in the west do we find the oak going far up the hill, hardly ever beyond 500 feet and usually not so high as that, but in the South-Eastern Highlands they reach 700 feet, and a few at Kenmore reach 1,000 feet. The oak fades off into birch and rowan.

Oak forest is a climax, but it is probably that the maintenance of oak-dominated stands of deciduous woodland in the Highlands will require management. The sparseness of oak in the Northern Highlands, and its marked preference for well drained soils on the north (and therefore sunward) shores of the lochs and sides of the glens, suggests that its habitat is of limited extent in the Highlands. Cessation of grazing by fencing may not be followed immediately by regeneration of oak within the undisturbed field layer of the wood. Shade from the close canopy of the pure oak woods, root and air-space competition with parent trees, scrub species and the field layer community, and consumption of acorns and seedlings by the fauna could all be limiting factors. Management may take the form of opening the canopy by felling (as would be done by gales in nature), by locally suppressing the field layer, delving and planting of acorns (as would be done by wild boar in nature), and if necessary reducing locally the numbers of animals known to damage the mast (as would be done by large predators in nature).

The species of oak in the Highlands include both *Quercus petraea* (=*sessiliflora*) and *Q. robur*. The status of both species and their identification is discussed by J. E. Cousens (1962), and apparent hybrids between the two by E. W. Jones (1959). An oak wood in the Highlands is very different in associated flora and in appearance from one in southern central England. The bark of Highland oaks is heavily bemossed, sometimes so much that the trunks are entirely green, and the branches festooned with lichens and polypody. Holly (*Ilex aquifolium*) is relatively common, gean (*Prunus avium*) occurs locally, but ivy (*Hedera helix*) is uncommon. The ground cover tends to moss and open bracken, with some bent grasses and *Holcus* grasses or even blaeberry. Bluebells (*Endymion non-scriptus*) grow under the oaks where the ubiquitous sheep do not graze them out. Bramble (*Rubus* spp.) is common in the Highlands but not as an undercover of oak woods as in the south. The yellow pimpernel (*Lysimachia nemorum*) and cow wheat (*Melampyrum pratense*) grow among the moss.

Though the northern oak woods may have more birds than the birch woods they are silent in comparison with those south of the Highland line. The chaffinch is probably the most numerous nesting species in the Southern, Central and Eastern Highlands while in the Northern Highlands the willow-warbler is probably more common. Pure oak and pure birch woods tend in our experience to possess a less diverse fauna than the mixed woodlands. Red deer, roe deer, red squirrel, rabbit, mountain and brown hare, brown rat, field-mouse, field-vole, bank-vole, common shrew, pigmy shrew, hedgehog, mole, stoat, weasel, pine marten, otter, badger and several species of bat occur in the mixed woodlands. Of all these only the badger (*Meles meles*) might be associated more with the oak than with other woods.

The oak woods are a favourite habitat of Daubenton's bat (*Myotis daubentoni*) when these are in close proximity to water. The pipistrelle (*Pipistrellus pipistrellus*) is common both in these oak woods and also throughout the Highlands even where there are no woods at all, as far afield as the Outer Hebrides. Natterer's bat (*Myotis nattereri*) is a local species occurring down the west shore of Loch Fyne and elsewhere in heavily timbered land near water. The whiskered bat (*Myotis mystacinus*) has been found in the open birch wood near Kinlochrannoch. The long-eared bat (*Plecotus auritus*) is common in many parts of the Highlands, and Millais (1904) records it at Balranald in North Uist.

The moths and butterflies of oak woods are generally richer in variety than those of birch, and in north Argyll and west Inverness-shire include the rare shade-loving butterfly, the speckled wood (*Pararge aegeria*). The insect apparently survived the last glaciation and has, since Pleistocene times, preserved a separate existence from the populations farther south.

The caterpillars of geometrid moths feed on sallow (*Salix cinerea*), birch, alder and oak, also on bedstraw and bent grasses. A few of these moths may be mentioned: the clouded border (*Lomaspilis marginata*), the common white wave (*Cabera pusaria*), the Galium carpet (*Xanthorhoe galiata*), the common carpet (*X. sociata*), the yellow shell (*Euphyia bilineata*), the green carpet (*Amoebe viridaria*), and the early thorn (*Selenia bilunaria*). The commonest geometrid occasionally causing severe defoliation of birches in the eastern Highlands is *Graunis aurantiasia* and these birches also have the bee beetle (*Trichius fasciatus*). The beating of the leaves of these trees in summer disturbs a large variety of diptera and of lacewings as well as froghoppers and pentatomid bugs. The lacewings prey on the aphids which swarm on the birch and sallow leaves. The oak leaf-roller (*Tortrix viridana*) does not occur in oak woods of the Highlands though it is so largely eaten in the larval stage by woodland birds in the southern oak woods. Its place is taken by the winter moth (*Cheimatobia brumata* and *C. boreata*), and in the more southerly area the mottled umber moth (*Hybernia defoliaria*).

The main constituents of the canopy of mixed deciduous woodlands are birch, oak, ash and alder ; each of these species may assume dominance. Dominance is usually shared by birch and oak, but occasionally one comes across a mature woodland which is predominantly ash or alder. Such is the case with the ash woods at Rassal, Loch Kishorn, the ash-hazel wood at Glasdrum, Loch Creran and the alder wood at Carnoch, Glencoe. There are, of course, many pure alder groves by rivers and lochsides.

McVean and Ratcliffe recognise the pure ash wood as a distinct vegetation type which they call *Brachypodium*-rich ash wood with six herb constants: common bent grass, slender false-brome, cock's-foot (*Dactylis glomerata*),

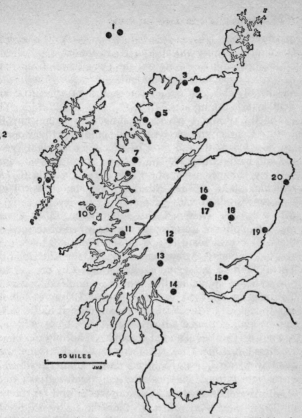

FIG. 10. National and Forest Nature Reserves in the Highlands and Islands: 1, North Rona and Sula Sgeir; 2, St. Kilda; 3, Invernaver; 4, Strathy Bog; 5, Inchnadamph; 6, Inverpolly; 7, Beinn Eighe; 8, Rassal Ash wood; 9, Loch Druidibeg; 10, Rhum; 11, Ariundle Oak wood; 12, Rannoch Moor; 13, Ben Lui; 14, Loch Lomond; 15, Loch Leven; 16, Craigellachie; 17, Cairngorms, 18, Caenlochan; 19, St. Cyrus; 20, Sands of Forvie
From 'Report of the Nature Conservancy, 1963' by permission of the Controller of H.M. Stationery Office
Since this figure was produced there are seven more NNR and FNR in the Highlands and Islands (*See Nature Conservancy Handbook 1968*, HMSO).

meadowsweet (*Filipendula ulmaria*), yellow pimpernel (*Lysimachia nemorum*) and violet. In the canopy, ash and birch predominate and on the woodland floor wild hyacinth, enchanter's nightshade (*Circaea intermedia*), melancholy thistle (*Cirsium heterophyllum*), primrose and slender false-brome.

The Rassal wood extends to about thirty acres and the Nature Conservancy have enclosed some of it to encourage regeneration of the ash most of which is over eighty years old and badly cankered. This is a most beautiful woodland in early summer when the early purple orchids (*Orchis mascula*) stand up majestically from the moss-covered limestone studded with primroses. There is bird song of willow-warbler, chaffinch, tits, robin, wren, mistle-thrush, meadow-pipit and treecreeper. The air is full of insects and the light brown loam is rich in earthworms. Moles, mice and voles abound. A few hundred yards from the wood the Torridonian sandstone moorland is a desert by comparison.

Another example of this type of woodland occurs at Tokavaig in Sleat, Skye. There is over 100 acres but the ash, though larger and with less canker than at Kishorn, are more widely scattered among birch, hazel, bird-cherry and hawthorn. This woodland also stands on an outcrop of the Durness limestone. The only other extensive limestone pavement with ash in the Highlands is on Lismore, which is intensively farmed and possesses only scattered ash trees.

All the woods cited by Professor Steven and Dr Carlisle in *The Native Pinewoods of Scotland* lie north of the narrow waist of Scotland between Forth and Clyde, and most of them are situated within the boundary of the North (Scotland) Conservancy of the Forestry Commission (Dickson and Innes, 1959). Most are in private ownership but the Commission do own some, the largest of which is in Glen Affric (plate 1). Much the same can be said for the native deciduous woodlands and the preservation of these is the responsibility of the landowners and the Forestry Commission. In some areas the native pines have been underplanted by exotic species.

The nation set the Forestry Commission the task of producing wholesome commercial timber, and a lot of it. They were not asked to rehabilitate relict native woodlands or resurrect the Wood of Caledon. Yet in many places the Commission are playing their part in the conservation of the old pine woods. Some woodlands in Glen Affric, Glen Loy, Glen Garry and in Glen More in the west Cairngorms have been fenced against grazing animals. The old pines were not underplanted and were left to regenerate naturally. These localities have at the same time escaped the ravages of fire. Apart from draining very little was done to assist regeneration and it is clearly not enough simply to protect the trees from grazing and fire ; a more direct type of management is required similar to that being done by the Nature Conservancy, and by the Forestry Commission itself in Glen Affric.

In the Affric Forest the Commission have set up a Caledonian pine reserve of about 2,100 acres to the south of Loch Benevean and Loch Affric. The native woodland which is a mixture of pine and birch is to be regenerated in pine by planting thirty acres a year of local stock over a period of seventy years. The existing stands are about 250 years old and it is thought that they will remain good for at least seventy years. The Commission are also establishing tree banks of good stock obtained from scions of elite trees in many of the native pinewoods throughout the Highlands.

The conservation of natural woodlands and of commercial forests have many common denominators. The basic studies of climate, geology, soils, hydrology and community ecology of plants and animals are the same, but the superstructure of aims, management and techniques are different. If Britain is spared from war, the nation's timber needs will be fully satisfied and more opportunity may be given for the diversification of the present monocultures of exotic conifers. An admixture of native hardwoods, particularly oak, would be desirable. At the same time the conservator of natural woodlands can draw upon the technical experience of the commercial forest officer, where conventional methods are appropriate. Such is now the case at Beinn Eighe, where the Reserve is being diversified with commercial plantation (managed by the Forestry Commission) and an ecological plantation of mixed woodland of pine, birch, oak, alder, rowan and bird-cherry (managed by the Nature Conservancy employing conventional treatments where necessary). Such areas of common interest between commercial forestry and nature conservation are now well established.

In earlier chapters, and also in this, we have mentioned points of interest in the status and ecology of plants and animals in the context of coniferous plantations. The increase in numbers of many animals is due mainly to the improved cover, and both the Forestry Commission and private owners have, in a very real sense, become conservators of wildlife. They can also be devastators, however, for in excluding deer and to a lesser extent sheep, from wide tracts of hill ground, they have, without giving due care to the matter, disrupted the deer forests and sheep runs. Deer and sheep have been deprived of ground to which they have been hefted for generations, and driven to areas in which they were unwanted. In most parts of the Highlands the deer returning from the high ground at the onset of winter are involved in an ever increasing complex of enclosures on forest and agricultural land. It is against this background of hastily erected fences in overstocked deer forests and marauding deer, that the Red Deer Commission now functions.

The natural forest is the most diverse of habitats. The mosaics of plants and animals which we have just described give an idea of the role which the Highland forests and woodland play in providing cover for animals. The most diverse cover is obtained in the idea of the climax forest where the trees are of mixed species and mixed age, with areas of extreme density grading into open woodland and natural clearings. The soil, field layer, scrub and canopy are all present in a discontinuous way and into this cover is laced the fauna. This is the rare impression which we have of the Wood of Caledon ; nothing we have described

above approaches near to the climax, but some good woodland remains with which we can work and from which we
can develop a modern natural type forest in the Highlands
—rich, diverse and beautiful.

Chapter Eight

The Summits of the Hills

It has been the theme of the last few chapters that the wild country of the Highlands is not untouched, virgin or unalterable country. Much the larger part of it shows the influence of man's hand and of the mouths of his domestic animals; yet the summits of the hills stand clear of man's influence, at least above 3,000 feet. In the first place, the herbage of this zone is scanty and dwarfed; secondly, it lies under snow and frost, or constant intermittency of these conditions, for several months of the year; and thirdly, man's domestic animals do not graze land over 3,000 feet except occasionally.

The country of the summits is, in the Highlands, an immense archipelago of biological islands holding relict communities of a past age—that of the last post-glacial epoch. When the topographical zones of the Highlands were being described in chapters 2 and 3, the northern area was said to have certain sub-arctic affinities in its vegetational complex. Some botanists might qualify this description and use boreal as a more correct adjective. But when we come to the summits, arctic affinities are obvious and beyond quibble. Even that common arctic phenomenon of frost action is found on the summits of several Highland hills—polygons and stripes of assorted sizes of stones, rock granules and sand in solifluction boils and terraces. The alpine flora as it appears on any one hilltop strikes the observer by its great stability and by its sharp exclusions, for example, only occasional dwarf pines at altitudes well above 3,000 feet in the Cairngorms and Lochnagar ranges. The nearest approach to tree cover above 3,000 feet is given by dwarf juniper (*Juniperus communis* ssp. *nana*) which spreads over exposed rocks at high altitudes.

What are the conditions obtaining on the summits which render them a distinct habitat? The tops are a country of extremes and of paradox. First in the matter of temperature: for a large part of the year the atmospheric and soil temperatures are low, frequently below freezing point. The soil temperature is particularly low when there is frost without snow—a common occurrence, especially on the western hills. The old Ben Nevis Observatory figures and those from Ben Macdhui (Baird, 1957) are, unfortunately, the only ones we have. Summer temperatures are also low as compared with those at sea level, but occasionally the temperatures of the barely-covered granular soil may be much higher than most places at sea level. Only in certain spots such as sandy beaches and cliff edges, upon which we shall have occasion to remark again, are similar conditions found of periodic unduly high soil temperatures. There are also sharp alternations between day and night temperature until July, when day and night temperatures on the tops come much closer to each other, and there may then be adiabatic inversion—that is, the temperatures at night on the higher levels may be higher than they are down below. Day temperatures on the tops are frequently higher than in the glens and on the coast. These differential and sometimes inverted temperature conditions and the related condition to relative humidity account for some of the most beautiful landscape effects ; for example, in late summer, when an observer who spends the night on a high summit may wake to see the sunlit tops brilliantly clear as blue islands in a level sea of white mist. Soil temperatures of the summits also vary widely according to the slope. A northern gully will be cold even when the southern face is extremely hot. There are climates within climates.

Second, there is the matter of moisture. The summits endure a higher precipitation in Scotland than any other part of the countryside. The plants and animals of the summits exist for days and weeks at a time in a supersaturated atmosphere. But at other times they suffer drought conditions from the action of several factors. A period of frost imposes a physiological drought, when the

soil is frozen and the plants transpire into a windy atmosphere. Except when under snow widespread damage occurs
to heather, crowberry (*Empetrum nigrum*), blaeberry,
cross-leaved heath, juniper and sometimes even pine, whin
and broom on lower ground. Bearberry (*Arctostaphylos
uva-ursi*) and bell heather are more resistant.

Straight drought conditions of lack of precipitation are
not at all uncommon in the Highlands, especially in the
second quarter of the year. There is also the factor of slope
which helps to drain water away rapidly, and the nature
of the soil. The peat, except in rare instances, has been
left behind a thousand feet or more below and the soil of
the tops is little more than a conglomeration of fairly large
granular particles of rock extremely deficient in humus.
This means that such soil has very little sponge-like quality
in retaining moisture.

Third, the plant communities of the tops have their
period of exposure to light and free air severely curtailed
by snow cover. On the Cairngorms and on Ben Nevis a
few places may have no more than a month of exposure,
and large tracts fail to get six months of light in the year.
Nevertheless, the extensive snowfields do provide much
shelter and are in themselves a factor ameliorating drought,
for their gradual melting in a hot dry time irrigates a
large surrounding area.

Fourth, we should not forget the frequent incidence of
high wind above the 3,000-foot contour. The 13-year
average of 261 gales per annum of more than 50 m.p.h.
was recorded on Ben Nevis.

Tansley (1939) warns the observer of arctic-alpine
vegetation against laying too much stress on zones of
altitude expressed numerically. He points to the fact that
on the Atlantic side of the country arctic-alpine species
are found at a much lower level. This brings us to the
point, which is developed in the discussion of coastal
vegetation which follows in chapter 9, that certain species
appearing on the higher plateaux and slopes have a habit
of cropping up again at sea level or on cliff faces. Sea
pink (*Armeria maritima*) and roseroot (*Sedum rosea*)

may be taken as examples. The intermediate zone does not carry these plants. Both of them are drought resisters in a high degree and both need good drainage from their root systems. Dahl (1951) has, however, drawn attention to the low average summer temperatures as an important factor in the occurrence of montane plants near sealevel in the north-west. The light-loving dwarf juniper is another member of the 250-foot stratum nearest the sea as for instance on the more northern Inner Hebrides, the Summer Isles, and the north Sutherland coast.

The arctic-alpine flora of the Highland summits is much affected by the geological formations which form them. The Torridonian and quartzite tops of the north-west have a poor flora compared with Ben Lawers, 3,984 feet, the latter being composed of a metamorphic schist rich in lime. The relatively flat granite tops of the Cairngorms also have some beautiful stretches of a heath composed of *Rhacomitrium,* mat grass (*Nardus stricta*) and three-leaved rush (*Juncus trifidus*). Though mountain tops in the North-West Highlands are generally poor, there are exceptionally fine prostrate shrub and moss heaths on Beinn Eighe.

The plant communities of the summits are a fascinating study and for many years the work of W. G. Smith (1911) and Tansley (1939) remained standard for the habitat. The recent monograph of McVean and Ratcliffe (1962) now provides a description of the Scottish montane vegetation such as has long been available to students of Scandinavian and central European plant sociology.

McVean and Ratcliffe have given a classified description of the mountain heath communities under three main headings: dwarf shrub, grass and moss. The natural dwarf shrub heaths (as opposed to the man-made heather moors) lie between 1,200 feet and 2,800 feet in the Northern Highlands and between 2,300 and 3,600 feet in the Cairngorms, and the vegetation falls roughly into two main types: a heather dominated heath which lies in the low alpine zone, and a blaeberry or crowberry (*Empetrum hermaphroditum*) dominated heath which occupies the low to mid-alpine zone. The transition between the sub-alpine shrub zone—already described in the grouse moors ⸺and the alpine dwarf shrub zone is represented by a dwarf juniper heath which has eight main species: bearberry, ling, dwarf juniper, wavy hair grass, deer grass (*Trichophorum caespitosum*), the mosses *Hypnum cupressiforme* and *Rhacomitrium lanuginosum* and the lichen *Cladonia uncialis*.

The heather dominated shrub-heath may take three main forms depending on the respective amounts of black bearberry (*Arctous alpinus*) and *Rhacomitrium* present. In addition to the plants already mentioned in the nearby shrub heath there is the mountain azalea (*Loiseleuria procumbens*) and a number of lichens. The blaeberry dominated heath also takes various forms depending on the relative abundance of crowberry, lichen and sheep's fescue grass (*Festuca ovina*). The heather is usually blown clear of snow in winter but there are tracts of the blaeberry heath which are covered for the whole winter and some of the spring. The main species in the snow-bed are blaeberry, crowberry, hard fern, wavy hair grass, heath bedstraw, mat grass, and the mosses *Pleurozium schreberi* and *Hylocomium splendens*. The stiff sedge (*Carex bigelowii*) is common in both the lichen-moss and fescue-rich blae-berry, the alpine lady's mantle (*Alchemilla alpina*) in the latter only.

From 2,200 feet to 3,600 feet there is a widespread distribution of crowberry heath, which is especially common in the north-west on block scree, bedrock or block detritus on summits and gently sloping ridges where it occurs below and sometimes in mosaic with the true moss (*Rhacomitrium*) heath. The main constituents are crowberry, blaeberry, stiff sedge, *Rhacomitrium* and the lichens *Cetraria islandica, Cladonia gracilis, C. sylvatica and C. uncialis*. On lime-rich ground there are tracts of *Dryas* heath which occurs from sea-level in the far north-west to about 3,000 feet in the south-east. McVean and Ratcliffe have classified those according to the presence or absence or relative amounts of carnation grass (*Carex flacca*),

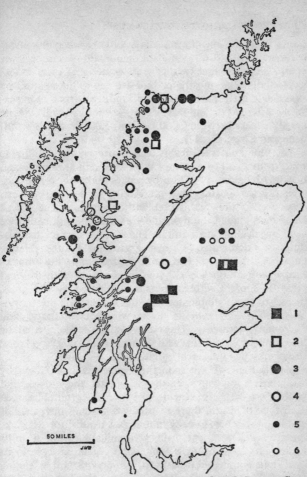

■	1
☐	2
●	3
○	4
•	5
○	6

50 MILES

FIG. 11. Dryas heaths: 1, *Dryas—Salix reticulata*; 2, *Dryas—Carex rupestris*; 3, *Dryas—Carex flacca*; 4, Unclassified *Dryas heaths*. (1-4 *from McVean and Ratcliffe 1962 by permission of the Controller of H.M. Stationery Office*); 5, Other sites with *Dryas* after 1930; 6, Other sites with *Dryas* before 1930. (5 and 6 *from 'Atlas of the British Flora', 1962 by permission of the Botanical Society of the British Isles and Thomas Nelson & Sons Ltd.*)

Carex rupestris and the reticulate willow (*Salix reticulata*) occurring with the dominant mountain avens (*Dryas octopetala*). The first two sedge types occur at or near sea-level, the *flacca* mostly on coastal sand dunes and the *rupestris* mostly on the Durness limestone from near sea-level to 1,700 feet on Ben More Assynt. The willow type mostly occurs between 2,300 feet and 3,000 feet in Breadalbane and Clova (fig. 11).

The montane grass heaths usually occupy the mid-alpine zone from 2,000 feet to over 4,000 feet and are dominated not only by grasses but also by sedges and rushes. The subalpine grass heaths and grasslands have already been described in the deer forest and sheep farm chapter as being man-made ; the montane grass heaths with which we are now dealing are natural. The lowest of these associations are dominated by mat grass and by the relative amounts of deer grass, stiff sedge and the mosses *Pleurozium* and *Rhacomitrium* in company with the mat grass. In mosaic with the heaths dominated by mat grass, there are those dominated by stiff sedge with the mosses *Polytrichum alpinum* in a snow-bed community, and *Dicranum fuscescens*. There are also those dominated by the three-leaved rush, with lichens (*Cladonia* spp.) and sheep's fescue in different situations.

The pattern of the montane vegetation is outlined by the relative dominance of shrubs, grasses and mosses, and while there can be a clear definition on the ground between each type of heath, there is usually a gradual mergence of one into the other. Locally in the grass heaths, for example, the mossy constituents will predominate giving a moss heath. The moss heath is a montane vegetation type in its own right dominated usually with *Rhacomitrium* extending from 2,000 feet to over 4,000 feet in the low and mid-alpine zones. The community also includes as constants blaeberry, wavy hair grass, stiff sedge, heath bedstraw and *Cladonia*. The moss campion (*Silene acaulis*) (plate XII), the mossy cyphel (*Cherleria sedoides*) and thrift occur locally within the community, especially in the north-west. Sometimes the *Rhacomitrium* heaths are species-rich when they occur on lime-rich rocks such as the dolomitic mudstones on Beinn Eighe. The community then also has alpine lady's mantle, thyme (*Thymus drucei*), least willow (*Salix herbacea*) and *Polygonum viviparum* together with mosses and lichens. In some snow-beds the *Rhacomitrium* is replaced by *Polytrichum* and *Rhytidiadelphus*.

The vegetation of the late snow-beds which linger on through spring to early summer is dominated by mosses, principally *Polytrichum, Rhacomitrium, Dicranum* and *Gymnomitrium*. Present also are wavy hair grass, starry saxifrage (*Saxifraga stellaris*), least willow, stiff sedge, dwarf cudweed (*Gnaphalium supinum*) and curved wood-rush (*Luzula arcuata*). The zonations of the snow-bed species depend on the relative periods of exposure to the air and sunshine and coverage by snow. Much also depends on the distribution on the slopes of melt water and the bedrock. We see an expression of the same principle when we look at the zonation of the seaweeds on the rocky shores.

Most of these alpine heaths are now included within National Nature Reserves in the Cairngorms, at Caen-lochan in Angus, Ben Lui, Beinn Eighe, Inchnadamph and Invernaver. They occur also on Ben Lawers with its rich montane meadows and on the mountains of Glencoe and Kintail all of which are properties of the National Trust for Scotland. Such beautiful flowers as the purple saxifrage (*Saxifraga oppositifolia*), starry saxifrage, alpine saxifrage (*S. nivalis*), alpine forget-me-not (*Myosotis alpestris*) and the rock speedwell (*Veronica fruticans*) occur sparingly or rarely. The mountain bladder fern (*Cystopteris montana*) occurs rarely among rocks and the club moss (*Lycopodium alpinum*) is common. Apart from the direct protection of the rarer species very little direct conservation of the vegetation of the mountain tops is required.

The herb-rich vegetation of the alpine grasslands and herb meadows is more restricted in its occurrence than the grasslands, and depends upon the presence of soft, lime-rich rock at the surface or at the emergence of water

which has percolated through calcareous rocks at a higher
level. The grasslands extend over a wide range of rocks
with a greater floristic variety on those of high lime
content. Such stretches of continuous green herbage play
their part in maintaining the deer from the end of June
until almost the end of the year, so long as the weather
remains good. Many of the plants mentioned above occur
in these meadows and richer grasslands, but the most
prominent are those which are also common at low levels
such as common bent grass, brown bent grass, sweet vernal
grass, wavy hair grass, sheep's fescue, meadow buttercup
(*Ranunculus acris*), water avens (*Geum rivale*), alpine
lady's mantle and tormentil. The stiff sedge, three-leaved
rush, spiked woodrush (*Luzula spicata*), alpine meadow
rue (*Thalictrum alpinum*) and alpine cinquefoil (*Potentilla
crantzii*) also occur.

Woody plants of the dwarf shrub heaths are also found
scattered through the grassland. The bog blaeberry
(*Vaccinium uliginosum*) is common but not so abundant
as common blaeberry, cowberry (whortleberry) or the
crowberries. The common bearberry and the black bearberry go fairly high but are rarely found in the grass heaths
and meadows at such an altitude. The leaves of the first
are obovate, evergreen and glossy and the berries bright
and smooth, while the leaves of the latter are thinner in
texture and turn bright purplish-red before being shed in
the autumn. The black bearberry is confined to dwarf
shrub heaths and a few bogs at intermediate elevations in
the Northern Highlands. That mountain member of the
bramble family, the cloudberry (*Rubus chamaemorus*) is
unlike the other mountain berry-bearing plants in having
large orbicular and reniform leaves. It is found mostly on
deep peat above 2,000 feet and the fruits are compound
like raspberries, large and yellow when ripe.

Few animals are confined to the high tops which may
only be visited by birds and mammals as an incidental or
seasonal event in their lives, and not because of any special
predilection for such altitudes ; among these are the red
deer. Common shrews (*Sorex araneus*) occur on the

highest hills in summer and are often seen in winter running over the snow even as high as 3,500 feet. Even the pigmy shrew (*Sorex minutus*) was observed at the top of Ben Nevis in the Observatory days. Short-tailed voles and field-mice have been trapped on the summits of the Cairngorms and sometimes voles are so abundant on alpine grassland at over 3,000 feet that most of the grass is eaten in spring. The highest fox den reported in Scotland was at 2,700 feet but foxes frequently cross high summits even in winter. The mountain hare (*Lepus timidus*) occurs much lower than 3,000 feet, of course, and ought more properly to be included in chapter 7, but in the West Highlands it is mainly confined to the area above 2,000 feet.

The food of mountain hares has been studied in Banff-shire by Raymond Hewson (1962). Analysis of stomach contents of forty-seven mountain hares, collected at monthly intervals for a year, showed that ling formed ninety per cent of the winter and half the summer diet; cotton grass about one tenth of the winter and a fifth of the summer; other grass species present as a trace in the winter increased to a quarter of the total food in summer. During the snows, gorse (*Ulex europaeus*), juniper, soft rush and young trees are also important.

Watson and Hewson (1963) describe how the white winter coat grows in a moult which lasts from October to December and is much denser and warmer than the grey-brown summer one. This moult is triggered off by a reduc-tion in day length, and conversely the lengthening days in spring cause the hares to moult to brown. The rate of moulting is influenced by temperature and those on the tops are white longer than those on the lower moors. A further moult takes place in late summer when the brown coat is replaced by another of the same colour. This is also found in the larger brown hare and the winter moult in mountain hares is additional as a form of protective colouring found also in stoats and ptarmigan.

The bucks are in breeding condition by the end of December and promiscuous mating takes place from then on. Mating is usually preceded by one or more males loping along for up to half a mile or more behind a female which, if approached too closely, may not stand but strike and chase the buck for up to fifty yards. The gestation period is likely to be about fifty days and litters which appear in March and April and later in August, at the end of the breeding season, are smaller than those born in June. Litters of up to four may occur, but usually one or two. Leverets are born fully furred and with eyes open in a form in the heather or in a peat burrow. During the breeding season mountain hares occasionally gather in groups of over a dozen and sit motionless. The groups disperse suddenly with a great burst of activity.

Desmond Nethersole-Thompson, author of *The Snow Bunting* and a forthcoming book on the dotterel, has kindly sent us this information, some of which is as yet unpublished.

The dotterel (*Charadrius morinellus*) breeds regularly in the Cairngorms, Grampians, Monadliadhs and several hills north of the Great Glen. Dotterels like long, hog-backed hills with rounded tops, gentle slopes and broad plateaux where they nest on alpine meadows, ridges covered with woolly fringe-moss, and on stony slopes, flats and plateaux with scanty vegetation. In the Grampians most dotterels nest from 2,700 feet upwards, but in the Cairngorms they seldom nest below 3,200 feet and sometimes well above 4,000 feet. In most years sixty to eighty pairs probably nest in Scotland; there may be more in exceptionally favourable seasons. They usually complete their clutches from the last week of May onwards on the lower hills and a little later on the high Cairngorms. On Scottish hills dotterels largely feed on craneflies, spiders, small flies, beetles and insect larvae.

Dotterels are remarkable birds with a "reversed courtship". The occasionally biandrous hens generally leave the duller-coloured cocks to brood the eggs and tend young. Some hens, however, do incubate; change-overs have been seen. When the cocks are sitting, hens associate in little groups. These "grass widows" then sometimes spar and display on rudimentary leks. In all, the Nethersole-

Thompsons have seen 108 nests in which 101 first clutches contained: 3 eggs in 91 and 2 in 10. Eggs normally hatch between the twenty-fifth and twenty-eighth day.

The snow bunting (*Plectrophenax nivalis*) is perhaps the most elusive Scottish breeding bird. From 1885 onwards, and almost certainly earlier, snow buntings regularly nested on mainland hills. In 1901 a pair nested on St. Kilda and between 1886 and 1910 nests or broods were found on some of the Sutherland hills, the Torridon hills in Ross, Ben Nevis, the east and west Cairngorms, and in Perthshire. Some were also seen and probably nested on other hills.

Nethersole-Thompson says that in the last thirty years there have been many good naturalists on the hills but, apart from the Cairngorms, there is only one acceptable record: a brood on Ben Nevis, in 1954. From the early nineteen thirties onwards, snow buntings have nested fairly regularly on the high Cairngorms. Numbers have varied from three pairs and seven or eight unmated cocks in 1947 to only a couple of cocks in 1956. There have also been some apparently blank years or years with perhaps only a single pair on the whole range, but the snow buntings nest in such wild and formidable country that no one can ever be certain of their status; it is remarkable that the Victorian naturalists, then so thin on the ground, so often saw them on many other hills. An amelioration of climate from the late 19th century onwards has probably adversely affected the small colonies.

The altitudinal nesting range of Scottish snow buntings has varied from 1,000 feet or less in St. Kilda to about 4,150 feet in the Cairngorms. They have nested on different kinds of screes including granite, Cambrian quartzite, Torridonian sandstone, gabbro and probably tertiary basalt. In early spring pairs often arrive together and then wander all over the hills before they take up territories. Some pairs almost certainly stay together for at least two years. Unmated cocks also often sing and display over great tracts of hill and scree and then end by pairing with mated hens whose chicks have left their nests. Leaving her first brood in charge of her original mate, the adulterous hen now builds a new nest and sometimes successfully rears a second brood. Apart from these triangles, Scottish snow buntings are quite often double-brooded as their pairs stay together and raise two families. In 1949, moreover, one cock was bigamous; his two mates had nests about 100 to 150 yards apart.

The Nethersole-Thompsons have found 37 snow buntings' nests in Scotland but it has taken 253 nights in small tents above the 3,000 foot contour to find them. *P. n. insulae,* the Iceland race, and the typical *P. n. nivalis* sometimes nest in the Cairngorms in the same season. The snowy owl (*Nyctea scandiaca*) has been seen on the Cairngorm plateau in the summers of 1952, 1953 and 1963-5 and bred in Fetlar, Shetland in 1967.

The lower breeding level of the ptarmigan in the Scottish Highlands rises from north-west to south-east. The average lower limit on Ben Loyal, Ben Hope and the Reay Forest in Sutherland is about 1,500 feet with a few down to about 1,000 feet or less in exposed places particularly on the hills near Cape Wrath. In Wester Ross the lower limit is about 1,800 feet and the road from Kishorn to Applecross is the only one in Britain which runs through ptarmigan breeding ground. In the Monadliadhs and Cairngorms the lower level is about 2,500 feet and in the Ballater district, 2,700 feet. The highest nest seen was at 4,025 feet in the Cairngorms where they became scarce above 3,800 feet and below 3,000 feet. The highest nesting grouse are found rarely at about 3,400 feet (Watson 1964; 1965 *a* and *b*).

During severe winter conditions ptarmigan often move right out of their summer range on the high tops down on to the lower slopes. On such occasions flocks of ptarmigan and grouse occupy the same ground, but do not mix except temporarily when disturbed. In the severe winter of 1951 the grouse completely deserted the slopes above 2,000 feet, while the ptarmigan remained between 2,000 and 2,500 feet. Ptarmigan have been seen after a January storm as low as 850 feet in a wood near Loch Laggan, and lower still on the coast near Aberdeen.

Ptarmigan feed mainly on the shrub heaths. The shoots of heather, crowberry and blaeberry are the staple diet with smaller amounts of cowberry, bog whortleberry and mountain azalea. The flowers and berries of these plants are also eaten and occasionally the shoots and seeds of least willow, heath bedstraw, rushes, saxifrages and others. Insects are taken in summer with traces of lichens and mosses. The amounts of the staple foods taken vary according to the altitude at which the birds are ranging; the higher birds take more blaeberry and crowberry, and the lower more heather. When forced down in winter the birds live almost entirely on heather tips.

The winter flocks break up between the end of February and mid-April during the first long spell of spring sunshine which melts the snows and exposes patches of the montane heaths. The cocks take up territories which they occupy all day in late spring and which they defend against all intruding cocks, of which there are many without territories. The hens move into the territories, but not all hens are successful in finding a mate. Different hens often associate with different cocks on different days, but as spring advances the bond between cock and hen becomes stronger, and the same hen is found with the same cock.

In the late afternoons in February and March the ptarmigan may leave their territories and flock; if the snow returns, flocking lasts all day, with the territory mosaic becoming re-established in the ensuing thaw. Hens return to the flocks more readily than cocks, but if good weather persists both birds remain continuously in the territory, leaving only during sexual displays, fights with intruders and when escaping from predators including man. The cocks have song-flights, rising up to about twenty-five feet with a rattling call which can be heard for a mile.

The average clutch is six or seven eggs, and there is a hatching success of about ninety per cent. The number of chicks successfully reared in a brood varies from one to ten, but many hens do not rear any. The nest is usually a scrape lined with moss in the shelter of a large stone. In summer they remain silent and move little except when leading an intruder away from their brood. On hot days they may move to snow beds and actually bathe in the snow. The families break up in September and October when there is again territorial and sexual display in the early morning. It is at this time that pairing of birds occurs. Ptarmigan turn white between October and January and are usually pure white in February. They turn dark again in March and April, but retain white bellies, legs and wings. These colour changes occur in complete body moults. Ptarmigan, hare and fox may dig holes in the snow for shelter and insulation, and the small mammals are snug in their tunnels under the snow all winter.

Invertebrates show much less specialization: for example, among the 52 recorded species collected by W. S. Bruce (1896), on the summit of Ben Nevis in the Observatory days, it was said that thirty could have been found in an English parish. Arctic species such as the weevil *Otiorrhynchus arcticus,* the sawfly *Amauronematus abnormis* and the dung bettle *Aphodius lapponum* occur in the high Cairngorms. The commonest flying insects of the warm summer's day are craneflies, green bottles, bibionids and empids.

There are a few species of Lepidoptera which seem disposed to inhabit the tops. The moths *Crambus ericellus* and *C. furcatellus* are mountain species though they are also found on moors at mid-height. They are daylight fliers. The larvae feed on mosses, particularly club moss (*Lycopodium* spp.). The eastern Cairngorm tops and Lochnagar are the only British stations of the mountain burnet moth (*Zygaena exulans*). The caterpillar feeds on such characteristic mountain plants as moss campion, cyphel and mountain azalea. The netted mountain moth (*Fidonia carbonaria*) has a much wider distribution, being found on many summits north of Perthshire ; it is nevertheless local in its appearances. The caterpillar of the netted mountain may be found on the highest-growing dwarfed birches, on blaeberry and on bearberry. The moth appears in April and May, when the summits are often still under snow. The broad-bordered white underwing (*Anarta melanopa*) and black mountain moth (*Psodos coracina*) are locally common between 2,500 and 4,000 feet on eastern hills. The northern dart (*Agrotis hyperborea*) occurs locally above 2,000 feet on the Monadliadhs and the hills of Angus. It feeds on the montane shrub heaths and has a biennial life-cycle with about two years spent as a caterpillar. It is valued by collectors and is in need of protection (Tod, 1953).

Spiders apparently show a slightly greater tendency to specialization. Like other invertebrates they live much of their lives under stone and snow escaping from cold and wind. Many do not have webs. Such species as the following occur at over 3,000 feet in the Cairngorms and other hills in the Central Highlands: *Lycosa amentata, Oligolophus morio* var. *alpinus, Trochosa biunguiculata, T. andrenivora,* and *Pedanstettius lividus.* Bristowe (1927) points to the great advantage in altitudinal range possessed by those spiders that live under stones and which walk abroad only when conditions are optimal. Such spiders are not subjected to anything like as wide a range of temperature as are web-making species.

The relict quality of the summits is emphasized once more by the discovery by Murray (1906) of specimens of both arctic and antarctic species of those primitive and microscopic arthropods, the Tardigrada or water bears, in moss on the summit cairn of Ben Lawers.

The deep and precipitous corries and the spiry summits may cause awe, but the high grasslands on a summer day have an idyllic quality. They are remote and quiet. They are green and kind to the eye. They are ease to the feet. The flowers have great variety and a new beauty, and the very pebbles among which they grow have a sparkle and show of colour. To climb to one of these alps of grass and descend again in a few hours is not enough. Take a little tent and remain in the quietness for a few days. It is magnificent to rise in the morning in such a place. The only sounds breaking the silence, if you get the best of the early July weather, will be the grackle of the ptarmigan, the flute-like pipe of the ring ouzel, and perhaps the plaint of a golden plover or a dotterel. See how the deer, now bright-red-coated lie at ease in the alpine grassland. Listen, if you have stalked near enough, to the sweet talkings of the calves who are like happy children. Of your charity disturb them not in their Arcadia.

Chapter Nine

The Sea-Coast and Coastal Seas

THE MARITIME GRASSLANDS

Not all Highland coasts are wild and forbidding. There is
an extreme contrast, however, between the western points
such as Ardnamurchan, Greenstone and Barra Head, and
the sheltered sea inlets of Loch Broom, Loch Sunart and
Loch Fyne. The one means ocean, the other little more
than salt water without storm. The winds on the outer coast
render the habitat comparable in many ways with the
summits of the hills. The outer coasts are washed with
driven spray and rain so that what little grass they grow
is sweet and clean. The main difference between the
summits and the outer coasts is in winter temperature and
the effects of snow-lie. Apart from that the rigour is
similar. Many plants of the shore, the cliff face and the
top of the cliff have to endure drought conditions caused
by lack of humus acting as a sponge for water and by
extreme paucity of soil for their roots. It is common, therefore, to find plants with tightly rolled leaves such as viviparous sheep's fescue which we found on the summits ; some
plants have deeply ribbed leaves, very stiff, such as marram
grass (*Ammophila arenaria*) ; some have polished glossy
leaves, like scurvy grass (*Cochlearia officinalis*) ; there are
little linear crowded leaves like those of sea pink, and fleshy
ones like roseroot and English stonecrop (*Sedum anglicum*)
(plate XII) which absorb moisture and store it. All these
plants are more or less drought-resistant. Then there are
others such as orache (*Atriplex patula*) and bladder
campion (*Silene maritima*) which are common on shore
and cliff and yet do not show any particular adaptation of
leaf to a droughty or salty habitat. The curled dock

(*Rimex crispus*) resists some measure of drought and occurs as far afield as North Rona, nearly fifty miles from other land.

Another common plant of the cliff edge and shingle beach which is highly salt- and drought-resistant is scentless mayweed (*Tripleurospermum maritimum*). Sea milkwort (*Glaux maritima*) is very closely linked with the spray zone, or the intertidal area of low shingle beaches where wave action is not too great to disturb its rooting. Its rather fleshy leaves turn bright yellow in conditions of extreme drought. Silverweed (*Potentilla anserina*) comes down as far as the high-tide mark but not below, and achieves its drought resistance by a fine hairiness which calms the air near the stomata of the leaves and thus slows transpiration. Such hard-surfaced, glossy-leaved xerophytes as sea holly (*Eryngium maritimum*) do not occur generally on West Highland coasts, though this species is recorded from South Argyll, Tiree and Skye.

Salt marshes are not of considerable extent in the Highlands and cannot be compared in richness of flora with those of England, which are generally on a much more muddy base. The estuarine salt marshes of the Highlands are on alluvial gravel and are always heavily grazed by sheep, which type of grazing in itself impoverishes the flora. There are other patches of salt marsh in the Outer Hebrides where the sea reaches into the ramifying inlets, mostly on the east coast. The salt-marsh zone stretches from the high-water mark of neap tides to just above the high-water mark of spring tides. Generally speaking it is a form of close grassland intersected by channels and studded with shallow holes where the grassland stops abruptly. The mud or gravel there is tinged with the green and blue-green algae, and quite different from the vegetational complex of the shingly and rocky shore line described below.

The occurrence of definite periodic immersion in salt water is, of course, the outstanding ecological factor in the life of the salt marsh. Immersion varies from regular twice-daily periods of some hours to fortnightly occasions of possibly less than an hour. The plants are halophytic, i.e. having a high tolerance of salt. Their cellular structure is similar in some respects to the xerophytic or drought-resistant plants. Salt-marsh plants are also subject to considerable fortuitous changes of salt concentration in their root medium, depending on such factors as rainfall and hot sunlight during periods of emersion.

The vegetational complex of a Highland salt marsh is commonly made up of the following plants. Dominant members of the flora are marked with an asterisk.

Marsh samphire	*Salicornia herbacea*
Sea pink or Thrift	*Armeria maritima**
Sea aster	*Aster tripolium*
Sea milkwort	*Glaux maritima*
Sand-spurrey	*Spergularia salina*
Sea plantain	*Plantago maritima*
Buck's-horn plantain	*P. coronopus**
Red fescue grass	*Festuca rubra**
Seablite	*Suaeda maritima*

On the high promontories of St. Kilda and other outlying islands there is a peculiar spray-washed grassland called *Plantago* sward. This is a favourite pasture for the Soay sheep which keep it smooth as a carpet. The important species in order of dominance are:

Red fescue	*Festuca rubra*
Sea plantain	*Plantago maritima*
Sea pink	*Armeria maritima*
Autumnal hawkbit	*Leontodon autumnalis*
Buck's-horn plantain	*Plantago coronopus*
Procumbent pearlwort	*Sagina procumbens*
Common mouse-ear chickweed	*Cerastium holosteoides*
Scurvy-grass	*Cochlearia officinalis*

Another peculiar salt marsh occurs on the low peninsula of Fianuis on North Rona. It is a *Stellaria* sward dominated by the common chickweed (*S. media*); it contains a large

H

number of annuals and covers most of the breeding grounds of the great seal herd. It is more fully described in chapter 10. The presence of nesting sea-birds and sheep greatly modify the constitution of spray-washed swards and good examples of this can be seen at St. Kilda. Hirta has *Plantago* sward heavily grazed by sheep but without puffins; Dun has "puffin" swards honey-combed with burrows but has no sheep; Boreray has both sheep and puffins. Dun and Hirta are separated by a narrow sea-filled chasm, and a comparison of the swards on either side is outstanding in island ecology.

There is another type of coastal vegetation in the West Highlands which must have special mention—namely, the *machair* already described in its physical features. The *machair* starts with the florally sterile tidal zone of shell-sand, then there is the bank of unstable dunes, on the sea-ward edge of which the marram grass begins to grow thinly. The marram is locally called "bent" but should not be confused with what is botanically referred to as bent (*Agrostis* spp.). The importance of the marram cannot be overestimated; it is one of the first plants to stabilize sand. The dune-*machair* vegetation has been described by Gimingham *et al.* (1948), MacLeod (1948) and Vose *et al.* (1957) in Harris, Barra and Tiree, respectively.

In the foreshore community marram is often accompanied by sand couch-grass (*Agropyron junceiforme*), sea rocket (*Cakile maritima*), sea sandwort (*Arenaria peploides*), ragwort (*Senecio jacobaea*), curled dock (*Rumex crispus*), Babington's orache (*Atriplex glabriuscula*) and bird's-foot trefoil (*Lotus corniculatus*). All those, with the possible exception of sand couch-grass, play little or no part in dune formation. Were it not for the spiky leaved marram grass there would be no rich *machair*.

The wind makes a throat in the sand and fans out on the landward side of it, removing sand and potential grass-land. There should never be grazing in the marram zone, nor should leave be given to cut "bents". This was done some years ago at Dunfanaghy, Co. Donegal, with alarming results. A large bridge on a main road and some houses were enveloped. Inspired work on the part of the Eire Department of Lands not only stopped further trouble, but caused the displaced sand to be blown back to where it had come from. Artificial dunes were started with whin bushes and wire fencing and bents were planted a few inches apart all over them. Landowners of the past knew the importance of marram and in the Tiree leases of the Dukedom of Argyll it was specially stated that if a man should find a hole in the dunes, he should fill it and plant a bent therein.

One of the most dramatic instances of collapses in the marram bondage is at the isthmus of Eoligarry, Barra, to the west of the great cockle-strand. With the combined effects of wind, rabbits, cattle, sheep, ponies and man himself, the fifty-foot rampart has crumbled, leaving isolated shaggy-headed dunes and causing a large-scale transgression of sand over the *machair*.

There is something almost purposive to the eyes of the human observer in the contemplation of a rhizome of sea sedge (*Carex arenaria*) moving forward and thrusting up a shoot every few inches. In a grossly overgrazed sandy grassland at Opinan in Ross-shire, the rhizomes of the sea sedge were thrusting out in all directions in the erosion scars. The conspicuous shoots were immediately pulled up and eaten by cattle and sheep at the same time exposing the rhizome which subsequently was caught in the hoofs and torn out of the sand.

The fully established dune pasture is a loose turf of red fescue grass, meadow grass (*Poa pratensis*), ladies' bedstraw (*Galium verum*), white clover (*Trifolium repens*) and ribwort (*Plantago lanceolata*) with scattered marram. It is when the marram has disappeared from the sward that the true *machair* is seen. The nitrogen-fixing legumes become more frequent as the marram becomes thinner and the soil more stable. These create conditions for other flowers and grasses to strike and soon a rich shell-sand grassland is developed, heavily grazed by crofters' stock. Most *machairs* are too heavily grazed and the sand is showing through the grass.

In bygone days great damage was caused on the *machair* by carts carrying heavy loads of seaweed to the kelp factories; when the ruts were inconveniently deep the cart took a new track. This is well shown at Balephuil, Tiree, where the *machair* is corrugated by the great expanse of tracks long since restored by nature to green meadowland. The advent of the pneumatic tyre has resulted in the reduction of damage. It is not often that one sees dunes increasing to give promise of new land but this is the case at Balevuilin, Tiree, where there are no rabbits to increase the grazing pressure and dig up the sand.

The flowers of the *machair* and the profusion of common blue butterflies (*Lycaena icarus*) make it a brilliant place in July, offset by the blue of sky and sea, the white edge of surf and the cream expanse of shell-sand beaches. Here is a list of flowers and mosses which one might expect to find in any *machair*:

Zigzag clover	*Trifolium medium*
Red clover	*T. pratense*
Wild white clover	*T. repens*
Hop trefoil	*T. campestre*
Red fescue	*Festuca rubra*
Cow parsnip	*Heracleum sphondylium*
Common storksbill	*Erodium cicutarium*
Ragwort	*Senecio jacobaea*
Tufted vetch	*Vicia cracca*
Lesser meadow rue	*Thalictrum minus* subsp. *arenarium* and *T. minus* subsp. *minus*
Yarrow	*Achillea millefolium*
Common dandelion	*Taraxacum officinale*
Lesser dandelion	*T. laevigatum*
Perennial rye-grass	*Lolium perenne*
Silverweed	*Potentilla anserina*
Mouse-ear hawkweed	*Hieracium pilosella*
Lesser knapweed	*Centaurea nigra*
Yellow-rattle	*Rhinanthus minor* and *R. m.* subsp. *stenophyllus*

Crested dog's tail	*Cynosurus cristatus*
Field woodrush	*Luzula campestris*
Sand couch-grass	*Agropyron junceiforme*
Fiorin	*Agrostis stolonifera*
Sand sedge	*Carex arenaria*
Carnation-grass	*C. flacca*
Crested hair-grass	*Koeleria gracilis*
Buck's-horn plantain	*Plantago coronopus*
Sea plantain	*P. maritima*
Ribwort	*P. lanceolata*
Yellow oat	*Trisetum flavescens*
Sea pearlwort	*Sagina maritima*
Marsh horsetail	*Equisetum palustre*
Harebell	*Campanula rotundifolia*
Hairy oat	*Helictotrichon pubescens*
Smooth hawk's beard	*Crepis capillaris*
Dove's foot cranesbill	*Geranium molle*
Blood-red cranesbill	*G. sanguineum*
Primrose	*Primula vulgaris*
Buttercup	*Ranunculus bulbosus*
Milkwort	*Polygala vulgaris*
Mouse-ear chickweeds	*Cerastium holosteoides* and
	C. semidecandrum and
	C. atrovirens
Wild pansy	*Viola tricolor*
Cathartic flax	*Linum catharticum*
Bird's-foot trefoil	*Lotus corniculatus*
Kidney vetch	*Anthyllis vulneraria*
Wild carrot	*Daucus carota*
Ladies' bedstraw	*Galium verum*
Daisy	*Bellis perennis*
Spear thistle	*Cirsium lanceolatum*
Creeping thistle	*C. arvense*
Cat's ear	*Hypochoeris radicata*
Sow thistle	*Sonchus oleraceus*
Forget-me-not	*Myosotis versicolor*
Germander speedwell	*Veronica chamaedrys*
Eyebright	*Euphrasia officinalis*
Red-rattle	*Bartsia odontites*

Wild thyme	*Thymus drucei*
Selfheal	*Prunella vulgaris*
Frog orchid	*Coeloglossum viride*
Marram	*Ammophila arenaria*
Yorkshire fog	*Holcus lanatus*
Meadow grasses	*Poa* spp.
Wooly fringe-moss	*Rhacomitrium lanuginosum*
Moss	*Rhytidiadelphus squarrosus* *Pohlia nutans* *Acrocladium cuspidatum* *Mnium cuspidatum*

In short, the *machair* is a natural grassland of calcareous type consolidating land initially stabilised by marram. Moderate grazing perpetuates it by adding organic matter and keeping a clear sward. The people of the country have in the past effected control of grazing by each crofter having a given "souming" of stock, and in the old days the township constable, appointed from among themselves, saw to it that the souming was not exceeded. The Crofter's Act of 1886 did away with the constable and there is evidence that the souming is now often exceeded, to the detriment of the *machair*.

The dunes, being still unstable and receiving the force of high prevailing winds, have an extremely scanty population of Mollusca; but as soon as the stable dune and *machair* vegetation is reached, full advantage is taken of the lime-rich conditions, and snails abound. *Helicella itala, Cochlicella acuta, Cochlicopa lubrica* and *Vitrina pelucida* are the main grassland species of snail; *Helix aspersa, Hygromia hispida*, and *Oxychilus alliarius* are common on rocks and fence posts and in rank vegetation. *Acanthinula aculeata, Vallonia agrestis, V. excentrica* and *Vertigo pygmaea* are also present. Slugs (*Agriolimax* spp. and *Arion* spp.) are common.

The inner edge of the *machair* bordering the moor is prolific in snails and slugs. Waterston (1936) found thirty species by the roadside between North Bay and Vaslain in Barra. On the moorland there are small green oases usually around a sheep fold or the ruins of crofts and shielings. These harbour island populations of snails and other invertebrates (often small mammals too) while the surrounding moorland supports slugs. Shell production limits the range of environment.

The invertebrate fauna of the *machair* has rarely been studied as a whole. Perhaps the most comprehensive was the work of the Biological Society of Edinburgh University in Barra in 1935. In another survey carried out over twelve months in 1955-56 in Tiree, Boyd (1961) examined the seasonal activity of the invertebrate community on the Reef *machair*. Only the larger forms were collected in pit-fall traps and small creatures such as Collembola and mites were neglected. Approximately 100 species were encountered. The close cropped grasslands teem with small chysomelid beetles (*Longitarsus* spp.), but the biomass of the catch was mainly composed of harvestmen (Phalangida). In September, nine 2-lb. jam jars sunk to the level of the soil caught 1,988 harvestmen and in the following month 2,552 chysomelids. A similar battery of traps set in the rank ungrazed *machair* of the airfield boundary caught 581 common black beetles (*Pterostichus vulgaris*) in one month. The most important species were: *Longitarsus luridus, L. pratensis, Pterostichus vulgaris, Staphylinus aeneocephalus, Calathus melanocephalus, Tachyporus* spp., *Myrmica rubra, M. scabrinodis,* Tipulidae (3 species), Homoptera (5 species), earthworms (10 species) and spiders (18 species). The little green grass-hopper (*Omocestus viridulus*) was common and the carrion beetles (*Necrophorus humator, N. invesitgator* and *N. vespillo*) were trapped with the corpses of field-mice which had been killed in the traps.

The untilled *machair* soil is usually a six-inch deep veneer of dark brown loam. It is a vegetable-mould/shell-sand mix, and is very uniform containing no stones. In most places this has been converted by cultivation into a loamy till of over a foot in depth and including stones brought up from the beach with seaweed manure. Undis-turbed *machair* soils are now scarce in the Hebrides, most having been ploughed or burrowed by rabbits or over-blown by sand transgressions. Some of the best examples of extensive undisturbed *machair* soils are to be found in Tiree where cultivation is restricted to enclosed ground and there are no rabbits. The pH of *machair* soils is about 7.9, the lime content about 50 per cent. The dung of sheep and cattle play an important part in the cycle of soil nutrients and the maintenance of a balanced flora and fauna. Many insects and earthworms would be unable to colonise the bare open close-cropped grassland without the presence of dung pats and stones. Stones are sometimes rare and the ants, earwigs (*Forficula auricularia*) and earthworms (*Lumbricus rubellus*) are found sharing the dung pat. The large iridescent dung beetle (*Geotrupes stercorarius*) tunnels below the fresh dung pats through to the sand, where a dung-filled cavity is excavated for egg-laying.

The vegetation of the sea-cliffs is closely related to that of the outlying islands mentioned in chapter 10. What is probably a typical Hebridean cliff community from Greian Head, Barra, is listed by Watson and Barlow (1936) as including:

Scurvy-grass	*Cochlearia officinalis*
Sea campion	*Silene maritima*
Dark-green mouse-ear chickweed	*Cerastium atrovirens=C. tetrandrum*
Procumbent pearlwort	*Sagina procumbens*
Rose-root	*Sedum rosea*
Scentless mayweed	*Tripleurospermum maritimum*
Field milk-thistle	*Sonchus arvensis*
Autumnal hawkbit	*Leontodon autumnalis*
Sea milkwort	*Glaux maritima*
Sea pink	*Armeria maritima*
Sea plantain	*Plantago maritima*
Buck's-horn plantain	*P. coronopus*
Ribwort	*P. lanceolata*
Curled dock	*Rumex crispus*

Sheep's sorrel	*Rumex acetosella*
Hastate orache	*Atriplex hastata*
Vernal squill	*Scilla verna*
Sea arrow-grass	*Triglochin maritima*
Mud rush	*Juncus gerardii*
Red fescue	*Festuca rubra*
Sea spleenwort	*Asplenium marinum*

THE SEA SHORE

The sea's edge is a frontier, a place of action whether it be on the quiet shores of the sheltered sea-lochs or the cliff faces of an outer coast (plate x). There are two tides a day, there are sudden changes through the action of heavy spates which occur so often in the West Highlands and there is a constant reaching up and reaching down of species from one element to the other. From top to bottom of the shore there is a wide range in habitat conditioned by the alternation of emersion and immersion ; there is also a wide range along the shore conditioned by exposure and shelter. What suits one organism does not suit another, and within the continuous graduation from the vertical walls of St. Kilda to the spacious tidal flats of Islay each is adapted for life within strict limits.

J. R. Lewis (1957) and in *The Ecology of Rocky Shores* (1964) has described the biological character of western Scottish coasts, comparing the zonations of the exposed shores with those of the sheltered lochs and kyles. The exposed coasts have a fringe of spray-washed rocks above the tidal zone, to which very few plants and animals are adapted ; it is neither land nor shore but a sort of no-man's-land between the two. Here grey, orange and black lichens are found in descending order (*Lecanora* spp., *Ramalina* spp., *Placodium* spp., and *Verrucaria maura*), intermittent growths of channelled wrack (*Pelvetia canaliculata*), flat wrack (*Fucus spiralis*), *Porphyra umbili-calis,* small periwinkles (*Littorina neritoides, L. rudis*) and barnacles (*Chthamalus stellatus*). The main shore populations occur below this fringe though in some localities where wave action is strong, intertidal animals such as the common limpet (*Patella vulgata*) and the mussel (*Mytilus edulis*) extend upwards. In similar situations *Chthamalus* extends downwards, and in very exposed situations such as at St. Kilda, is found at low water mark. T. B. Bagenal (1957) found some unusual molluscs and crustaceans well above sea-level there.

The main intertidal zone of the exposed rocky shore is dominated by the acorn barnacle (*Balanus balanoides*), the mussel, the limpets, and the bladderless form of the bladder wrack (*F. vesiculosus*). The barnacle line is usually well developed but is broken by heavy wave action. The bladder wrack occurs about a foot below the barnacles and is also broken up by strong surf ; there is seldom full local coverage of the shore as often occurs with the dominant weeds on the loch shores. The mussel usually occupies the lower half of the barnacle zone to the fringe of surf-washed rocks below low-tide mark. Where wave action is great the pink encrusting *Lithothamnion*, the large brown tangle-weed (*Alaria esculenta*) and the limpet (*Patella aspera*) which usually inhabit the lower shore extend well upwards and a gradual depletion of the mid-shore populations takes place at mid-tide level. Red algae are common on the lower shore, *Rhodymenia, Laurencia, Corallina* and *Gigartina* being the most widely distributed. In sheltered localities *Alaria* is replaced by *Laminaria digitata*.

The exposed coasts have superlative beaches which delight the eye. The west coast of the Outer Hebrides from south Harris to Barra is one great sweep of surf-swept shell-sand. The churning sand and pebbles is a desert of alternating calm and storm where no plant or animal can find a footing. Such well-known inhabitants of the sheltered sandy shore as the lugworm (*Arenicola marina*), the cockle (*Cardium edule*) and the small bivalve (*Tellina tenuis*) are rare, but where the strand runs from the exposed beach to the sand-silted inlets, these animals become abundant.

There is an interesting exposed beach at Village Bay,

St. Kilda. It is a two hundred yard stretch of pale yellow sand of low shell content. In winter the sand recedes, exposing more of the huge boulder beach upon which it rests, and at low spring tide the sand may not be exposed. In summer the sand returns to the level of the high spring tides and about 50 yards may appear above low-water mark. Movements of sand occur on all beaches, but an annual beach of this type must be comparatively rare. Gauld, Bagenal and Connell (1953) surveyed the fauna of this beach and remarked that there was only one resident species, the isopod (*Eurydice pulchra*).

Though the sheltered shores of the firths and sea-lochs are a different type from those of the open coasts, there is a continuous gradation of conditions from one to the other. The open coast communities on Ardnamurchan (plate x), for example, change gradually as the coastlines run eastward to the shelter of Lochs Sunart and Moidart. The reduced wave action in turn reduces the effects of spray above the tidal shore and results in stable beaches varying from boulders of mud. The wide band of lichens which may extend for over a hundred feet on the bastions of St. Kilda are compressed to a few feet on the loch shores and are replaced locally by salt marsh. The black and orange lichens are usually submerged at high spring tide. The fauna is poor on the upper limit of the shore consisting mostly of amphipods, isopods and fly larvae living mostly in deposits of rotting seaweed.

The shores of the sea-lochs are dominated by fucoid weeds with densest growths on the tracts of large stable boulders or wide exposures of bedrock. The characteristic expanse of knotted wrack (*Ascophyllum nodosum*) stretches from the band of channelled and flat-wracks immediately below the lichens of the spray-fringe, to the laminarian forest below low-water mark. Some flat wrack usually occurs on the shore above the *Ascophyllum* and sandwiched between it and the Laminariae is another thin band of serrated wrack (*Fucus serratus*). The bladder wrack is seldom conspicuous except in strong tidal currents which occur at the entrances to many sea-lochs and through kyles and on mid-shore shingle. In places where boulders are graded to shingle the knotted wrack is replaced by the flat wrack, and in estuaries, where sharp changes of salinity occur, the bladder wrack, *Fucus ceranoides,* and a specialised form of knotted wrack (*A.n.*f. *mackaii*) occur with the fresh green *Enteromorpha* spp.

Among the neatly zoned fucoids there is a rich and less conspicuously zoned fauna. Below the lichens and salt marsh there are stratified communities of acorn barnacles, periwinkles (*Littorina rudis, L. littorea* and *L. littoralis* in descending order), the purple top shell (*Gibbula umbilicalis*), limpets, mussels and amphipods occupying the middle and upper levels of the boulder shore. On the lower shore, most mid-shore communities die out as the knotted and serrated wracks are replaced by laminarians. *Laminaria digitata* gives way to *L. cloustonii* and *L. saccharina* at greater depth. *Saccorhiza bulbosa* has a large frond and a large hollow, papillated swelling which looks rather like coarse tripe turned inside out.

The laminarian fauna is rich and includes limpets, tube-worms (*Pomatoceros*), a barnacle (*Balanus crenatus*), hydrozoa, polyzoa, sponges (Porifera) and sea-squirts (Ascidiacea). On the shingle shore the laminarians are accompanied by *Chorda filum, Halidrys siliquosa, Codium fragile* and the beautifully glossy thong-weed (*Himanthalia elongata*) though the thong-weed occurs more commonly on the exposed coasts. Where the bed is muddy, eel-grass (*Zostera marina*) often occurs. Among the laminarians are also found the silver tommies (*Gibbula cineraria*) and the sea urchins (*Echinus esculentus* and *Psammechinus miliaris*).

The fucoid weeds may contribute to the gradual change of the shore. Bladder wrack and knotted wrack will each grow on a stone and when the plants get large enough they will almost float the stone. At least, wind and wave action are sufficient then to make the weed carry the stone ashore and deposit it near the high-tide mark.

The wrack weeds are used by man for manure especially on the eastern side of the Outer Hebrides where sheltered inlets are numerous, but laminarians are of much greater importance for this purpose, especially on the Atlantic side where immense quantities of the weed are torn from the sea floor and washed ashore in spring. The tractors and trailers are kept going busily in spring carrying the torn weed from the tide mark on to the ploughed ground and meadow land from the crofts. Most of this ground on the west side of the Outer Hebrides and some of the Inner Hebrides is *machair*. When ploughed, this limy, sandy soil must be kept well supplied with organic matter if it is to remain in good heart. The annual casts of laminarians provide exactly what is needed.

The luxuriant growth of wrack from the east coast inlets of the Uists is cropped and taken by lorry to the processing stations near Lochboisdale and Lochmaddy. This provides employment for a few full-time and a large number of part-time workers throughout the length and breadth of the Isles. Most of the Atlantic crofting townships have shore where, after the winter gales, great quantities of "tangle" are cast. The stipes of the laminarians are gathered by young and old and dried on stone emplacements made for the purpose. Work sometimes goes on all winter and by the early summer an ablebodied schoolboy may have gathered and dried several tons. As much as 26 tons of dried tangle has been gathered by one family there in one season and the amount of weed handled to achieve this may be more than ten times as much. A recent rate of payment was £8 per dry ton.

At low spring-tide level, carragheen (*Chondrus crispus*) and dulse (*Rhodymenia*) are found. Both are red algae, gelatinous and edible. Though still in general domestic use in western Ireland, they are not used much now in the Hebrides. The older generation of housewives used carragheen as a thickener for milk to make a blancmange-like mould. It is a pity that Hebridean and West Highland hotels and boarding houses do not include this regularly on the menu.

There can be few amusements so absorbing as wandering among the Laminariae at the equinoctial low spring tide, where marine life has amazing variety and beauty. There are squat lobsters (*Galathea* spp.), pellucid limpets (*Patina pellucida*) and hundreds more forms of invertebrate animal life: molluscs, crustacea, echinoderms, sea-squirts, hydroids, worms, sponges and teeming protozoans. One should look under boulders for there may rest a lobster (*Homarus vulgaris*) waiting for the tide to return. Perhaps a "hen" lobster may be found to be "berried" carrying her 10,000 eggs glued among her legs. The lobster fishermen are supposed by law to return berried hens to the sea as an obvious measure of conservation. Sometimes a fisherman may pluck a few hundred of the tiny black eggs between his thumb and first finger, for the taste of these eggs is most strange and delicate.

The ship's barnacle (*Lepas anatifera*) with shells almost an inch across and with fleshy stalks up to a foot long, is not seen unless a floating log comes along from far out in the Atlantic. Acorn barnacles have free-swimming larvae in April, when a fisherman is wise not to bring his boat into a harbour where it will dry out with the tides, because on such a surface alternately wet and dry the barnacle takes its stance, and by the end of the summer has grown crusty enough to retard the speed of the boat. And yet, if the boat remains always in the water at a mooring, young *Laminaria* and other algae will begin to grow on the bottom of her so that she must be dried out, scraped and dressed with anti-fouling paint. Floating timber is frequently infested by the wood-boring mussels (*Teredo norvegica, T. megotara* and *Xylophaga dorsalis*) and drift wood cast up on the shore is usually burrowed both by these and the wood-boring crustacea *Limnoria lignorum* and *Chelura terebrans*. *Teredo* burrows can usually be identified by their lining of lime.

The artistry of the sea upon the shore rocks produces an infinite variety of shapes, some of which might have been the work of a human sculptor. The sculpture is moulded together into the whole grotesque architecture of the cliff-faces and the rolling shores. The graven surface is embellished with an equally infinite variety of rock pools, each of which is a unique and vividly coloured gem; each has its own quota of plants and animals. The position of the pool on the shore will determine the creatures which live in it. Exposure to surf is, of course, an important factor, but there are also the effects of temperature and the depletion of oxygen. Competition for living space and food is heavy.

One of the first things which catches the eye is the darting shadow of a shanny (*Blennius pholis*), or butter fish (*Centronotus gunnellus*), or black goby (*Gobius niger*), or spotted goby (*G. ruthensparri*), or bullhead (*Cottus* spp.). Most pools are lined with coralline weeds which are lime encrusted and bright pinky purple in colour. *Lithophyllum* lies flat on the rock and completely covers it, as if an artist had worked it on with a palette knife. Semi-transparent shrimps (Mysids, *Praunus* spp., Gammarids) flit to and from the cover of the red and green weeds which are also the refuge of shore crabs (*Carcinus maenas*) and hermit crabs (*Eupagurus bernhardus*). The catalogue of life in these pools is long and the reward which comes to those who potter among them in warm summer days is unending.

The three-spined stickleback (*Gasterosteus aculeatus*) is a fish with a wide range of salt tolerance. It probably occurs in most of the estuaries in the West Highlands and the Islands. The orange-red belly of the male may be seen to advantage if he has just chased another away from the nest which he guards so assiduously. At other times it is paler. This highly adaptable fish is common in shore, stream and loch in Tiree and is even present in the tidal pools on Leac Mor Fianuis on remote North Rona. This extreme accommodation to a varying range of salinity is a character probably developed by natural selection in an environment which through storms, splash, rainfall and drought is peculiarly liable to sudden and continual changes in salinity. The larger fifteen-spined stickleback (*Spinachia vulgaris*) is more truly a salt-water species and may be found in the larger and lower rock pools.

An organism which emphasizes the phenomenon of the Atlantic Drift lapping some of our British shores and not others is the sea urchin (*Echinus esculentus*). D. M. Reid (1935) explained its present distribution as an almost direct consequence of the Atlantic Drift. The sea urchin is absent from the Channel coast and from Wales and the east coast of Ireland as a littoral form but is very plentiful on the west coast and off the West Highland coasts. It occurs in Shetland and all the way up the Norwegian coast as far as the North Cape. The brilliant red and purple sea urchins are found attached to the rocks and cliff face at about low spring-tide level, or on large leaves of tangle below the surface of the sea, for they are vegetable feeders. Their temperature range is very great—from 4.2°C. in deep waters to 17-18°C. off the north of Spain.

The influence of the Atlantic Drift is again apparent in the occurrence of crabs of western Irish distribution in the Inner Hebrides. These were discovered by Nicol (1939) and Campbell (1948) in waters off Muck and Canna, and were *Xantho incisus, Pilumnus hirtellus, Pirimela denticulata* and *Corystes cassivelaunus*. The strong current from the west of Ireland evidently carries the planktonic larval stages of these crabs north-eastwards. The intertidal zones of the most isolated islands of St Kilda (Gauld, *et al.,* 1953) and Rockall (Fisher, 1956) have been described. That of Rockall would seem to have four main zones:

1. The *Alaria* zone. 9-12 feet above l.w.m., is the lowest dominated by *A. esculenta.*
2. Zone of red (and some green) algae. Above the *Alaria* for about 8-10 feet the rock is apparently sparsely covered with algae: *Ceramium rubrum, Polysiphonia ureolata, Porphyra leucosticta, Ulva lactuca, Enteromorpha compressa, Rhizoclonium riparium, Rhodochorton rothii, Bangia fuscopurpurea.*
3. The black zone. 10-12 feet vertically, dominated by blue-green algae *Pleurocapsa kerneri, Plectonema battersii* and *Phormidium fragile,* small green algae, *Blidingia minima, Prasiola stipitata* and the black lichen *Verrucaria. Calothrix scopulorum* is also

present here. The effective tidal range seems to be of the order of 20 feet

4. The green zone. From 40-50 feet to the top has much of the rock bare but there are patches of the green algae *Prasiola stipitata* and *Blidingia minima* with the blue-greens of the black stone. *Hildenbrandia proto-typus* (red), *Enteromorpha prolifera* and *Phor-midium fragile* were also present.

Six species of animal only have been found on Rockall: the rough periwinkle (*Littorina rudis*) an amphipod (*Hyale nilssoni*), an unidentified orange rotifer, a flat-worm (Trematode) larva parasite in the rough periwinkle and two mites *Hyadesia fusca* and *Ameronothrus* sp.

The fauna of the protected sandy shores is very different from the rocky ones which we have so far discussed. Where a sandy shore is well sheltered from wave action and possibly much surrounded by land, it tends to become muddy in the intertidal zone through the accumulation of terrestrial detritus, and by the decomposition of fucoid seaweeds and possible casts of laminarians. In such places the sand may appear black if the condition is extreme, but quite often the blackness appears at a depth of two or three inches and there is an objectionable smell. The smell and the blackness are caused by the generation of hydrogen sulphide (H_2S) from the decay of the sulphur-rich weed, and the deposition of iron sulphide in the sand. Wave action causes aeration and dispersal of the products of decomposition.

The intertidal zones of the Outer Isles are extensive and of exceptional interest. The earliest investigator on scientific lines was W. C. McIntosh (1866), though that observant man Martin Martin (1703) made special mention of the interesting phenomena of the brackish-water lochs of North Uist, which he visited in 1695. For example, Martin records that these lochs, reached by spring tides only and therefore but slightly brackish, may contain cod, ling and mackerel. A. C. Stephen (1935) investigated the North Uist sands and found that the shores subjected to considerable wave action contained no molluscs and but few polychaete worms. Where the strands were sheltered, several molluscs were found, *Macoma baltica* and *Cardium edule* in great abundance. The well-known lug worm (*Arenicola marina*) was common on the Vallay strand and other places but not everywhere. The most remarkable of Stephen's findings in North Uist was the virtual absence of that other very common bivalve mollusc *Tellina tenuis*. The usual co-occupant of the zone, *Donax vittatus,* was also absent. These molluscs usually occur at and lower than the lower end of the *Cardium-Macoma* stratum which reaches three-quarters of the way down to low-water mark. The cockle, though maintaining itself, does not thrive specially well judging by the size of the shell, though in Barra it is large and abundant. *M. baltica* flourishes; the specimens taken in North Uist being the finest met during the survey of Scottish coasts.

On the eastern side of North Uist there is a maze of interconnected lochs and inlets which receive the sea at different states of the tide; each is more or less influenced by the salt water than the next. There are those which are only identifiable at low-water springs and others which are only overwhelmed at high-water springs; the gradation from the mildly brackish to full-strength sea-water is stepped out along the chain and the effects on the flora and fauna are outstanding. Dr. Edith Nicol (1936) made a survey of these lochs. Salinity varies from 30 per thousand down to 2-3 per thousand. The fresh water of the island has a hydrogen-ion concentration of about 5.4 but in the brackish-water lochs it rises to 7.8-9.9. They are rich in species compared with other brackish-water areas in Britain.

The nature of the sub-stratum is in itself of equal importance to salinity in controlling distribution. The burrowing amphipod *Corophium volutator,* for example, occurs only where the bottom is of sandy mud or mud. Nicol's numbers of species found in the lochs were 59 marine, 24 freshwater and 25 brackish-water forms, as well as five euryhaline forms, such as the salmon, which are at home in any salinity. The distribution of the polychaete worms and the bottom fauna of Lochs Nevis and Creran have been investigated by A. D. McIntyre (1956, 1961) and of the littoral sea squirts of Argyll by R. H. Millar (1952).

The sands of the shore are derived from the erosion of the rocks of the district and from materials taken from solution in the sea by plants and animals. All contain different proportions of those derivatives; those containing a high marine fraction are called shell-sands to distinguish them from silica and basalt sands which have a high erosion fraction, though there is no hard and fast line between the two. Analysis of sands from 18 different localities on West Highland and Island coasts showed a range of shell content from 38 to 86 per cent. There is wide local variation; within a single deposit at Durness the range was from 40 to 70, and South Uist sand gave a value of 66 compared with 39 in Benbecula.

The most extensive sandy shores are those immense stretches of shell-sand on the western sides of the Outer Isles and some of the Inner Hebrides—Barra, the Uists, Harris, Tiree and Coll. Their influence on the adjacent land is benign and their cream expanse is a joy to the eye. The dead-white silica sands of Morar are not to us nearly so pleasing, nor do thcy so profoundly influence the coastal strip. At Gruinard Bay, near Loch Broom, and at several places further north on the west coast of Sutherland there are bays of Torridonian sand under rocky shores and cliffs of Archaean gneiss. The sands of Islay and Broad Bay, Lewis, are also Torridonian sands derived from the local rocks. The Kentra sands between Ardnamurchan and Moidart are calcareous to an extent of about 50 per cent, but at low spring-tide level there is a bank of almost pure shell-sand. On the west coast of Mull there are dark grey sands derived from the basalt behind them. There is a particularly fine beach of this type at Carsaig.

Lastly, there are the few banks of coral sand which are exposed for but a short time at low spring-tides. They occur at Tanera Beag of the Summer Isles; at Claggan and near Staffin, Skye; at Duncraig Island, Loch Carron; and at

Erbusaig near the Kyle of Lochalsh. Few people know that such sands exist in these parts and that beautiful pieces of pink and purple-tinted coral, much branched, can be picked up in quantity. This coral is not made by the coral polyps of sub-tropical seas but by a plant, a seaweed called *Lithothamnion* related to the pink encrustations in pools. This weed lives a few fathoms out from shore in not too exposed places and forms a semi-circular cushion about four inches across. Bits break off and are cast up to make the coral sand beach.

The old folk knew the fertilizing value of the shell and coral sands, and thousands of tons were carted to the crofting townships for the inbye land. Today an observer may see black and green ground sharply demarcated here and there, and it was the sand that made this possible. In recent years there has been a large-scale revival of this practice in the Outer Hebrides, where hundreds of acres of moorland are being reclaimed for grazing. With financial and technical assistance from the Government and the North of Scotland College of Agriculture, the crofters are applying ten tons of shell-sand to the acre, together with a few hundredweight of combined fertilizer before sowing a seed mixture containing rye-grasses, clovers and timothy. The full regenerative effect, however, cannot be accomplished in a single treatment and a boost of shell-sand may be required every five years or so. The green flush of the maritime grasslands is, with man's help, invading the moors and changing the complexion of the countryside. The senior author used this method on Isle Tanera in 1940.

Boyd (1958) shows the diversity of birds on the shores of Tiree and Coll, which are fairly typical of the Hebrides as a whole. One of the most striking things about the bird life of the littoral zone on the West Highland and Hebridean coasts is its paucity compared with that of the east coast and its muddy firths. Clean sand is beautiful, but if you want numbers and variety of shore birds there is nothing like mud. Mud means a rich invertebrate fauna, the staple food of so many of the waders and ducks.

The oystercatcher (*Haematopus ostralegus*) (plate XI)

is found almost everywhere, whether the shore is sheltered or open, and has a liking for the cleanest sand though its food consists substantially of molluscs and crustacea which are not found on the surf-swept beaches. The curlew (*Numenius arquata*) is common on the shore whatever the characteristics. It feeds in the intertidal zone on sand, or among rocks and weeds and many are present throughout the year on the estuaries and saltings. The whimbrel (*N. phaeopus*) is seen on spring passage most years, but is not a shore bird. The redshank (*Tringa totanus*) is a bird of curiously local distribution in the West Highlands and Islands, and it too is not a shore bird though it is so commonly found there. Several estuaries of the western mainland are devoid of redshanks, others have several all the year round. Islands which have stagnant gull-haunted pools or places where a great mass of decaying seaweed accumulates are likely to hold a few redshanks. This bird is a scarce breeder in the Outer Hebrides. The high-pitched yelp of the oystercatcher and redshank and the rippling trill of the curlew are dominant voices on most shores.

The ringed plover (*Charadrius hiaticula*) (plate XI) agrees almost equally with the oystercatcher in its ubiquity. It occupies precisely the same habitat, but lives on rather different food than the oystercatcher—smaller mollusca and crustacea, insects and annelids. The oystercatcher and ring plover will nest in very close proximity on shingly shores. The dunlin (*Calidris alpina*) occurs commonly, sometimes in vast flocks, but it is nothing like so common as on the east coast, where it is considered the most plentiful of all shore birds. The winter dunlins on the west may be of the races (*C. a.*) *alpina, shinzii* and *arctica*. Large numbers of turnstones (*Arenaria interpres*) come to both sandy and rocky shores in winter, and small numbers stay all year round. Birds in breeding plumage have been seen in late spring, but no nests have been found. In winter they work assiduously on the patches of mud churned up by the seals on North Rona. The sanderling (*Crocethia alba*) is rarely seen on the western main-

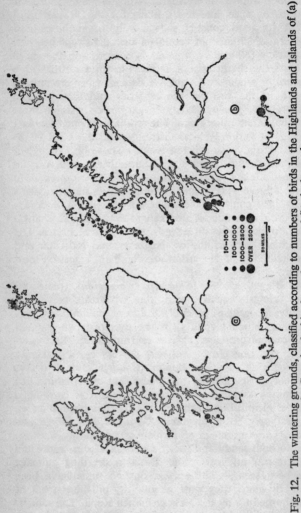

Fig. 12. The wintering grounds, classified according to numbers of birds in the Highlands and Islands of (a) the Greenland white-fronted goose and (b) the barnacle goose

From 'Wildfowl in Britain', Wildfowl Trust, 1963 by permission of the Controller of H.M. Stationary Office land shore, but is common in winter on the strands of the outer fringe of isles from Islay to Lewis.

The greenshank (*Tringa nebularia*) appears on the coasts during migration, and when its breeding haunt is within reasonable distance of an estuary the greenshank comes out of the deer forest or island heath and down to the estuarine flats in the very early morning and late at night. The knot (*Calidris canutus*) is not nearly so common on the west coast as it is on the east. Small flocks find their way to the island beaches in passage. Birds in breeding plumage are occasionally seen on the oceanic islands. The bar-tailed godwit (*Limosa lapponica*) is common in small flocks in winter, and is present throughout the year on the west coast where summer flocks are present notably in Islay, at Loch Indaal, but no nests have been found. The black-tailed godwit (*L. limosa*) occurs on passage; the spring cocks are among the most colourful birds to visit the Hebrides. The little stint (*Calidris minuta*) occurs in small flocks and often singly on passage. The purple sandpiper (*C. maritima*) is a constant winter resident on the exposed rocky shores of the West. It works singly or in small parties, often in company with turnstones and redshanks, following the wash of the waves down the barnacled rocks.

So much for the waders. There is a long list of other species of birds frequenting western shores the status of which is given by Baxter and Rintoul (1953) or Bannerman (1953 *et seq.*). The list includes gulls (*Larus* spp.), terns (*Sterna* spp.), diving and dabbling ducks, shelduck (*Tadorna tadorna*), saw-billed ducks (*Mergus* spp.) cormorants (*Phalacrocorax* spp.), the black guillemot (*Cepphus grylle*), the heron (*Ardea cinera*), crows (*Corvus* spp.), the starling (*Sturnus vulgaris*), the rock dove (*Columba livia*) and the rock pipit (*Anthus spinoletta*). The coastal grasslands have large nesting populations of lapwings (*Vanellus vanellus*), ring plovers, oystercatchers, terns, skylarks (*Alauda arvensis*) and meadow pipits (*Anthus pratensis*). In autumn and spring enormous flocks of lapwings and golden plover (*Charadrius apricarius*) frequent the *machairs,* and in winter flocks of grey lag geese (*Anser anser*), white-fronted geese (*A. albifrons*) and barnacle geese (*Branta leucopsis*), especially in Islay (fig. 12). Snipe (*Capella gallinago*) and dunlin breed plentifully on the marshes beside the *machair* lochs and in late autumn very heavy passages of snipe occur in Tiree and Islay. In an analysis of the game record of Tiree, Boyd (1956) found that 37,247 common snipe and 2,355 jack snipe had been shot between 1929 and 1955. The petrels, auks, kittiwakes and gannets are mentioned in chapter 10.

THE COASTAL SEAS

If one wishes to have a full list of the fishes found in the inshore waters of the Highlands and Islands reference should be made to the *Vertebrate Fauna of Scotland* series, particularly the Outer Hebrides volume (Harvie-Brown and Buckley, 1888) and the Shetland volume (Buckley and Evans, 1899). Further additions have been recorded in such publications as *Scottish Naturalist* and *Annals of Scottish Natural History,* and in *Reports of the Fishery Board of Scotland.*

The sharks, skates and rays are common. The small blue shark (*Carcharinus glaucus*) though once common is now rare. A hundred years ago their livers were an important source of oil in the Hebrides, but the subsequent fickle appearance of the shoals caused the fishery to die out. The same applies to some extent to the basking shark (*Cetorhinus maximus*). In Scottish waters the greatest numbers occur in the Minch but it is common also in the Firth of Clyde and in Shetland. It is not known where they come from; one view is that they come north on migration, but the spontaneous appearance of the fish in Irish, Hebridean and Shetland waters indicates that they most probably move in from deep water along a wide front. The basking shark is the second largest fish in the sea, and may grow to about 30 feet.

After the last war a shark fishery was carried on by Major Gavin Maxwell from the Isle of Soay (Skye) where his factory was based. The project was started in 1946 but the fall in demand for shark oil saw the company in trouble by 1949 and Maxwell gave up. Tex Geddes was a harpooner with Maxwell and continued the shark fishery from Soay and Mallaig in 1950-51 on a reduced scale, in competition with a more highly equipped fleet of Norwegian shark-boats. In these few years with the sharks Maxwell and Geddes learnt much of the habits of these great fish which are described in their books *Harpoon at a Venture* (Maxwell, 1952) and *Hebridean Sharker* (Geddes, 1960). The sharks rid themselves of a superficially-planted harpoon by sounding to the bottom—in possibly 70 fathoms—rolling over and over and either twisting or snapping the shank of the instrument until free. They are strong enough to pull a 76-foot fishing boat for several hours. With all this power the basking shark or sailfish is yet quiet and inoffensive.

The basking shark comes very close inshore. We have seen one rub his side along the end of the stone quay on Tanera and within a stone's throw of the quay of St. Kilda in about twelve feet of water. It progresses at about two knots, feeding on the small surface life of the sea. The water containing the food is taken in through the almost constantly open mouth and is sieved out through the multiple gill slits, the relatively small quantities of food being left behind to be swallowed. Maxwell says that the contents of the stomach are almost indistinguishable in scent and appearance from shrimp paste!

The thrasher or fox shark (*Alopias vulpes*) occur sporadically. They were unusually and unpleasantly common in the summer of 1938. The thrasher is about 12 to 14 feet long, extremely savage in attack on such a creature as a whale, and given to suddenly springing out of the water and coming down a "belly-flopper" which resounds like a gunshot. In attack it uses the tail as a weapon of offence. Harvie-Brown (1888) cites a battle between thrashers and swordfish on the one hand and a large whale on the other. The whale apparently had much the worst of it and the sea was red with blood.

The following is a list of some rare and exotic fishes found off the west coast of Scotland between 1953 and 1967 (Rae and Wilson, 1953-61 ; Rae and Lamont, 1961-64 ; Rae, pers, comm.).

Blue shark	*Carcharinus glaucus*
Six-gilled shark	*Hexanchus griseus*
*Frilled shark	*Chlamydoselachus anguineus*
*Long-finned bream	*Taractes longipinnis*
Black bream	*Spondyliosoma cantharus*
Deal fish	*Trachypterus articus*
Boar fish	*Capros aper*
Sturgeon	*Acipenser sturio*
Pilchard	*Sardina pilchardus*
Ray's bream	*Brama raii*
Red mullet	*Mullus surmuletus*
Trumpet fish	*Macroramphosus scolopax*
Black scabbard-fish	*Aphanopus carbo*
Red band fish	*Cepola rubescens*
Sunfish	*Mola mola*
Electric ray	*Torpedo nobiliana*
Pilot fish	*Naucrates ductor*
Opah or moon-fish	*Lampris guttatus*
Angel-fish	*Squatina squatina*
Eagle-ray	*Myliobatis aquila*
*Bogue	*Box boops*
Stone-basse	*Polyprion americanum*
Pearl-fish	*Fierasfer dentatus*
Black fish	*Centrolophus niger*
Blue-mouth	*Scorpaena dactyloptera*
Tunny	*Thunnus thynnus*

*First record from Scottish waters.

Most of these fish are not strictly speaking coastal. Many have been taken at the fishing banks around Dhuheartach, Skerryvore, St. Kilda and Rockall.

The herring (*Clupea harengus*) is a fish about which a

great deal is known; and yet not enough to predict its behaviour and numbers. The volume of research has been great and the echo-sounders have helped, but fishing for herring is still a gamble. The fish feed largely on the free-swimming copepod *Calanus,* which itself is feeding on the microscopic diatoms. It must be that the micro-plankton differs much in quality, depending on the nutritive salts in the water. It is reckoned that the best herring come from outside the West Hebridean coast.

Herring fisheries take place round the British Isles apparently following a clockwise movement of the shoals. It was once thought that this signified a migration of a single vast stock, but it is now thought that these fisheries arise from different concentrations of herring in different places at different times. Herring may be caught at almost any time of the year on the west coasts of Scotland. The herring fisheries of the west coast are at present prolific and the only real problem is the lack of outlets for the catch. There is no doubt that the herring stocks there at present are under-exploited. There are two methods in use in the west fishery: (i) mid-water pair-trawling which started in 1965-66 and had become a sizeable catching method in 1967 and (ii) purse-seining started in 1966-67. Local stocks in the lochs, which seemed numberless in the 18th century, have, however, almost disappeared. Loch Fyne has still a considerable herring fishery, but not nearly equal to what it was. Lochs Sunart, Creran, Nevis, Broom and Glendhu of the mainland and Lochs Roag and Maddy were all prolific of herring years ago. Take Loch Broom for example: the rich herring fishery was discovered by the Dutch. They fished from sailing busses and apparently made quite long trips to and from the netting grounds. The English speculative companies which were so common in the Highlands in the latter half of the 18th century, saw that the sailing trips to and fro could be shortened by building herring-curing stations on the outer coasts, such as at Isle Martin and Isle Tanera. Now the busses had to go less than a mile into the mouth of Loch Broom or into the Badentarbet Sound, through which the herring came

into the loch. The boats came back inside an hour sometimes to clear their catch. But from 1820 onward the herring
were scarcer and by 1880 the Loch Broom fishery was
finished as a serious pursuit.

Though young herring may grow to their first spawning
season a good way out in the Atlantic or as far north as
the Arctic Ocean, spawning takes place in parts of the
Minch and Firth of Clyde, for example on the Ballantrae
Bank. The Firth of Clyde and many west coast lochs have
stocks of young immature fish which move out as they
become mature, to join the shoals of adult fish in the
Minch and grounds between Barra Head and Ireland.

Mackerel (*Scomber scombrus*) occur in vast shoals in
summer penetrating to the heads of the long sea-lochs such
as Sunart, Hourn and Broom. They are abundant throughout the isles as far west as St. Kilda where they appear in
Village Bay in July and August. Their play of breaking
water with their dorsal fins and making a tiny crackling
concert, is a sound far-reaching and characteristic of calm
summer days in these parts.

The common cod (*Gadus morrhua*) is of general
occurrence on suitable banks between ten and forty
fathoms such as round Dhuheartach, Skerryvore, off
Gairloch and Coigach, near Canna and at several places
off the Outer Hebridean coast. After a century of commercial fishing, stocks in coastal waters are greatly reduced ;
many local fishermen take very few cod on inshore lines.
Virgin stocks of cod no longer exist in coastal waters. This
does not necessarily mean however, that the fish have been
poached from inshore grounds ; it is more likely that the
dearth has arisen from operations on off-shore feeding
grounds. The haddock (*Gadus aeglefinus*) forms the greater
part of the fishery at the St. Kilda Bank at present and is
also obtained at Rockall by trawl. It is not generally
common on inshore lines. Localities do occur, however,
where good catches can be obtained ; the narrows between
Scalpay and the Harris mainland have recently yielded
good haddock for the local boats in summer. Probably
the most common fish on the inshore lines of the local

fishermen are whiting (*Gadus merlangus*) saithe or coal-fish (*Gadus virens*), and lythe or pollack (*Gadus pollachius*).

The coal-fish is exceedingly numerous throughout the region. The young ones, about a foot long, are called cuddies and frequent the inshore skerries in summer. In the late dusk the shoals begin to jump and the sound is like heavy rain. At this time of year and time of night, West Highlanders who never go to the sea at any other season may be seen rowing little boats out to the skerries, where the cuddies are fished with a darrow. This is a line with perhaps as many as eight hooks let down from a stationary boat. Whiting and haddock are also taken in this way, and lythe and mackerel with a rubber eel trailed behind the boat. In autumn coalfish and lythe move in shoals around the coastal rocks and are fished on headlands with bamboo rods, limpet bait or a home-made fly. These are social occasions; vast catches are taken and much time is spent giving away fish to friends in the township. The lythe and coalfish are salted and dried for winter use.

Hake (*Merluccius merluccius*) is fished along the entire west coast of Scotland, mainly in deep water and by trawl, though in the Firth of Clyde and the Minch it is also taken by seine-net, and line. It is now less plentiful on banks to the west of Barra Head and near St. Kilda where it was heavily fished by English and Spanish trawlers in the first half of this century. Ling (*Molva molva*) used to be an important fish in the local economy of the Highlands and Islands, but it is now almost unknown in most crofting townships. Local boats are not big enough to reach the grounds. Ling is plentiful on the edge of the continental shelf north-west of St. Kilda and west of North Rona. It is the special harvest of Scandinavian boats. Incidentally, the filleting of the fish ashore in the Atlantic townships of the Hebrides meant the accumulation of a large quantity of backbones which were ground to meal for cows by women who were known as the *cosnaiche cnamh*—bone labourers.

The halibut (*Hippoglossus hippoglossus*) occurs with the ling on the edge of the continental shelf, and like the ling, tusk (*Brosmius brosme*) and skate (*Raja batis*) is taken on great lines.

The trawlers and seine-netters working out of Oban, Mallaig, Kyle, Gairloch, Kinlochbervie and Stornoway to the banks scattered from Skerryvore through the Sea of the Hebrides, and the Minches to Butt of Lewis catch a wide assortment of other white-fish and Elasmobranchs, the most important being: plaice (*Pleuronectes platessa*), lemon sole (*Microstomus kitt*), dab (*Limanda limanda*), eel (*Anguilla anguilla*), conger (*Conger conger*) and dogfish (*Squalus acanthias*).

The Atlantic loggerhead turtle (*Caretta caretta*) appears occasionally; it has been reported from Vallay (North Uist), Lochinver, Girvan, Unst and Dunvegan. Kemp's loggerhead turtle (*Lepidochelys kempii*) has been found at Kinlochbervie and Troon. Other unidentified turtles have been reported from Loch Seaforth, Firth of Clyde and Loch Lomond. A large squid (*Stenoteuthis caroli*) was stranded at Portree in January 1952, and the giant squid *Architeuthis* may grow to a length of sixty feet and is one of the foods of killer and sperm whales. It is an occasional vagrant to the coastal seas of north-west Britain and is probably the cause of many legends of sea-monsters. The jelly-fish, the by-the-wind-sailor (*Velella spirans*), which is a tropical form brought to Britain by the Atlantic Drift, has been reported from Canna. The common jelly-fishes are *Aurelia aurita* with its four pale violet circlets on the bell and *Cyanea lamarcki,* with its wonderful shades of blue and violet and nematocysts which can sting a man. The spiny lobster or crawfish (*Palinurus vulgaris*) is reported very rarely in Scottish waters, mostly off the north-west coast. J. L. Campbell says that most lobstermen have encountered it once or twice in a lifetime, but the ordinary creel is not very suitable for its capture; it may be less rare than is commonly supposed. The Norway lobster (*Nephrops norvegicus*) has been the subject of a fishery on western coasts since the early 1950's. The development of this fishery posed some difficult problems regarding conservation of the associated white-fish. The use of small-mesh nets led to certain abuses and the regulations concerning *Nephrops* fishing were revised in 1962 so that such nets could no longer be used by British vessels in Scottish waters.

MARINE MAMMALS

Whales are common migrants along the West Highland and Hebridean coasts. It would be better to say migrants rather than residents, for whales are seen more often in summer than in winter, and there seems little evidence of any stationary population. Happily there is now no whaling regularly prosecuted from British shores, but well into this century there were whaling stations in Harris and the Shetland Isles. Indeed, the station at Rona's Voe, Shetland, was not established until 1903.

The common whale of our Atlantic coasts which was the original quarry of the Basque hunters was the Atlantic right whale or nordcaper (*Balaena glacialis*). It was so heavily fished that it disappeared for years and was thought to be extinct. But a few were seen again in the '80's of last century and thereafter more frequently. Almost immediately they became the object of a fishery, along with fin whales (*Balaenoptera physalus*) and other whales, by Norwegian ships from the British bases mentioned. The catches were recorded by Haldane (1905 and annually to 1910) in *Ann. Scot. Nat. Hist.* and *Scot. Naturalist* and the whole data was reviewed by D'Arcy Thompson in 1918. The catches of this whale were commonest to the west and south-west of the Hebrides and were round about a score a year, but after a few years numbers fell rapidly again and it is good to know that such of these whales as now exist are left alone. They probably do not frequent the Minch; Maxwell did not see any during his shark fishing between 1946 and 1949. They are inhabitants of the temperate seas, ranging from the Spanish and North African coasts in winter to Norway in high summer.

The sperm whale (*Physeter catodon*) is another global

oceanic species which appears west of the Hebrides in August, mainly north and west of St. Kilda and out to Rockall. It also has made a northern migration but those seen off the western Scottish coasts in summer are almost invariably males.

The blue whale (*Balaenoptera musculus*) is also found in all the oceans of the world and one stock appears west of the Hebrides in summer. It formed a large part of the early 20th century whale fishery from Harris, but its pursuit later became more profitable in Antarctic waters, and those visiting Hebridean seas have been left in peace. Once more we may be thankful that though the profit motive starts exploitation, there is hope for an animal's continued existence when profit sinks to a low level and the species is a widely dispersed one. During the North Atlantic fishery many of the blue whales killed were pregnant females. As is well known this is the world's largest animal.

The common rorqual (*B. physalus*) is indeed common. Maxwell says that a week has rarely passed without him seeing one or more, usually singly. It is one of the finners and, seen end on, appears to have a very stiff dorsal ridge. It is greyish in colour with white underparts which are "plaited" or furrowed longitudinally. The whale formed the largest part of recent temperate North Atlantic fishery. Though a baleen whale, it does not live on small crustacea, exclusively, as does its larger relative the blue whale. It is known as a fish eater as well and is often called the herring whale. Though Rudolphi's rorqual (*B. borealis*) is thought to be common, Maxwell did not identify any during three years of fishing. It is between the common and the lesser rorqual and may be distinguished by its high dorsal fin. It is an active fast-moving whale and when coming up for air may blow no more than once or twice. The lesser rorqual (*B. acutorostrata*) is probably the most common inshore whale of the north-west seaboard, though Maxwell did not see any. It has little fear of man and quite often comes nosing about the cliffs. We have had several chances of closely observing it. Looked at from above they are dark-coloured, of course, but they are immediately recognisable by the broad band of white across each flipper on its upper surface.

The porpoise (*Phocaena phocaena*) is very common, especially so when food in the shape of small fish is plentiful. Large schools will stay in a bay or sea-loch for a week or more and then disappear. They are often ready to play with a boat; they leap from the water round the boat, crossing and recrossing its path at incredible speed. The best opportunity for watching the speed is a flat-calm night in a wide anchorage, when the bottom can be seen clearly several fathoms down. Sometimes in such flat-calm weather the explosive sound of blowing porpoises may be the continual and only noise to be heard. For a week or two in 1939 when a very large school of two or three hundred porpoises were in Badentarbet Sound, there was one which made a barking sound every time it blew. Unfortunately the senior author wasted an opportunity which a good naturalist should not have lost—of measuring the intervals between blows of the distinctive beast.

The killer whale (*Orchinus orca*), or grampus, is really the largest of the dolphins, reaching to rather over twenty feet in length. The adult male has an extremely long dorsal fin which is not very broad at the base. A full-grown bull's fin appears to stick up about six feet when the whale rises to blow. Usually four to six travel together, and as they all blow at the same time the very long fin of the bull is quite clear.

The killers are the terror of the sea. They will attack the largest whales, porpoises and seals. We have occasionally seen grey seals jump right out of the water in obvious fear, not up and down, but in a low forward parabola, obviously trying to get away from something behind. We have also seen killers visit the seal inhabited coast of North Rona, but when they were about the base of the cliffs all the seals in the locality were in the innermost recesses of the geos and caves. Killer whales frequent the waters around Eigg, Rhum and Canna hunting seals. J. L. Campbell thinks that many of the accidents to small boats attributed to basking

I

sharks, are in fact the work of killers. There is no evidence to show that they are resident in any locality for more than a few weeks. Maxwell saw small packs of killers, about a dozen times in a season. He estimated that some were travelling at thirty knots. H. I. C. MacLean (1961) witnessed what may have been mating killers in Loch Fyne— a very rare sight!

The pilot whale (*Globicephala malaena*) is a small toothed cetacean which moves in large herds, usually of hundreds of animals together. When a herd of these whales entered a Shetland voe, everyone turned out in small boats to hunt them, but the terrain is not so good in the Hebrides and the pilot whales were let go. The bottle-nosed dolphin (*Tursiops truncatas*) is probably more numerous than the common dolphin (*Delphinus delphis*) which tends to keep farther out in the ocean. Maxwell rarely saw common dolphins: a dozen times in three seasons' fishing. He saw white-sided dolphins (*Lagenorhynchus acutus*) at Rodel. Risso's grampus or dolphin (*Grampus griseus*) is reported from Canna and Maxwell saw them in most weeks of fishing.

The grey seal might well come in this section but is described later in chapter 11. The common seal (*Phoca vitulina*) should be mentioned, for it occurs particularly in sheltered and inshore waters, and to a great extent lives a life separate and different from that of the larger grey seal. It is sometimes called the firth seal and that would be a good name for it anywhere in Scotland. It is in the Firth of Clyde and at all points northwards, frequenting favourite rocks in most sea-lochs and island coasts of both Inner and Outer Hebrides. They are particularly numerous in the Firth of Lorne-Loch Linnhe-Sound of Mull area and south to Kintyre and Islay. In the Outer Hebrides the common seal frequents the east coast lochs and the sheltered tidal rocks in the sounds of Harris and Barra. It is also found on the shallow sandy tidal firths which transect the Long Island and is frequently seen in the protected tidal inlets of the Atlantic coasts of the Uists. They can be viewed in many places from main roads, particularly at Ronachan (Kintyre), seal rocks in Loch Feochan and Loch Linnhe, and from pleasure boats at the north end of Lismore. At the other end of the Caledonian Canal they can be seen on the banks of the Beauly and Dornoch Firths.

It is difficult to distinguish our two British seals in the water. Here are some points: the common seal has a much shorter face than the grey seal, as well as being much smaller. As the common seal raises itself to look at you its nose seems almost to turn upwards. The grey seal never gives that impression for in both male and female the face is distinctly roman-nosed. The nostrils are different; the common seal's nostrils are closely spaced at the base forming a wide V-shape, and those of the grey seal are widely spaced and limbs of the V are parted. Common seals vary a good deal in colour from black, through blue and brown to fawn and greeny cream, but the mottling overall is small and uniform, quite unlike the harlequin splashings of the throat and belly of the female grey seal. Juvenile grey seals are, at a distance, hardly distinguishable from adult common seals. It was thought that the habitat was a discriminating factor but now that grey seals have been identified from the inner reaches of most firths and in many sea-lochs, this ecological character cannot be relied upon. If the seals are present in strength then the variation in size and colour patterns of the beasts will usually betray their identity. The species seldom mix even though frequenting the same rock; each species usually hauls out separately and there are often full-grown adults in each assembly at some time of the day.

The common seal is hunted with vigour in the Firths of Tay, Beauly, Dornoch, Lorne and at places along the coasts near fixed salmon nets, but it is quite alive to the hazards and successfully overcomes them. Contrarily, the common seal will become tame, like the one that used to come out on the wharf at Ayr and accept fish from anyone who cared to offer them. A young common seal is easily tamed and will live quite happily about the house if it can also get down to the water for a swim and to feed. The grey seal calf is not easy to quieten. It has its own stock reactions of hostility which it repeats time after time, and while nursing you cannot offer it anything better than to be left alone. When weaned the calf can be tamed and successfully fed on fish.

The fundamental difference in the life history of the two British seals is that the common seal is better adapted to a wholly aquatic life than the grey. The young are born ashore on a rock or a sandbank. The calf sometimes has a white coat of a sort at birth but usually it has not, having shed the white coat *in utero*. The young common seal is most often born with a short sea-going coat. If it is clothed in a white coat, this is short and no handicap to the young animal in the water. The young common seal will happily go off with its mother as the tide envelops it, but the grey seal calf does not last more than two or three hours in the open sea if it should be so unfortunate as to be there on its first day. The young common seal with its mother forms one of the most delightful sights in a June day's watching among the islands. The disposition of these seals is much more amiable than that of the grey species, so they are not constantly bickering and fighting among themselves when hauled out on their usual rock. The babies will play and when the mother and pup go into the water, the pup may be often seen sitting on the back of the old one and holding on by its hands. The lactation period in the common seal seems to be at least two months and this fact alone is important in creating family cohesion. The oil of this seal is used in the Hebrides and elsewhere for cattle food, but much less so than formerly. There are some pedigree cattle breeders who use seal oil particularly because it gives such a bloom to the cattle for spring sales.

The migrations of the common seal have not so far been investigated. Tagging sufficient numbers of these seals is an immense task. They are thought to migrate only locally, but nothing is known of the dispersal of young. The common seal will often move considerable distances solitarily, in an exploring sort of way ; for example, it may go some miles up a river into fresh water, or it may go up a burn from the sea and enter a fresh-water loch to fish sea trout or salmon. They are reported from Tangland Bridge on the Ythan about twelve miles from the sea, in the River Lochy five miles from Loch Linnhe, in Loch Awe, Loch Shiel and Loch Hope.

Chapter Ten

The Outlying Islands

Oceanic islands are remote, but they may be a key in a communication system, they may be a means of yielding some of the resources of the ocean, they may provide the biologist with a field for the study of evolution if they carry isolated populations of animals and plants; and what should not be disregarded, they fill the romantic heart of man.

Oceanic islands have an immense significance for animal and plant life. There are those animals which cannot recruit their numbers from outside, such as the wrens and long-tailed field mice of St. Kilda. We in Britain have no flightless birds, but the wren of St. Kilda is an anchored species whose population seems unaugmented from elsewhere. It is in the nature of islands that they tend to be poorer in numbers of species of animals and plants than are larger masses of land nearest to them. The islands do gain a few species not to be found on the mainland, but they lose more. This is shown by a comparison of the birds which breed regularly in a series of National Nature Reserves from the Grampians to the farthest Hebrides, a distance of 200 miles (fig. 10, p. 191). The Reserves are Cairngorms (with Craigellachie), Rhum (Evans and Flower, 1967) and St. Kilda, and represent a change of habitat from the high afforested interior of the Highlands through the inner islands to the outliers. The birds which breed only sporadically in these Reserves carry an asterisk.

In the first edition similar lists for Wester Ross mainland, Isle Tanera and Priest Island (Summer Isles) had 85, 44 and 29 species respectively.

TABLE OF BREEDING BIRDS

CAIRNGORMS (with Craigellachie) 59,000 acres	RHUM 26,400 acres	ST. KILDA 2,100 acres
MAINLAND	INNER ISLAND	OUTLYING ISLAND
–	Red-throated diver	–
–	–	Leach's petrel
–	–	Storm-petrel
–	Manx shearwater	Manx shearwater
–	Fulmar	Fulmar
–	–	Gannet
–	Shag	Shag
Little grebe	–	–
Mallard	Mallard*	–
Teal	–	–
Wigeon	–	–
–	Eider	Eider
Red-breasted merganser	Red-breasted merganser	–
Goosander	–	–
Golden eagle	Golden eagle	–
Buzzard	–	–
Sparrow-hawk	Sparrow-hawk*	–
Hen-harrier*	–	–
Peregrine	Peregrine	Peregrine*
Merlin	Merlin	–
Kestrel	Kestrel	Kestrel*
Red grouse	Red grouse	–
Ptarmigan	–	–
Black grouse	–	–
Capercaillie	–	–
–	Corncrake	–
Moorhen	–	–
Coot	–	–
Oystercatcher	Oystercatcher	Oystercatcher
Lapwing	Lapwing*	–
–	Ringed plover	–
Golden plover	Golden plover	Golden plover*
Dotterel	–	–
Snipe	Snipe	Snipe
Woodcock	Woodcock	–
Curlew	–	–
Common sandpiper	Common sandpiper	–

CAIRNGORMS	RHUM	ST. KILDA
Greenshank	-	-
Dunlin	-	-
-	-	Great skua*
-	Greater black-backed gull	Greater black-backed gull
-	Lesser black-backed gull	Lesser black-backed gull
-	Herring gull	Herring gull
Common gull	Common gull	Common gullb
Black-headed gull	-	-
-	Kittiwake	Kittiwake
-	Arctic tern	
-	Razorbill	Razorbill
-	Guillemot	Guillemot
-	Black guillemot	Black guillemot
-	Puffin	Puffin
-	Rock-dove	-
Woodpigeon	Woodpigeon	-
Cuckoo	Cuckoo	-
Tawny owl	-	-
Long-eared owl	Long-eared owl*	-
Great spotted woodpecker	-	-
Skylark	Skylark	-
Raven	Raven	Raven
Carrion crow	-	-
Hooded crow	Hooded crow	Hooded crow
Jackdaw	-	-
Great tit	Great tit*	-
Blue tit	Blue tit	-
Coal-tit	Coal-tit*	-
Crested tit	-	-
Tree-creeper	Tree-creeper*	-
Wren	Wren	Wren
Dipper	Dipper	-
Mistle-thrush	Mistle-thrush*	-
Song-thrush	Song-thrush	-
Ring-ouzel	Ring-ouzel	-
Blackbird	Blackbird	-
Wheatear	Wheatear	Wheatear
-	Stonechat	-
Whinchat	Whinchat	-
Redstart	-	-
Robin	Robin	-

CAIRNGORMS	RHUM	ST. KILDA
Whitethroat	Whitethroat	-
Willow-warbler	Willow-warbler	-
Goldcrest	Goldcrest	-
	Spotted flycatcher	-
Hedge-sparrow	Hedge-sparrow	-
Meadow-pipit	Meadow-pipit	Meadow-pipit
	Rock-pipit	Rock-pipit
Pied wagtail	Pied wagtail	-
Grey wagtail	Grey wagtail	-
Starling	Starling*	Starling
Greenfinch	-	-
Siskin	-	-
	Twite	Twite
Bullfinch	-	-
Crossbill	-	-
Chaffinch	-	-
Snow-bunting	-	-
	House-sparrow*	
		Tree-sparrow*

Regular breeders

69 species	54 species	25 species

The outlying islands are most remarkable for their place in the lives of migrant and semi-migrant fauna. The numbers of species may not be large but there is often an amazing density of those which are present. Sometimes, as at Fair Isle, there may be waves of small land birds passing through in spring and autumn ; or a species may come to an island to breed in such numbers that, as at St. Kilda, the island becomes the biggest gannetry in the world. So also with North Rona where the grey seal congregates in larger numbers than on any other island, and which holds about one seventh of the world's population of the species. It is necessary to keep in mind that an oceanic island is a metropolis in the animal world and usually an important port of call in the systems of communications which animals establish. Every naturalist at some time of his life wishes to visit such places, and fortunately there are some in reach of all, but others, like Sule Stack, are so inaccessible that this rock has had but four naturalists ashore in the last half-century.

The number of such islands and rocks off West Highland and Hebridean coasts is not very large, and it is difficult to set a criterion for what is and what is not an oceanic island in the sense they are being considered in this chapter. The biologist's general meaning of the term oceanic island refers to those islands far from any continental land, such as the Galapagos, the Azores and Tristan da Cunha, where evolution has gone its own way for a long period of time. The outlying islands (fig. 13) listed below are not oceanic in the strict sense, though St. Kilda with its endemic fauna and geographical position on the edge of the Continental Shelf comes close to fulfilling the terms of the definition.

1 St. Kilda lies 45 miles west of Griminish Point, North Uist, Outer Hebrides. The name alludes to a group of islands and stacks, seven in all, of which Hirta is the largest, extending to 1,575 acres and reaching a height of 1,397 feet. The other small islands are Soay, 244 acres, 1,225 feet; Boreray, 189 acres, 1,245 feet; Dun, 79 acres, 576 feet; Stac an Armin, 13 acres, 627 feet; Stac Lee, 6 acres, 544 feet; Levenish, 1 acre, 185 feet. There were no permanent human settlements on St. Kilda between September 1930 and April 1957. Hirta now has an Army garrison of about 40 men who man the radar station which is part of the Guided Weapons Range in the Outer Hebrides. It is the premier sea-bird breeding station in Britain with the world's largest gannetry, and has populations of wrens, mice and Soay sheep which are of outstanding scientific interest. There are many sites of archaeological interest. The group is owned by the National Trust for Scotland and is leased to the Nature Conservancy as a National Nature Reserve.

2 Rockall is just a rock protruding 70 feet above a reef, lying 184 miles almost due west of St. Kilda. It has been landed on in 1810, possibly in 1887 and 1888, and in 1921,

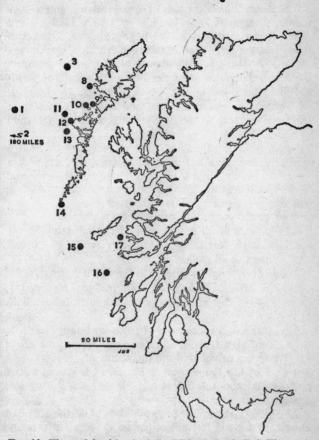

FIG. 13. The outlying islands: 1, St. Kilda; 2, Rockall; 3, Flannans;
4, North Rona; 5, Sula Sgeir; 6, Sule Stack; 7, Sule Skerry; 8,
Gasker; 9, Coppay; 10, Shillay; 11, Haskeir; 12, Causamul; 13,
Monachs; 14, Berneray and Mingulay; 15, Skerryvore; 16, Dhuheartach; 17, Treshnish

A party landed on 18th September 1955 by helicopter from H.M.S. *Vidal*.

3 Flannan Islands consists of 7 main islands and a large number of skerries lying 17 miles WNW. of Gallan Head on the west side of Lewis. The largest island is 39 acres in extent and the highest point 288 feet. All the Flannans are cliff-bound. There is a lighthouse on the largest island.

4 North Rona lies 47 miles NE. of the Butt of Lewis and the same distance NW. of Cape Wrath. The island is 300 acres in extent, cliff-bound and reaching a height of 355 feet. Together with Sula Sgeir it is a National Nature Reserve.

5 Sula Sgeir is 12½-miles WSW. of North Rona. It is a rock of a few acres but is about half a mile long with a sea cave running right through it in the middle, and reaching a height of 229 feet. It has a gannetry which is visited most years by fishermen from Ness who take a crop of young gannets or *gugas* in September.

6 Sule Stack is 30 miles north of Loch Eriboll off the north coast of Scotland; about 6 acres and 120-130 feet high. It has a small gannetry.

7 Sule Skerry rightly belongs to the Orkney group. It lies a few miles NE. of Sule Stack, a flat reef half a mile long, 35 acres, 40-45 feet at highest point with a lighthouse, built in 1895.

8 Gasker lies 5 miles west of Husinish Point, Harris. It is 105 feet high and 57 acres in area. At present about 1,000 grey seal calves are born there annually.

9 Coppay lies 2 miles west of Toa Head, is about 30 acres in area and about 100 feet high. It is a small grey seal nursery.

10 Shillay (Harris) is the most western isle in the Sound of Harris. It is a 265 feet hill of gneiss, 113 acres in area and a small grey seal nursery.

11 Haskeir is a group of islets and stacks lying about 7 miles west of Griminish Point, North Uist. Small grey seal nursery.

12 Causamul is a group of islets and skerries rising to 33 feet above high-water mark 2 miles west of Hougarry, North Uist. It is a small grey seal nursery.

13 Monarch Isles lie 8 miles SW. of Hougarry Point, North Uist. They became uninhabited in 1942 when the lighthouse on the most westerly isle (Shillay) was closed. There are two main islands, extending to about 600 acres in all. They are inhabited in summer by lobster-fishermen from North Uist. There are small grey seal nurseries on Stockay and Deasker.

14 Berneray is the southernmost island of the Hebridean chain, 400 acres and 580 feet in height. The only habitation is the lighthouse at the southern end, Barra Head.

15 Skerryvore and *16 Dhuheartach* are just small rocks lying 10-20 miles SW. of Tiree and Mull. Both have lighthouses.

17 The Treshnish Isles lying between Tiree and Mull are open to the ocean on the SW. They are of basalt and the two largest islands, Lunga and Dutchman's Cap, have basalt cones, the latter having such a marked one as to give the island its name, and to make it a well-known landmark. The largest island, Lunga, is 170 acres and 337 feet high. The cone of the Dutchman set on a level platform of rock is 284 feet high. They are a grey seal nursery.

The natural history of the small outlying islands off the outer west coasts of Scotland was worked out by few men before the last war. Since then, however, an increasing number have been attracted to work there, particularly at St. Kilda and North Rona which are now both National Nature Reserves managed by the Nature Conservancy. The first who left a record of his work was Sir Donald Munro, High Dean of the Isles, who wrote in 1549. We need not let ourselves be irritated by his vagaries of size and distance, but can be grateful instead that he wrote at all considering the difficulties of the subject. Next came Martin Martin, Gent., whose book was first published in 1703. His work was detailed, accurate, and is invaluable to the student of today. In our opinion his work is much to be preferred to that of Pennant (1774), and Martin is also more reliable in some branches of natural history than John MacCulloch (1824).

The 19th century was remarkable for the number of men of independent means who were both interested in natural history and were keen to visit the out-of-the-way corners of their own country. There was the group which centred round Harvie-Brown and which is perpetuated in the *Vertebrate Fauna of Scotland* series: T. E. Buckley, J. Swinburne, R. M. Barrington and M. E. Heddle. Their description of remote islands in the series are sometimes a little breathless but no less valuable for that. The late Duchess of Bedford was a pioneer; she made good use of her steam yacht in exploring the outlying isles. The bird migration work of Eagle Clarke on these remote islands is well known, that of J. Wilson Dougal in geology and general observation much less so.

The next move came after the first German war, consisting of young men who worked alone or in pairs. The late A. M. Cockburn (1935) set a wonderful standard in the late 1920's with his geological survey of St. Kilda. Thirty years have seen very few amendments to his outstanding paper. Malcolm Stewart visited most of the remoter islands and rocks and recorded his observations in a book (1933) and in papers.

Following the pioneer work of Eagle Clarke on bird migration at Fair Isle at the beginning of the century there came George Waterston's original idea of a bird observatory on the island, on the lines of that already established on the Isle of May. In 1948 the Fair Isle Bird Observatory Trust was formed and Kenneth Williamson began his work on the island which lasted till 1956. In the eight years Williamson produced many scientific papers and started long-term work which is being continued by his successors Peter Davies and Roy Dennis.

Following the evacuation of St. Kilda in 1930 there arose a new spirit of adventure and discovery in the Hebrides which is still fresh today. The Oxford-Cambridge expedition to St. Kilda in 1931 included Tom Harrisson who travelled widely among the isles contemporarily with James

Fisher, Robert Atkinson, Fraser Darling and others. Darling spent a year on Priest Island, four months on the Treshnish Isles and six months on North Rona between 1936 and 1939.

After the war, Fisher, Atkinson and Darling started where they had left off in 1939, but to join them and later to carry on from them came a stream of naturalists. The declaration of St. Kilda and North Rona as National Nature Reserves opened up unprecedented opportunities for research. Kenneth Williamson left Fair Isle and went to St. Kilda as Warden for six months in 1957. In the same year Morton Boyd was appointed as the Nature Conservancy's Regional Officer for West Scotland including the Reserves at St. Kilda and North Rona. The facilities at St. Kilda, where the Conservancy have a house with domestic and laboratory accommodation, have been used by a large number of naturalists, many of whom have made a lasting contribution to the records of the Reserve.

THE VEGETATION OF THE OUTLIERS

The number of habitats in these remote islands is severely limited, but even so the reduction of species is striking. At North Rona, however, Atkinson (1940) claimed that four of Barrington's thirty-five species had dropped out by 1939, and he found only twelve more than Barrington had seen. McVean (1961) states that eight species of vascular plant recorded by Barrington and Atkinson could not be found in 1958, and four new species were recorded. The latest tally, therefore, is 43 species. The Glasgow Univerity party of 1958 claimed to have found 44 (including *Plantago major*). On Sula Sgeir, a much smaller island a dozen miles away, Atkinson found 7 species. St. Kilda has about 140 species, but here the size and number of habitats are much greater. The Flannans have 22 species of flowering plants.

The vegetational complex of the outlying islands is a distinctive one as shown in a list of plants from North

Rona compiled from those of Barrington, Atkinson and McVean.

Lesser spearwort	*Ranunculus flammula*
Creeping buttercup	*R. repens*
Scurvy-grass	*Cochlearia officinalis*
Sea pearlwort	*Sagina maritima*
Mouse-ear chickweed	*Cerastium triviale*
Chickweed	*Stellaria media*
Blinks	*Montia fontana*
Wild white clover	*Trifolium repens*
Silverweed	*Potentilla anserina*
Marsh-pennywort	*Hydrocotyle vulgaris*
Lovage	*Ligusticum scoticum*
Angelica	*Angelica sylvestris*
Daisy	*Bellis perennis*
	(2 plants only, 1939)
Scentless mayweed	*Tripleurospermum maritimum*
Autumnal hawkbit	*Leontodon autumnalis*
Sea milkwort	*Glaux maritima*
Buck's-horn plantain	*Plantago coronopus*
Sea pink	*Armeria maritima*
Sand-spurrey	*Spergularia salina*
Babington's orache	*Atriplex glabriuscula*
Orache	*A. patula*
Sorrel	*Rumex acetosa*
Curled dock	*R. crispus*
Broad-leaved dock	*R. obtusifolius*
Cotton-grass	*Eriophorum angustifolium*
Common spike-rush	*Eleocharis palustris*
Tufted sedge	*Carex nigra*
Yorkshire fog	*Holcus lanatus*
Mat grass	*Nardus stricta*
Red fescue	*Festuca rubra*
Creeping bent-grass	*Agrostis stolonifera*
Meadow-grass	*Poa pratensis*
Annual meadow-grass	*P. annua*
Heath-grass	*Sieglingia decumbens*
Adder's tongue	*Ophioglossum vulgatum*

McVean's new finds were:

Marsh violet	*Viola palustris*
Rough meadow-grass	*Poa trivialis*
Common bent-grass	*Agrostis tenuis*
Procumbent pearlwort	*Sagina procumbens*

Losses from Barrington's and Atkinson's lists were:

Slender spike-rush	*Eleocharis acicularis*
Milfoil or Yarrow	*Achillea millefolium*
Eyebright	*Euphrasia* sp.
Knotgrass	*Polygonum aviculare*
Jointed rush	*Juncus articulatus*
Toad-rush	*J. bufonius*
Great plantain	*Plantago major**
Sheep's sorrel	*Rumex acetosella*
Field woodrush	*Luzula campestris*
Early hair-grass	*Aira praecox*

Found by Glasgow University Party in 1958

The grassy mound of Rona is made up of *Holcus, Nardus, Eriophorum* and *Carex,* but as the sea is approached, *Plantago, Armeria, Festuca* and *Cerastium* increase. *Trifolium* is abundant on what was once the arable ground. *Potentilla* forms exclusive beds about the village site. *Stellaria* forms acres of luxuriant growth on Fianuis, where *Poa annua* and *Atriplex* also occur in thick patches. The cliffs show such plants as *Rumex, Ligusticum* and *Angelica* in among the luxuriant *Festuca. Cochlearia* grows to enormous size near where kittiwakes are breeding. It is interesting that plants of cultivation such as *Plantago major, Bellis perennis* and *Ranunculus repens* seem to be dropping out.

Two weeds of cultivation, *Stellaria* and *Atriplex*—and *Poa annua* as well—are thriving on Fianuis and have apparently greatly extended their range over ground previously covered with *Armeria.*

During the centuries of human occupation a considerable part of North Rona must have been stripped of turf for domestic fuel. Fianuis has numerous alignments of stones and piles of stones which may have been associated with the cutting of turf. The northern tip of Fianuis still retains a luxuriant mat of *Armeria* and probably escaped the cutter's blade. If the human population cleared the *Armeria* turf from most of Fianuis, the seals have probably prevented it from becoming re-established, and have contributed to the maintenance of the *Stellaria*. Locally the thin pockets of soil are churned to mud by the wallowing seals in October and November, and in summer the whole area is given a manurial dressing by the flocks of fulmars and greater black-backed gulls which nest there. Now that the *Armeria* mat has gone, erosion from wind, rain and sea-spray is rapid ; this is accelerated by the seal concourse which, having wallowed, return mud-covered to the sea. There is less mud on the seal ground now than 25 years ago.

Donald McVean has published (1961) vegetation maps of both North Rona and Hirta, St. Kilda. The peninsulas of Fianuis and Sceapull on Rona are mainly occupied by a *Stellaria* sward, the principal species of which are :

Sea pearlwort (annual)   *Sagina maritima*
Sea milkwort (perennial)   *Glaux maritima*
Orache (annual)   *Atriplex* spp.
Scurvy-grass (annual)   *Cochlearia officinalis*
Scentless mayweed (either)   *Tripleurospermum maritimum*
Lesser sea-spurrey (annual)   *Spergularia salina*
Common chickweed (annual)   *Stellaria medida*
Blue-green algae

The vegetation of those portions of the cliffs which are not sheer and have some covering of soil is much affected by the birds which use those portions for nesting purposes. They are the puffin principally, and the fulmar to a lesser extent. The puffins burrow into the *Armeria-Festuca* sward and the fescue is not usually grazed in these situations and grows long and luxuriant under the heavy manuring from the birds. The "puffin" vegetation on the ungrazed island of Dun at St. Kilda has been described in detail (Petch, 1933).

The principal species are:

Creeping soft-grass	*Holcus mollis*
Annual meadow-grass	*Poa annua*
Chickweed	*Stellaria media*
Sorrel	*Rumex acetosa*
Common mouse-ear chickweed	*Cerastium holosteoides*
Sheep's fescue	*Festuca ovina*
Meadow buttercup	*Ranunculus acris*
Wild angelica	*Angelica sylvestris*
Blinks	*Montia fontana*

Hirta, St. Kilda, has greater diversity of habitat than North Rona and has room for many of the plants characteristic of Highland hill ground, such as sheep's fescue (*Festuca ovina*), heather (*Calluna vulgaris*), bell heather (*Erica cinerea*), blaeberry (*Vaccinium myrtillus*) and even cowberry (*V. vitis-idaea*), though McVean says that cowberry appears to be on its way out. Since 1932 the grazings of Hirta have been freely used by a population of Soay sheep founded by the late Marquis of Bute, the then owner of St. Kilda.

The flora of the high cliffs of St. Kilda is much richer than that of Rona and contains such additional species as:

Common polypody	*Polypodium vulgare*
Lady fern	*Athyrium filix-femina*
Dwarf willow	*Salix herbacea*
Broad shield fern	*Dryopteris dilatata*
Bladder campion	*Silene maritima*
Moss campion	*S. acaulis*
Primrose	*Primula vulgaris*
Honeysuckle	*Lonicera periclymenum*
Dandelion	*Taraxacum palustre*
Purple saxifrage	*Saxifraga oppositifolia*

The flora of the Flannans grows on a well-drained table-land on top of the sheer cliffs and is predominantly green and grassy. The cliff edges give the same complex as has been given already for other islands (Traill 1905, Bennet 1907). The vegetation of the Monachs (Clarke, 1939), Gasker (Atkinson and Roberts, 1952), and Haskeir (Roberts and Atkinson, 1955) have been briefly described.

THE BIRDS OF THE OUTLIERS

The avifauna of our outlying islands is dominated by the oceanic seabirds, and many species not confined to islands breed only on fairly remote sea-cliffs on the mainland. Only a few breeding species such as the meadow-pipit and rock-pipit, are common to a mainland habitat near the sea. It is as well to distinguish the truly oceanic birds which find a breeding stronghold in the remote isles, and the coastal birds which are widespread along the main western sea-board. The former are petrels, gannets, auks and kittiwakes; the latter are the cormorants, eiders, gulls and terns.

The gannet or solan (*Sula bassana*) is a species about which we know a good deal, thanks to the interest the bird has aroused in man for centuries, and to James Fisher and Gwynne Vevers who organised the world gannet census of 1939. The species is confined to the north Atlantic. Of the twenty-one extant colonies, thirteen are in waters of the British Isles, eight of these in Scotland, and of the Scottish gannetries three fall within the scope of this volume if we exclude Ailsa Craig. This last station has been studied by Dr. J. A. Gibson who has shown that between 1949 and 1955 the gannetry increased from 5,000 to 10,000 pairs approximately. Since 1955 numbers have fluctuated: 1957, 7,700; 1958, 9,500; 1959, 9,400; 1960, 13,500; 1961, 8,500; 1962, 9,600; 1963, 11,700.

The bird has been used as a source of food since time immemorial, the toll being taken when the young birds have reached their limit of growth on the nest and are being deserted by the parents. Sula Sgeir is the only British gannetry from which young birds are taken nowadays.

Over 3,000 young birds or *gugas* are taken after 31st
August under the Wild Birds (Gannets on Sula Sgeir)
Order, 1955. The species as a whole endured a period
of vast reduction during the 19th century which is directly
referable to the activities of man. Evans and Flower
(1967) doubt the accuracy of Munro's (1549) statement
about gannet hunting on Rhum and Eigg, suggesting
that gannets were mistaken for Manx shearwaters. Fisher
and Vevers say in their papers (1943-44): ". . . the
twentieth-century recovery is largely due to the relaxation
of his (man's) predation, to the control of it, or to positive
protective measures. In the story of the gannet man
appears in the different roles of mass-destroyer, harvester,
conservator and protector. . . . By mass destruction man
reduced the gannet population by about two-thirds in 60
years."

We are back again on an old truth, that a species which
has a very wide feeding distribution, but which must gather
for reproduction to a few isolated places, is in very great
danger. These gatherings are very large, or at least they
seem so to predatory man, and he goes hard at them,
wastefully and thoughtlessly, and suddenly the resource
disappears. The gannet must come to a cliff face to reproduce, risking casualty of broken wing should the wind
change to an onshore airt, and above all the approach of
predators from which it is free for the rest of the year ; and
for the same reason the Atlantic seal climbs out of the
buoyancy and freedom of water to move very laboriously
on land. These oceanic birds and mammals are harvesters
of fish and small crustacean life which they convert into
fat either in their own bodies or in that of their young.
The fat is in effect stored there against an inevitable period
of starvation. The seal mother leaves her calf, the hen
gannet leaves her chick, so does the puffin and all the petrel
family. Man is always wanting fat and when its presence
coincides with the period of the animal being out of its
element, man finds it hard to keep his fingers off it.

To quote Fisher and Vevers further: "At certain
colonies, however, man has continuously harvested gannets

for his own use, apparently without endangering the population. This applies to Ailsa Craig up to about 1880; to the Bass Rock up to 1885; to St. Kilda up to 1910 to Sule Stack up to 1932; and to Sula Sgeir, Myggenaes (Faeroe Islands), the Westmann Islands and Eldey (Iceland) up to the present day.

"There is no doubt that at the majority of these colonies man has acted as an unconscious conservator. Indeed, at Myggenaes the inhabitants carefully plan their takes of birds in each year, and set an upper limit to their bag before they start killing; here we can justifiably call them conscious conservators. At the other colonies it is perhaps the physical circumstances that have prevented man from taking too many, and there is still a danger that improved methods of transport or a greater demand for gannet flesh, may materially alter the situation in Lewis or Iceland."

Sula Sgeir provides a good opportunity for the study of optimal yield of the gannetry, and of whether the recruitment of new nesting birds comes only from birds hatched on Sula Sgeir, or from other colonies. It is remarkable that this gannetry has survived the ravages of centuries, and that in recent times more than half and sometimes about three-quarters of the year's production of young have been cropped. Only a minor fraction of gannets leaving the rock as fledglings will survive to breed. Recruitment is probably slow, and ground already colonised will probably be fully stocked before any new sites are annexed. By filling up the gaps of years of recession in a gannetry it might become half or more as large without spread to new ground. If recruitment of birds hatched in other colonies takes place on a large scale on Sula Sgeir little damage is likely from a large crop of young in September; were adult birds taken instead of young, however, the situation might be very different. Sula Sgeir is part of a National Nature Reserve and the Nature Conservancy are co-operating with the men of Ness in the management of the gannetry on the basis of a sustained optimal yield without damage to stocks.

The St. Kilda gannetry is the largest in the world and is situated on Boreray and its two great stacks, Stac an Armin and Stac Lee. A detailed study of this colony has been made recently by Boyd (1961) by analysis of aerial photographs. An index of 44,526 pairs was obtained for the whole; Boreray had 54.2 per cent; Stac and Armin 21.6 and Stac Lee 24.2. An index of 16,900 was obtained in 1939 by a less exhaustive method, and though there is evidence of an increase it cannot be measured by a direct comparison of the two totals.

Birds of the auk tribe, the guillemot, razorbill and puffin are inseparably associated in the mind with the outlying islands. The puffin probably is the most numerous, then the guillemot and the razorbill far behind. Puffins (*Fratercula arctica*) were extremely abundant in 1959 at the Flannan Islands (Anderson *et al.*, 1961), much more so than at North Rona. At Sula Sgeir there are comparatively few puffins; there is no suitable ground in which to burrow or extensive talus in which to find nesting niches. There is none on Sule Stack. They are present in vast numbers at St. Kilda where they find wide tracts of suitable ground on all islands in which to burrow, and extensive talus slopes. It is by far the commonest bird at St. Kilda, and breeds also at Gasker and Haskeir in small numbers. Berneray and Mingulay at the southern end of the Outer Isles also have large puffinries, as have the Shiant Islands, but from there one must go as far as Handa, Sutherland, before reaching the next colony. There are a number of stations around Cape Wrath, Sutherland, the most important being at Clo Mor. They breed on Eilean Bulgach, 7 miles south-west of Cape Wrath. The basalt islands: Shiants, Fladda Chuain, the Ascrib Islands (Skye), Eigg, Canna and the Treshnish group provide perfect conditions for large colonies of puffins. There are small colonies on Rhum, but none on Mull, Coll and Tiree.

The guillemot (*Uria aalge*) is a ledge breeder and wherever it is found one is almost certain to find the kittiwakes as well, nesting on smaller ledges or projections above or below the wider and slightly inward-sloping ledge which the guillemot prefers. The guillemot is a very social bird on the nesting ledges, and yet likes room to move about even if the ledge is crowded. It breeds on the outliers where there are suitable sites ; the Monachs have no cliffs, and although there are colonies on Haskeir the low cliffs of Gasker are not tenanted. It seems to be the only bird which may breed on Rockall where there is only one ledge near the summit. At St. Kilda the guillemot is rarer than might be expected of such a great breeding station, but wide areas of the cliff seem to be unsuitable. Boyd (1960) counted 13,850 pairs (+ up to 20 per cent) of guillemots in the St. Kilda group, distributed in 112 colonies. It is surprising how scarce is the guillemot on the Torridonian cliffs of Sutherland and Ross-shire. Indeed, there is no mainland colony between Cape Wrath and Ardnamurchan. Clo Mor, just east of Cape Wrath, is a magnificent station (Pennie, 1951).

The guillemot population can be divided into those which possess a "bridle" around and behind the eyes, and those which do not (plate XIII). The bridle is like a white monocle with cord appearing on the black breeding plumage. This area of feathering which is white in winter anyway, fails to go black with the rest of the head and throat in summer. The character is thought to be referable to a simple genetic allelomorph, but whether dominant or recessive is unknown. H. N. Southern has investigated the incidence of the bridled variety in three surveys at ten-yearly intervals since 1938-39. The second survey in 1948-49 showed that the proportion of bridled birds had changed in some colonies, but in 1958-59 proportions had returned approximately to the values of twenty years earlier. The latest survey shows that there is a cline in the bridled type stepped from west to east in ascending latitudes across the Atlantic, along the line of influence of the North Atlantic Drift. The figures on page 271, some of which show the range of percentages, are given by Southern (1962):

The razorbill (*Alca torda*) tends to choose a place of its own rather than a ledge with others, in a little hole in the cliff face, under an overhanging ledge, or even on a tuft of sea pink growing in the face of a high cliff. Although it is much less numerous than the guillemot, it is found almost everywhere where guillemots occur and much farther afield than this bird: for example, there were a few razorbills nesting far up Loch Broom, and a few more in parts of the Summer Isles in the late 1930's. Razorbills were abundant at North Rona in 1938 and 1958 but much less so than the guillemot. At St. Kilda the razorbill is an abundant breeder, nesting under boulders and in cliff crevices on all the islands and large stacks. The main colonies are on Dun and Soay. The razorbill and guillemot seldom select the same type of nesting ledges and the species do not appear to compete for space. At Ceann a' Mhara, Tiree, however, the guillemot has been completely replaced by the razorbill as a breeding species in the last half century.

Station	Percentage bridled	Station	Percentage bridled
Ailsa Craig	1	Handa	10–9
Islay	8–5	Bulgach Isle	9
Mingulay	12	Sula Sgeir	19
Canna	10	North Rona	13
St. Kilda	17–10	Fair Isle	9–16
Haskeir	12	Foula	24–29
Shiant Isles	10	Noss	25–24
Flannan Isles	16	Hermaness	24–17

The black guillemot or tystie (*Cepphus grylle*) is present on the outlying islands, but there are never many of them. It is a sociable bird on the sea, but nests privately in some deep cracks in the rocks. Two eggs are laid instead of the one characteristic of the auks, and these are roundish-oval in form, not pointed at one end like those of the guillemot and razorbill. The following is a summary of what is known of the status of the tystie in some of the outliers:

St. Kilda: "minimum of 7 or 8 pairs" (Williamson and Boyd, 1960)

Flannans: "minimum of 10 birds" (Anderson *et al.*, 1961)

N. Rona: "15-20 pairs" (Bagenal and Baird, 1959)

Sula Sgeir: No mention (Bagenal and Baird, 1959)

Haskeir: No mention (Roberts and Atkinson, 1955)

Gasker: No mention (Atkinson and Roberts, 1952)

Monachs: "nested Shillay July 1949" (Allan, 1955)

Mingulay and Berneray: "breeding" (Sergeant and Whidborne, 1951)

Though the presence of all members of the gull family in outlying islands is taken for granted, the kittiwake (*Rissa tridactyla*) is the only one which breeds there in large numbers and the common gull (*Larus canus*) and black-headed gull (*L. ridibundus*) do so in very small numbers. These species are at the two ends of a gradient from oceanic, as seen in the kittiwake, to continental situation, as seen in the black-headed gull which nests far from the sea. The other members of the family fall between the extremes; they are coastal birds and followers of ships. The herring gull (*L. argentatus*), lesser black-backed gull (*L. fuscus*) and greater black-backed gull (*L. marinus*) are scavengers on the human scene of which the Army camp at St. Kilda is the only outpost in our group of outliers.

All three species of large gull were rare at St. Kilda before the human evacuation in 1930; the natives had little refuse or carrion for scavenging gulls and less sympathy for them in pillaging the colonies of puffins, fulmars and gannets which were the staple food for both gull and man at St. Kilda. After the islands became a sanctuary small colonies of herring gulls and greater black-backs became established, but it was not until 1947 that lesser black-backs were found breeding in any strength. In recent years all three species have increased in numbers and this they have done at the expense of the puffin and the eider duck (*Somataria mollissima*).

The kittiwake is always a bird of the cliff face. The coasts of St. Kilda are suitable for them, but are untenanted over wide areas. As usual they mostly occur with the guillemot. There may be some subtle connection between the two species in their respective choice of nest sites. There were 7,660 pairs (± 20 per cent) in 85 colonies (Boyd, 1960); 3,388 occupied nests on North Rona (Bagenal and Baird, 1959); at least 1,400 pairs at Flannans (Anderson *et al.*,

1961) ; it nests at Berneray, Mingulay and Haskeir, but not at Gasker, Shillay or the Monachs.

The herring gull nests in small numbers in most of the outliers, usually in company with the greater black-back which it treats with respect. Greater black-backs may in fact cause serious damage to the eggs and young of both herring gulls and lesser black-backs which are nesting in the same locality. The lesser black-back is a common breeder at St. Kilda with usually about 100 pairs in Gleann Mor. In 1958 there were only two or three pairs left on North Rona; in 1959 there was none breeding on the Flannans, but nests were found by Atkinson and Roberts (1952 and 1955) on Gasker and Haskeir.

The greater black-backed gull is a big, handsome, ugly-faced creature and a very important bird in every way ; it exerts a strong predatory pressure on the vast colonies of oceanic sea fowl and has a widespread manurial effect on the soil and vegetation of these small islands. In the St. Kilda group (2,107 acres), there are probably less than 100 pairs ; North Rona upwards to 500 pairs, Glas Leac Beag, Summer Isles (34 acres), 120 pairs. The puffin is the great gull's meat and manna for the months of May, June and July, and a few days of August. Then the puffins go from the islands in a night, and so do the guillemots and razorbills, and a lot of the kittiwakes leave the ledges. The island seems a desolate place and the greater black-backed gulls are momentarily hungry. If a thousand black-backs on Rona eat each a puffin every other day, the toll must be heavy in a season. The gull stands above a puffin's burrow ; the little bird comes out quite cautiously, but not cautiously enough, for there is a sudden lunge and he finds his whole head in the black-back's mouth. The puffin is skinned and the skin is left as clean as a glove drawn inside out. Both adults and young are taken, but the resilience to predation is such that the vast population of puffins must be capable of overcoming the losses inflicted by gulls and others, just as the rabbit has shown itself capable of withstanding all the depredations of man and natural predators. But what controls the number of gulls? We have such a disparity between St. Kilda with its myriad of puffins on the one hand and Glas Leac Beag with no puffins on the other!

The manurial effect of gull flocks in the outlying islands is well seen on the *Plantago* swards of St. Kilda and the *Stellaria* swards of North Rona, but nowhere is it better demonstrated than on Glas Leac Beag. The other Summer Isles are heather-covered, but Glas Leac is brilliantly green and heather is found only at the south end, growing small and low in the grass. In August the island is bright with sheep's-bit scabious (*Jasione montana*) truly a heavenly blue. This remarkable enrichment of the herbage by the gulls draws a flock of 300 barnacle geese in winter, and the island is the chosen sanctuary of the whole Summer Isles race of grey lag geese in their flightless time in July and August.

The greater black-back meets his match in the adult fulmar which has a deterrent from which most birds would shrink: a squirt of oil from the mouth. We found a fulmar on North Rona in 1958 incubating a clutch of greater black-back eggs. The gull was standing about 10 yards away looking very sorry for itself; its size, guile and powerful bill could not overcome the immovable fulmar. Greater black-backs also take toll of Manx shearwaters and small petrels as they visit their nesting sites in the dark.

The sudden departure of the auks and kittiwakes from the outliers causes the black-back population to migrate in search of food. But with islands like North Rona, Gasker and the other nurseries of the grey seal an autumn build-up in numbers takes place. The observer will see many tugs-of-war between the gulls, the rope being the elastic afterbirth of the cow seal. The slack is down the birds' gullets, but in the course of the strife it may not stay there. After the feast of afterbirths there is a crop of dead calves which, at Rona, reaches about 20 per cent of the number born. The gulls find it difficult, however, to penetrate the thick hide of the newly dead, prodding with limited success at the eye sockets, umbilicus (if exposed) and wounds. We once dissected a young bull which was found freshly dead at Sceapull and left the eviscerated carcass to the gulls; they made a poor impression on the tough masculature. As the carcasses putrify, however, the skin yields and in time they are turned inside out, just like puffins in summer.

The large gulls of the outlying islands are joined in winter and spring by a few non-breeding Iceland gulls (*Larus glaucoides*) and glaucous gulls (*L. hyperboreus*). Both have been so recorded at St. Kilda.

Arctic terns (*Sterna macrura*) nest on most of the outlying islands but not on St. Kilda or the Flannans. North Rona has a few small colonies nesting directly among the gneiss pebbles which their eggs so much resemble. Haskeir and Gasker both have colonies, and they probably nest also on Shillay (Harris) and Causamul. The Monachs like Tiree, are well suited for terns but there are only small colonies of Arctics, common terns (*Sterna hirundo*), and little terns (*Sterna albifrons*). A thriving population of feral cats may be the reason for this.

Most of the outlying islands of the West carry members of the petrel and shearwater family, the Procellariidae. St. Kilda heads the list with four species—the fulmar, Leach's petrel, the storm-petrel and the Manx shearwater. Rona, Sula Sgeir, and the Flannans lack the shearwater, but are among the few other British stations of Leach's petrel. Manx shearwaters (*Puffinus puffinus*) are found on Rhum, Eigg, Canna and the Treshnish Isles but not so far from the Shiants or south Barra Isles. The Treshnish also have the fulmar and the storm-petrel. Only small islands seem to hold the storm-petrel (*Hydrobates pelagicus*). They are probably present on most of the outliers where there is talus, high storm beaches or dry stoneworks. It nests in large numbers in the storm beach on Rona and also in the boulder strewn localities on Shillay (Harris). In many places such as the Carn Mor, St. Kilda, it nests side by side with the larger Leach's petrel. Carn Mor has all four species of tube-nose nesting, and is one of the most popular ornithological sites in the Hebrides. We have heard stormies calling from burrows on Carn Mor in April, and have found young on Rona in late October. There is no record of breeding on the Scottish mainland. The Manx shearwater will go farther inland ; on Rhum it nests on the summits of Hallival and Askival at over 2,000 feet.

The Leach's petrel (*Oceanodroma leucorhoa*) is probably our rarest breeding petrel though it is not a rare bird in the outlying islands. The main breeding station in Britain is at St. Kilda where large colonies are present on all the main islands and on some of the stacks. Ainslie and Atkinson (1940) claimed that there were 377 pairs on North Rona in 1936. Bagenal and Baird (1959) claimed that the population was about ten times that number in 1958. The estimates, as with those of the gannets at St. Kilda, were obtained by different methods and are not directly comparable. Sula Sgeir and Flannans (breeding on at least six islands of the group) are other breeding stations of the Leach's which have recently been visited by Bagenal and Baird. The rareness then is in the remoteness of the few oceanic stations, and in the restricted area in which these lie. A search for new colonies has been made in recent years for breeding Leach's in the other outlying isles of the west and north but all have drawn blank except for Atkinson and Roberts (1955) who found two adults in a burrow on Bearasay off Loch Roag, Lewis ; see also Robson (1963).

These two petrels—we say it consciously and without shame—are dear little birds. The churring noise of the stormie is one of the comforting things to the human visitor in his nights on a lonely island. Leach's petrel is larger, about the size of a thrush, and has lovely grey shading round the full black eye. The birds in the air have an arresting, wild staccato calling, but those in the burrows make an exceptionally sweet ascending trill which is not so often heard. Both species have an aerial display in the summer darkness but that of the Leach's is much more impressive and takes place with larger numbers of birds. To be present at one of the breeding stations on a fine July night is one of the great experiences for a naturalist. Recent work by Donald Baird points to most of the aerial display in the Leach's petrel being caused by immature birds which have no burrows.

The Manx shearwater's cry when indulging in somewhat similar aerial evolutions is anything but comforting. It is an unearthly shriek which has to be heard to be believed. The breeding habits of these birds have been closely studied and described by Lockley (1941) wherein he shows that the south Wales shearwaters feed in the Bay of Biscay area during the nesting season. It still remains for someone to pin a flag on a map showing where the West Highland shearwaters feed and to find out exactly how long each sex is away from the nesting burrow in this area.

Lastly among the sea-birds there is the fulmar (*Fulmarus glacialis*) (plate XIII). It is truly oceanic, feeding on plankton and coming ashore only to breed usually on high sea-cliffs. There they occupy the upper parts, nesting on broad ledges, scree slopes or broken grassy slopes; recently, however, they have adopted other nest sites. In Sutherland, for instance, they nest on the roof of Dunrobin Castle, and go about 5 miles or more inland up Strath Fleet, near Rogart, nesting on rocky outcrops on the hillside. On uninhabited islands in Orkney and at North Rona they nest beside ruined walls, on old buildings, even on the open hillside, and in the Monach Isles in sand dunes.

St. Kilda has been a fulmar stronghold since before 1698. Indeed it was the only breeding colony south of Iceland until the early nineteenth century. Since then the species has increased dramatically in number and spread all round the British coast and in Brittany and Western Norway. This spectacular spread has been exhaustively documented by James Fisher (1952) who has sifted the great mass of published information and pieced together the details. Fulmars colonised Foula from Faroe in 1878, and arrived at North Rona, Sula Sgeir and Hoy between 1890 and 1900. Fisher thinks that the Orkney birds colonised the north coast of Scotland and eventually the east coast and the North Rona birds colonised the west mainland, starting at Clo Mor about 1897 and moving from there to Cape Wrath and the North Minch area, and

Handa. From these northern stations the species soon spread south, usually colonising the main headlands first and slowly colonising the gaps as time went on. The St. Kilda population, which Fisher estimated at about 38,000 pairs in 1949, with about half the total on Hirta, is thought to have contributed only slightly to the main spread. In 1956 and 1961 Sandy Anderson (1957 and 1962) repeated the census of fulmars on Hirta and found the numbers to be between 19,400-19,700.

The fulmar is an enigmatic bird. Fisher states, though the evidence is not very precise, that young fulmars do not return to land until four years old and "prospect" at nesting colonies for a further three years or so before breeding. Then they lay only one egg each year, and it cannot be replaced if lost. This very slow breeding rate can be reconciled with their great increase in numbers and breeding range in the last 80 years or so, only if the birds live for a very long time.

Recently Robert Carrick and George Dunnet, and later Dunnet and Anderson of the University of Aberdeen have been studying the fulmars on Eynhallow in Orkney. By having each bird marked with a different combination of coloured rings, and observing them year after year (continuously since 1950) they have shed some light on the breeding biology and longevity of this species.

Breeding adults return to the island as early as November and numbers gradually increase during winter and spring. Just why an oceanic bird should spend so many months at its nest-site before egg-laying (in May) is still a mystery. None of the other petrels is known to do this. Nobody has yet studied fulmar behaviour during these stormy months and it may well be then that pairing takes place and nest-sites are secured. By late April all the breeders are back, paired, on their sites, but before egg-laying they take part in a spectacular exodus from the colonies. Both male and female leave the island and are absent for 10-14 days. In 1962 a breeding adult male from Eynhallow was recovered 230 miles to the east, just off the Norwegian coast, on 10th May, at the height of

8

the exodus, and this is our only indication of where they go. The male usually returns a day or two before the female, and she normally scrapes out the cup of the nest and lays the egg within a few hours of arriving back. She incubates for only a few hours before flying away again leaving the male in attendance. By contrast, his first incubation stint is long, averaging about 7 and sometimes lasting 11 days. After this marathon effort the sexes share the incubation more or less equally in stints of about 4 days till hatching approaches when the parents change over more frequently. Incubation lasts about 7 weeks, and the chick fledges after a further 7 weeks.

Recoveries of British fulmars ringed at nests mainly on Sula Sgeir, St. Kilda and in Orkney, have revealed a very large-scale dispersal. Though some thousands have now been ringed, there are only about 40 recoveries of British-ringed birds, but these show clearly that it is the youngest birds which usually go as far as the Grand Banks off Newfoundland, the Barents Sea and the Bay of Biscay while old birds are recovered mainly in Western European waters. No fulmar, ringed as a nestling, has yet been found breeding.

By ringing adult fulmars with individually distinctive rings and observing them each year over a period of 13 years, Dunnet and Anderson have been estimating the average length of *adult life* as $15\frac{1}{2} \pm 2$ years. Assuming Fisher's estimate of an average of 7 years as pre-breeders to be correct the average age of an adult fulmar is about 23 years. Breeding fulmars show great tenacity of mate and nesting site and some Eynhallow birds have bred together at the same nest for 13 years up to 1962.

So much for the sea-birds. Some of the outlying islands have a rather surprising list of other birds nesting, and none is more interesting than St. Kilda. There are breeding in some years: peregrine, kestrel, oystercatcher, golden plover, snipe, raven, hooded crow, wren, wheatear, meadow-pipit, rock-pipit, starling, twite, tree sparrow. Some, like the snipe, wren and starling breed in astonishingly high numbers for such a habitat. In historical times St. Kilda

K

has lost as breeders the white-tailed eagle, corncrake, rock-dove, skylark, song-thrush and corn-bunting. The breeding snipe are of the Faeroe race *Capella gallinago faeroeensis* nesting in dry situations in the meadows and hillsides. The wrens are the resident St. Kildan sub-species *Troglodytes t. hirtensis*.

There are no breeding wrens on Rona or the Flannans, but St. Kilda is like a little country unto itself and has its own wren. The bird was noticed by Martin Martin in his famous visit to St. Kilda in 1697 and by Macaulay in 1764. The history of the bird is reviewed by Williamson and Boyd in *St. Kilda Summer* (1960). The St. Kilda wren is slightly larger than the mainland wren, its nape and head are uniform grey-brown, and the rest of the upper parts a brighter brown than *T. t. troglodytes*. The under parts are paler. Barring is more pronounced and there is a whiter superciliary eye stripe. Its song is similar to that of the mainland form with slightly different phrasing, and is sweeter.

"The uniqueness of the St. Kilda wren" write Williamson and Boyd, "was recognised by the ornithologist Henry Seebohm in 1884, when he described it as a species new to science, under the name *Troglodytes hirtensis*. Many thought it would have been better by far had science remained in ignorance of its special virtues, for in those bad old days of the egg-blowpipe and the walking-stick gun a new name for a new bird was sufficient to set the ornithological underworld aflame with lust and greed. One dealer, indeed, is known to have advertised widely, soliciting subscriptions to finance two special trips to St. Kilda in quest of skins and eggs of this bird and Leach's petrel. A letter from a St. Kildan accompanying skins to the Norwich Castle Museum in 1904, asks for one guinea each for adults and 12s 6d each for young. So greatly was this persecution feared by bird-lovers of the day that an Act of Parliament was passed for the protection of both species in 1904, by which time collecting by and for the dealers had led that fine literary naturalist W. H. Hudson to throw up his hands in despair and declare that the St. Kilda wren was already extinct!

"But it was not extinct, and indeed this craze for cabinet-specimens could never have reduced its numbers significantly, for even after the collectors had got to St. Kilda (no mean achievement in those days) the difficulties awaiting them in their task were well-nigh insuperable. For although a few pairs nested annually in the dry-stone *cleitean* where the St. Kildans stored their peats and hay (and these village wrens were always vulnerable), the bird's chief haunts are inaccessible places along the towering, awe-inspiring range of the island's cliffs. There they are, and always will be, secure from human depredation ; and there, as it turns out, they are far from scarce."

The estimate of 68 pairs on St. Kilda in 1931 was followed by estimates from Hirta: 31 in 1939 and 48 in 1948. But a thorough investigation of the population of wrens on Hirta was not carried out until Williamson's protracted stay in 1957. In May he painstakingly went round all the cliff edges and the landward areas where they were likely to occur, and, on consecutive mornings plotted the positions of singing cocks in the dawn chorus. There were 117. With the population of the other islands of the group added, the total number of singing males at St. Kilda for that year was probably well over 150. Even Stac an Armin had its quota of at least three males in May 1959. The bird has now complete protection and is thriving in the absence of predators. The fact that Williamson's results were so much in excess of the earlier estimates is not conclusive evidence of a substantial increase ; it is more likely to be connected with the different methods used by the various observers.

Bagenal (1959) watched two pairs of wrens feeding young in the village area over periods of about twelve hours ; there were between 20 and 30 feeds per hour with both parents sharing the task equally. In recent years small numbers of St. Kilda wrens have had rings attached but so far none have been observed elsewhere than on the island. Waters (1964) has made further observations.

Before leaving the birds we must mention the exceptional interest of the outlying islands as ports of call for migrant birds. Some are better placed than others. Modern techniques using radar show that the bulk of migration takes place at heights of up to five thousand feet or more and that only "grounded" passage or migration which has gone astray is seen on the islands. What we see on the ground is, however, fascinating and does not always give the impression of birds gone astray.

This is particularly so with geese and swans: St. Kilda is not nearly so well placed in the ocean as Fair Isle to catch the wind-blown continental migrants but it does catch migrants to and from Iceland and Greenland. Flocks of barnacle geese (*Branta leucopsis*), and pink-footed geese (*Anser brachyrhynchus*) pass through St. Kilda in late April and disappear to the north-west. On 29th October, 1960, there was a wonderful passage of geese and swans at North Rona. Great skeins were seen flying southwards towards the Sutherland seaboard. Some flew over the island but it presented no inducements to the passing flocks and we saw a very small proportion of numbers passing through the sea area that day. Most of those identified were pinkfeet, but a few grey lag were also seen and no barnacles. Of all the outliers, the Monachs are the only ones where geese are present in strength all winter.

We have no record for St. Kilda, the Flannans and North Rona to compare with Fair Isle and the Isle of May. At St. Kilda many extraordinary birds have been recorded such as the nightingale (*Luscinia megarhynchos*), the short-toed lark (*Calandrella cinerea*), the red-headed bunting (*Emberiza bruniceps*) and the rock-thrush (*Monticola saxatilis*). Summer and autumn expeditions to North Rona have also recorded the red-headed bunting and the red-breasted flycatcher (*Muscicapa parva*). Unless an ornithologist is resident on the islands most of the rarities are missed.

THE MAMMALS

The tendency of remote islands to produce differentiated races of animals is further exhibited on St. Kilda by its mice, of which there were two species: the St. Kilda house-mouse (*Mus musculus muralis*) and the long-tailed field-mouse of St. Kilda (*Apodemus sylvaticus hirtensis*). It is probable that the house-mouse was introduced within historical time, but *Apodemus* is considered to be a true relict. If *Apodemus* becomes isolated for centuries in a small island it will begin to show differences from the main stock through the process of natural selection of the best surviving form of the animal. This is what has happened in a large number of the stocks of the Scottish islands, which show size and coat colour differences from the mainland mice. The scale of change varies greatly, but is nowhere greater than with the St. Kilda field-mouse. The differences in size, coat colour and behaviour, to mention only the more obvious characteristics, point to its being isolated from the mainland stock for a much longer period than all other Hebridean populations. Its origin on St. Kilda may go back thousands of years to before the Quaternary Ice Age, when St. Kilda was the only island lying outside the ice sheet.

M. m. muralis was a robust type of *M. musculus* and had a lighter underside than the normal mainland house-mouse. When St. Kilda was inhabited this house-mouse was numerous in the cottages and byres. Its fate after the evacuation in 1930 was one of the especial interests of the 1931 expedition. Harrisson and Moy-Thomas (1933) carried out a good deal of live trapping and marking, and finally assessed the total population at between 12 and 25 head. That was the last seen of the house-mouse. The new situation would demand that the house-mouse went into the open to forage for food. A reciprocal movement probably occurred in the field-mouse, which could now venture into the deserted byres and cottages. The subsequent extinction of the domesticated dog and cat populations at and after the evacuation released both mice from what was a heavy predatory pressure. The field-mouse, a much more powerful animal than the house-mouse, invaded the habitat of the latter and probably contributed to its downfall.

Darling's experience of house-mice on Lunga of the Treshnish Isles in 1937 seems relevant here. The expedition was housed in a bell tent situated on a ledge about 80 feet above the sea and about 20-50 yards from the ruins of the old roofless houses. The island had not been inhabited for about 80 years. Within a few nights of arrival, mice of some sort were coming into the tents and tackling the stores, and it was not long before they were showing all the cheekiness of the house-mouse and indulging in games. Trapping was essential and we caught 75 individuals. It was obvious that there were always more coming in, so the total population present must have been considerable. These mice showed no different features from type specimens, and after the years of successful living as field-mice, and in the absence of *Apodemus,* this island race was ready immediately to take up the traditional existence of house-mice again, granted without the dry warmth of the hearth stone. The St. Kilda field-mouse reacted in a similar way when the Army camp was established in 1957, and the population in the village glen now lead a commensal existence.

In 1957 when St. Kilda was declared a National Nature Reserve the field-mouse population on Hirta was large. It was present all over the island, densest in the village area and rarest on the ridges. Dry-stone works (*cleitean,* walls, and ruined houses) and talus play a highly significant part in the widespread colonisation of Hirta. The great upheaval in the habitat of Hirta which occurred with "Operation Hardrock" (Williamson and Boyd, 1960) did not adversely affect the mouse. During the summers of 1957 and 1958 up to 300 men were present on the island, employed in building a radar station, but subsequent surveys in 1959 and 1960 showed that the distribution and size of the population had not changed. Investigations showed that if St. Kilda mice are displaced at midnight from the pastures beside the beach to the top of the island—a distance of up to half a mile—some were back in the same traps before dawn.

The work on the field-mouse, like that in all departments of natural science at St. Kilda, is being developed by the Nature Conservancy (Boyd, 1959 and 1963). The great danger to the future of the mice is the introduction of rats, cats and dogs and this is being watched.

The food of the St. Kilda field-mouse has not so far been studied, but is likely to include grasses, seeds, insects, and carrion. In the invasion of the camp buildings by the mice they have consumed a wide variety of materials including food-stuffs, soap, plastics, and rubber! They harbour several species of fleas and mites and have a heavy intestinal worm burden. The round worm *Rictularia cristata*, which almost occludes the bowel of the St. Kilda mouse, has not been recorded from mice elsewhere in Britain. The tape worm is *Hymenolepis diminuta* which has its bladder-worm in the flea. It is interesting to note that the bladder-worm (*Cysticercus fasciolaris*) of the cat tape-worm (*Taenia crassicolis*) found in the mice in 1931 is not now recorded in the mice.

Rona and the smaller outlying islands listed at the beginning of this chapter have no mice or voles. The Shiants are, unfortunately, full of brown rats (*Rattus norvegicus*). Several of the Inner Hebrides and even a small island like Priest Island of 300 acres have pygmy shrews (*Sorex minutus*). Elton (1938) found both *Apodemus* and *Sorex minutus* on Pabbay, one of the outermost islands in the Sound of Harris. He remarks also on the submerged forests of this region and on the remains of birch and hazel in the peat. McVean (1961) says that St. Kilda once possessed a partial cover of birch-hazel scrub. This evidently is the point; the St. Kilda field-mouse and the pygmy shrew may be relics of an era of woodland now long gone.

During the summers of 1957 to 1960 M. J. Delany travelled widely in the North-West Highlands and Islands studying small mammals. The species caught were: common shrew, pygmy shrew, water-shrew, field-mouse, house-mouse, bank-vole and field-vole. Delany (1961) examined as many habitats as possible on South Rona, Raasay, Rhum Mull, Colonsay, Lewis, North and South Uist and Barra. On all the islands except Mull there were fewer species than on the mainland. He found: the pygmy shrew and the field-mouse on all the islands; the bank-vole and water-shrew on Raasay and Mull only; the field-vole on Mull and the Uists; and the common-shrew on South Rona, Raasay and Mull. The common-shrew and water-shrew were not taken on Colonsay and the Outer Hebrides. The house-mouse was found in fields in the Outer Hebrides and Colonsay with the field-mouse. Collections of bank-voles from the mainland, Mull and Raasay showed the Raasay animals with larger skulls, but they differed little in colour (Delany, 1960).

The Soay sheep of St. Kilda may have been brought there a thousand years ago by the Vikings; or by the pre-Viking inhabitants, since archaeological evidence points to the islands being inhabited before the Viking conquest. An exactly similar sheep did not survive elsewhere. It is not unlike the wild moufflon, though smaller. From archaeological evidence throughout Britain, the Soay is similar to the sheep possessed by the Neolithic farmers and they closely resembled those which survived in a smiliar habitat in the Faeroe Islands, until the middle of last century. The colour is brown. There are two types; most are of a dark brown almost black in autumn but fading somewhat by spring, but others are light brown. The horns of the Soay ram are of the same type as in the moufflon, i.e. lifting well up from the head and taking a full, wide curve. Soay sheep grow a short, fine coat of soft wool mixed with hair. The mane and hair on the throat which is characteristic of the moufflon also appears on the Soay.

They were formerly confined to the islands of Soay and Dun at St. Kilda, and were the property of the proprietor and not of the St. Kildans, who, through the centuries, had a succession of breeds including the Soay, the four-horned, Blackface and mixed breeds of modern sheep. The Soays remained intact as a breed on the isle of Soay throughout the improvement of the islands' sheep, and now they also occupy Hirta. Sheep from Soay have been taken for park collections and small populations have been established on other islands.

These little sheep have a great deal of native common-sense which the domesticated sheep have lost. It is practically impossible to work them with a dog; if an attempt is made to catch them they retreat to the cliffs. A good sheep dog was taken to Hirta to gather Soays for scientific inspection in 1956. It failed completely to herd them; they scattered in all directions. The St. Kildan dogs were trained to run individual sheep to earth and hold them.

The lambs vary in colour. Some are a very dark brown and a few are jet black, but others are a warm ginger colour. They have a thick coat of curly wool and hair. They are usually born in March and April, and by the tupping season in October many of the ram lambs are already running with the older rams in the nuptial chases. A large proportion of ewes carry a lamb in their first winter. Bottle-fed lambs become very tame, but later in the year usually join their kinsfolk for the rut. The flocks which graze through the village glen have become accustomed to the presence of men on the island and move about among the buildings without concern. They are, however, very watchful and will not allow an approach of closer than ten yards without cantering off. Those in Gleann Mor and on the cliff terraces have not become accustomed in the same way. They have acute senses and as soon as a man makes his appearance several hundred yards distant, they stream along their airy cliff-edge paths to safety.

The Hirta population was founded in 1932 by the introduction of a sexually balanced flock of 107 Soays from the isle of Soay to the vacant pastures of Hirta. Since 1955 a census of this flock has been made annually:

1952: 1,114	1957: 971
1955: 710	1958: 1,089
1956: 755	1959: 1,344

1960: 610   1962: 1,056
1961: 910   1963: 1,589

This demonstrates an abundance cycle in the population which is at present being studied together with age structure, flock sociology, food and parasites of the population. Work is also in progress on the anatomy of the Soay. In this work the Nature Conservancy has co-operated with the Hill Farming Research Organisation and the Wellcome Institute, London Zoo.

The "crash" which occurred in the population in the winter of 1959-60 was caused by an epidemic of enterotoxaemia following a winter of extreme poverty. Malnutrition was probably acute because of the large number of mouths to feed. Ironically enough, it was not the starvation that killed the animals, but the spring flush of maritime vegetation upon which they gorged themselves. This created excellent conditions for the enterotoxaemia germ, which halved the population during March 1960. There are clearly years of good increase and those of bad; the bad ones of 1955-56 and 1961-62 were the result of a very late spring flush of pasture and high mortality of lambs. Since 1959 about ten per cent of the island population had been caught, measured and tagged annually. Over 100 can usually be counted annually on the isle of Soay, from the neighbouring cliffs of Hirta.

On Boreray of St. Kilda there is a feral flock of Blackface type sheep: the descendants of those left behind by the St. Kildans in 1930. The island is 190 acres about half of which is bare rock. The other half is rich pasture supporting more than three sheep to the acre in most years. The inaccessibility of the island protects this flock against interference by man and also hampers investigation. The following counts have been made:

1951: 340   1960: 340
1956: 360   1962: 250
1959: 440   1963: 413

The success of this domesticated flock in maintaining itself on this rich but savage island is of great interest. No less interesting are its origins, evolution after thirty years by natural selection of this Blackface strain and its comparison with mainland strains.

The invertebrates of the outlying islands are not nearly so well known as the birds and mammals. Here again, Hirta has been much more thoroughly worked than all the others put together. The important features of the associations lie as much in what is absent as in what is present. Races of invertebrates peculiar to the outliers may yet be discovered. The following surveys have been carried out at St. Kilda: insects (Lack, 1931, 1932 and 1933 ; Smith, 1963), spiders (Pickard-Cambridge, 1905, Duffey, 1959), terrestrial molluscs (Waterston and Taylor, 1905 ; Hunter and Hamilton, 1958 ; Smith, 1963), earthworms (Boyd, 1956), freshwater fauna (Hamilton, 1963). Other invertebrate studies on the outliers include the collembola of Shillay (Goto, 1955 and 1957) and invertebrates on Pabbay (Elton, 1938).

The natural history of the Scottish outliers could provide material for a large volume. This inadequate attempt to summarize some of the happiness and good thinking which naturalists through a hundred years have gained and given, can hope to do little more than focus attention on what will still remain a joyous pursuit for all young folk under eighty years. Read Robert Atkinson's *Island Going* for an outstanding appreciation of this. The only thing which can dull enthusiasm for reaching the little islands must be the stiffening joints which may prevent one from getting ashore in the moment of the slackening swell. Seasickness is certainly not enough to keep the naturalist on the mainland, and neither is the thought of holing the boat and swimming for one's life. There are certain situations which arouse a tremendous feeling of exhilaration and physical well-being. We know of none to beat the approach in a launch to a remote and uninhabited island where the swell is whitening the foot of the cliff. Whether you are at the engine and tiller with a kedge anchor rope running astern, or poised on the forepeak, ready to jump ashore with the forrard rope, it is all the same. You are trying to beat the elements but you are also working with them as the boat lifts and falls in the broken water at the foot of the cliffs. The island is still remote till your foot touches down, and sometimes the swell will beat you, and you will make the long journey home unrewarded, only to return again.

The islands are a paradox: bare and remote to our eyes, they are nevertheless among the most heavily populated areas in the kingdom and contain forms not found elsewhere; on the one hand there is the falling off in the number of species, and on the other the immense numbers of those there are; the vast comings and goings of the creatures, and at the same time the irrevocable isolation of the tiny mammals. We tend to think of these northern islands as storm-bound and mist-wrapped, yet nowhere can there be greater brilliance of colour, the sea so blue, the grass so green, the rocks so vivid with saffron lichen. A meadow of sea pink in June contains all colours between white and deep purple, and the predominantly white plumages of most of the birds reflect the boreal intensity of the summer light.

Chapter Eleven

The Life of the Grey Seal

J. G. Millais wrote in 1904 that excepting the whales the grey seal (*Halichoerus grypus*) was the British mammal of whose life history we were most ignorant. This lack of knowledge was more surprising when it is remembered how man regularly hunted the animal. The hunters of the grey seal, however, contributed little to our knowledge of it and the ardent naturalists of the 19th century seemed unable to find the truth. The animal did not receive its specific name until 1791, when Fabricius distinguished it from the bearded seal (*Erignathus barbatus*). *E. barbatus* is a northern species that does not come south into British waters at all, but breeds on the edge of the ice in early spring in Northern Eurasia and America. The grey seal on the other hand, is a beast of temperate North Atlantic and Baltic waters.

The largest part of the world population of grey seals to our present knowledge breeds in the waters of the British Isles. This alone should make the British people proud of their heritage of wild life and determined to uphold it by proper measures of conservation of one of the world's rarer seals. They are numerous off the west coast of Scotland, but despite their increase during the last century, there are only one or two places, the caves of Loch Eriboll and east Caithness, where they breed on the mainland of Scotland. All the other breeding stations are on uninhabited oceanic islands or on the most remote parts of larger inhabited ones. In Ireland there are small colonies on Clare Island off Galway, Lambey Island off Dublin, Great Saltee off Wexford and perhaps a few in Bantry Bay and Donegal. In Cardiganshire and the islands of Pembrokeshire in south-west Wales the grey seal often breeds in sea caves as they do in western Ireland and at Loch Eriboll. A few grey seals breed on Lundy, on the north coast of Cornwall in sea caves and there are a very few left in the Isles of Scilly.

The grey seal does not breed in the English Channel but on the east coast a small colony has become established in recent years on Scroby Sands near Yarmouth, and there is a large colony on the Farne Islands, Northumberland. At the Farnes about 800 calves are born each year. The seals are protected throughout the year by the owners, the National Trust, and have been studied by the local naturalists for many years (Hickling, 1962).

This seal probably breeds in very small numbers on the east coast of Scotland. One has been born on the Isle of May on three occasions since 1956 and breeding probably takes place in the remotest parts of the east coast of Caithness. It occurs in Orkney and Shetland in large numbers; about 3,000 calves are born annually in Orkney alone. It is found on the Norwegian coasts in small numbers. In the Baltic it breeds in the Gulf of Bothnia in spring as do the grey seals of the Canadian coast. These two groups, whose numbers are not accurately known, have had a separate history from the British grey seal, since the last interglacial period. It has another station in the Faeroes where again it is a cave breeder. There are colonies of moderate size in Iceland. It has been shot in Greenland, though its breeding is not confirmed there, but it breeds down the Labrador and Newfoundland coasts in the Gulf of St. Lawrence and south as far as Nova Scotia, but no farther. H. D. Fisher put the west Atlantic stocks as at least 5,000 (Davies, 1957). Fraser Darling was told by American biologists in Alaska that the grey seal is found singly and very rarely along the Arctic coast, as if occasional individuals make the North-West Passage. The eastern and western Atlantic stocks were united in the last inter-glacial period and parted company in the last glacial period.

North Rona has the largest colony of any individual island, possibly about one seventh of the world population (fig. 14). About 2,500 calves are born there annually, and about 7,500 seals of all ages are present there (not all at once) between September and December. The grey seal breeds at St. Kilda, but did not do so before the St. Kildans were evacuated in 1930. Now there is a colony of about 300-400 producing possibly about 100 calves annually. Breeding has not been recorded at the Flannans. We sailed within 200 yards of Sula Sgeir on 20th October 1962 and saw a score or more of seals above high-water mark. We could not see white calves which may have been hidden by boulders. The presence of about 50 greater black-backed gulls overlooking the seals was evidence that calving may have been in progress. Breeding took place on Sule Skerry before the building of the lighthouse in 1895 ; five calves were born there in 1962, the first recent record of such. It is unlikely that any are born at Sule Stack which is much too steep and exposed.

The islands off the west end of the Sound of Harris are favourite breeding grounds: Gasker (1,000 calves), Shillay (160), Coppay (100), Haskeir (60), Monachs (50), Causamul (50). A few calves are born each year on Kearstay, off Scarp. We visited the Taransay Glorigs on 7th October, 1961, taking the deep channel through the reefs. Only one seal was seen in the water and none ashore. These skerries would be untenable by seals in autumn gales. Grey seals are born each year on Flodday in the south Barra isles.

There is a colony of over 1,000 of all ages, with about 350 calves born annually on Eilean nan Ron and Eilean Ghaoideamal off Oronsay south of Colonsay (Hewer, 1960). They are well protected and breed on the beaches of the flat islands intersected by numerous sheltered channels which give the seals sea ways. There is another colony on the Treshnish Isles producing more than 100 calves. There has apparently been a reduction in the size of this colony in recent years. Gunna, between Coll and Tiree, has usually over 50 and Oidhsgeir off Canna has a smaller colony. A few calves have recently been found on an islet in the Sound of Jura south-east of the Gulf of Corryvreckan. Solitary grey seal calves have been reported in autumn 1958-61 from many parts of the coast. These have probably been swept from their breeding station by heavy seas and

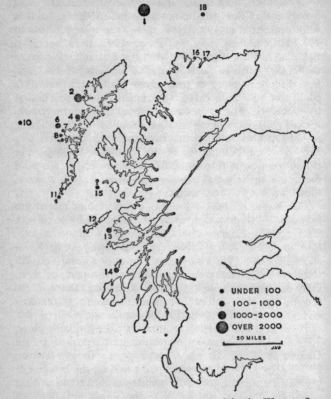

FIG. 14. Breeding stations of the grey seal in the Western Isles classified according to the number of calves born annually: 1, North Rona; 2, Gasker; 3, Kearstay (Scarp); 4, Shillay; 5, Coppay; 6, Haskeir; 7, Causamul; 8, Deasker; 9, Stockay (Monachs); 10, St. Kilda; 11, Flodday (Barra); 12, Gunna; 13, Treshnish Isles; 14, islets off Oronsay; 15, Oidhsgeir; 16, caves at Loch Eriboll; 17, Eilean nan Ron; 18, Sule Skerry

carried many miles before again reaching a safe landing. Colonies of one or two cows and a bull are not enough to keep up a regular population.

The grey seal varies greatly in colour. It has been so called for a hundred years or more but this is not the general colour. The bulls are often olive-brown, the same colour as seaweed, or they may be dark iron-grey with little or no light markings on the throat and chest. All tend to show a bright steel grey cap on top of the head when new-moulted, but this is faded by the breeding season.

The most common colour for the cows is pale slate-grey on the whole of the upper surface. The underside is lighter: lemon or deep yellow splashed with black. Occasionally one of these normally-coloured cows will have chestnut shading on the throat which makes them handsomer still. Then there are light and dark mole-coloured cows not showing the black-and-white underside. When the young cows cast their first coat and go to sea, they have a short blue-grey coat and a lighter underside not very different from the adult cows. The young bulls are darker, sometimes jet black, velvety and beautiful in their first sea-going coat, and resemble the adult bulls. During the first two years these juvenile coats usually fade to a buff, fawn or ginger colour before the adult-like coat is gained.

Many seals are migratory, some markedly so, such as the fur seal (*Callorhinus ursinus*) from California, to the Pribilov Islands, Alaska, the hooded seal (*Cystophora cristata*) and the Greenland or harp seal (*Phoca groenlandica*) southward in October from Baffin Land and East Greenland to the Grand Banks of Newfoundland. The grey seal is not generally considered a migratory species, but in fact it is. We know that dispersal migrations of animals in their first year are considerable. Calves tagged on North Rona in October, 1960, were reported in the following December from points as far apart as east Iceland and Co. Donegal, 480 and 300 miles respectively from the island (Boyd and Laws, 1962). Movements of grey seals to breeding stations from their summer feeding range are apparent in August and September and in the opposite direction January to June (Boyd, 1962). The return to the summer range is much slower than the departure from it and the post-breeding moult has probably an influence on dispersal, affecting both sexes differently.

What is not known is how far adult grey seals go from each breeding station and whether the areas of diffusion overlap. We now know a great deal about the dispersal of young seals from nurseries in Pembrokeshire, Farne Islands, Orkney, Shillay and North Rona. From Pembrokeshire they mainly go south as far as Santander in Spain; from the Farnes northward and eastward to Scandinavia and Scotland; from Orkney to Scandinavia, Faeroes and north-east coasts of Scotland; from North Rona to points between Iceland and Ireland including Faeroes, Norway, Shetland, Orkney and the Hebrides (one was found off Gourdon, Kincardineshire). The overlap in these ranges is substantial. Due to the life of the present tags being only a few years we have so far been unable to trace the movements of adults, which may be different from juveniles.

This diffusion and reconvergence to the breeding islands probably means that the social system of the seals concerned with reproduction is biologically isolating the several groups. Grey seals may return to the place of their birth to breed and this may be involved in the maintenance of breeding assemblies in localities like the ridge of North Rona where the bulls haul themselves laboriously up a steep hill to hold territories about 300 feet above sea-level, followed later by cows which go there to deliver calves. On the way to this high-level nursery the animals pass over ground which is suitable for breeding and which holds large numbers of seals, but undeterred they climb to seek a favourite spot, unseen but well remembered. Evidence from the Sound of Harris and Orkney where there are groups of colonies, suggests that there is little or no isolation between individual colonies. Full-term cows, for example, may find it impossible to get ashore on Haskeir because of the weather at the time of imminent delivery and may proceed to Shillay. The presence of the camp on Shillay in October 1961 occasioned a sharp drop in the birth rate of calves; the cows may simply have been scared off and have calved on Coppay. The same has been seen on the Greenholms in Orkney by E. A. Smith.

In 1960 the Nature Conservancy began an investigation of the social structure of the Scottish population of grey seals by branding the young before they take to the sea. The key to the problem is the placing of a permanent mark on the seals, peculiar to the colony of their birth. The only known method is by branding, and in the past few years juveniles have been branded on North Rona, Orkney, Shillay, Harris and on the Farne Islands. By continuing this over a period of years it is hoped to show whether or not adult seals return to their birth place to breed, or whether they settle in other colonies. This will take some time for the first branded cows will reproduce in 1965, but the bull may not do so till 1970. Parallel to the field-work of branding there has been an investigation of techniques to overcome infection of the brand scar. The brand mark for Rona is "Y", Hebrides "H" and Orkney "V".

The grey seal has a very varied diet which has been listed by Rae (1960) from the analysis of stomach contents: cod, saithe, ling, conger, "toad-fish", lump suckers, salmon, sea trout, halibut, herring, cuttlefish, squid, octopus, lythe, flat-fishes, crustaceans, mackerel, skate, ray, dog-fish, mussels, bib, haddock, whiting, pilchard, brill, wrasse, bass, grey mullet and seaweed.

The feeding haul-outs of the grey seal in summer are on skerries scattered all along the coast from the exposed rocks of Skerryvore and the Flannans to the shores of the firths and sea-lochs; there is an assembly as far up the Firth of Clyde as Innellan. The overlap with the range of the common seal is considerable. It is reasonable to suppose that most of their food will be drawn from local stocks of rock fish, but we have yet to appreciate how far they travel to feed each day, or over a period of days. It might not be the close proximity of food in the form of rock fish at a skerry which decides its adoption by the seals; it could be its being a conveniently sheltered place within an hour or more of a fishing bank, or a resting place for seals in transit between distant feeding grounds.

If the observer is at a breeding station, say Rona, Gasker or Treshnish in mid-summer, he will find the place almost deserted. Only a very small fraction of the breeding population remains to harvest the fish in the immediate neighbourhood of the breeding island ; there may be two or three hundred at Rona from May to July out of a population of about 7,500 of all ages associated with the island from September to December. Very few yearling seals return to the breeding islands, although there are numbers of juveniles approaching adulthood, which are usually assembled on rocks on the fringes of the breeding grounds.

From July onwards the number of seals gradually increases on the nursery islands. They begin to spend more time out of the water in August, but not on the main mass of the island as yet. Both at the Treshnish Isles and at North Rona there are certain skerries favoured by the gathering seals, and it is on these they assemble in close groups. The adult bulls tend to have a rock of their own where they almost overlap each other in their slumbers and many are seen as well among the increasing numbers of cows on resting rocks. Cows are more quarrelsome and more vocal than the bulls, but they pack close all the same. A few yellow yearlings haul out on some rocks, but other skerries are frequented wholly by adults, and it is at these latter places that it is possible to make accurate counts of the increase of numbers through late summer. Here is a typical example from Rona, the counts being made during the afternoon each day: August 14th—56 ; 15th—72 ; 16th—103 ; spell of rough weather during which the rocks were untenable ; 26th—170.

The movements of grey seals from a wide summer range to the breeding islands takes place in August and September ; movement back takes place from January to June (Boyd, 1962). There is a peak of seal movements (both common seals and grey seals) in May and June which is synchronous with the periods of finest weather on the north-west coasts.

The resting period before breeding, finishes at the end of August and now the adult bulls (plate XIV) begin to come ashore to the breeding grounds. It is amazing to see the climbing power of those 8-foot and 5-cwt. animals. There is great gripping strength in their hands, which hold on while their belly muscles contract and expand as they heave themselves upward and forwards. The bulls take up their chosen places and lie quiet there. Often they lie by a shallow pool which becomes more or less the centre of their territory and is the place where coition occurs later. Now the Treshnish Isles are volcanic in origin, with sheer cliffs falling to erosion platforms at approximately sea-level. These shelves of lava are the breeding ground of the seals. The animals cannot get far away from the sea and we find the territories of the bulls set in linear fashion along the coast. Rona is cliff-bound and an immense swell makes the sea's edge a dangerous place. The seals of Rona come farther inland and stay there without frequent return to the sea. The bulls come as far as 200 yards from the sea, and over 250 feet above it. The plan of the territories, therefore, is not linear as on the Treshnish, but a mosaic. This is also the pattern at Gasker, Shillay and Coppay.

Cows come into the territories from the sea 2-5 days before calving and the number of seals ashore increases throughout September and October. On one strip of shore on the Treshnish Isles where the first bull took up his territory on August 28, numbers grew from two on that date to 78 on September 15. The cows leave the breeding ground after the calves have been weaned and they have mated, rarely more than three weeks after calving and more commonly 14 days. Once more the seals are found on the resting rocks, lying in close masses. Bulls disperse quickly after breeding but the cows linger longer on or near the grounds, returning to the breeding islands in some cases to moult. The moulting of cows probably takes place in January and February and is followed by the bull moult in March and April, perhaps also on or near the breeding islands.

Let us look closer at the life of the seals on the breeding

territories and nursery grounds. There are many more adult bulls on Rona than can immediately take up territories. These animals lie on the rocks at a place where there is most traffic up to the territories, and this traffic tends to be up accepted tracks which give easiest access. This bull rock may be called the reservoir, for 500 bulls can be seen there, and the cows stay among them a short time before going up to calve. No challenging behaviour is to be seen at the reservoir, which is strictly neutral ground.

The bulls inland in possession of territories will not trouble each other much either. Challenge comes from fresh bulls emerging from the sea and working their way up from the reservoir. Sometimes the sight of the possessor is enough to deter the new bull, but if not, the two will indulge in a primary display of weapons which, though so different, is comparable with the challenging display of stags. The bull rolls over from side to side, turning his head sideways in the direction of the roll, opening his mouth and raising his hand. The canine teeth on each side and the powerful claws are thus shown to the opponent. A bull grey seal has such a large development of muzzle and foreface that the canines cannot be seen head on. The limit of challenging behaviour is when the bulls come muzzle to muzzle, heads raised. If that does not suffice there is a fight with teeth and claws. Great rips may be made in the hide and once started the fight goes on for some minutes until both appear seriously wounded, and one retreats. Defeated bulls or spent ones return to the reservoir and there all challenging behaviour is set aside. Here again the comparison with the stags is close. A detailed description of bull *versus* bull is given by H. R. Hewer (1960).

When a bull comes ashore for the breeding season he is very fat, extraordinarily fat, yet he looks what he is, as fit as a fiddle. He has had his last meal for a month or two, or perhaps longer. He now begins to live on his blubber and gradually loses condition. This is the position on Rona and other colonies with mosaics of territories, but on Treshnish and Oronsay where the animals are nearer the sea and can enter the water direct from their territory without infringing another, the bull will spend many hours in the water opposite his territory, gently patrolling the length of it. It is unlikely that they feed though they are in the water ; to be in the water is perhaps the best method of defence of the ground. The inland territories of Rona are not kept by the same bull from beginning to end of the breeding season ; there is always the traffic up of fresh bulls and down of spent ones, and it is probable that some territories are occupied by several bulls in the course of the breeding season. On Rona some of the principal bulls were in territory from about 4th to 24th October 1959. It is uncertain whether or not some of these spent bulls return to the territories again, but it is unlikely. Grey seals are polygamous, bulls usually having about ten cows in the denser colonies. The sex ratio on Rona is nearly equal at birth and there is a slightly heavier mortality in the males thereafter ; the adult stock of a polygamous species, therefore, still appears to have a large excess of males. Males mature a year or two after females and the oldest ones die several years before the oldest females. Almost all the adult bull population may be in service some time or other during the season, and the apparent excess of males at any one time is no true indicator of the situation for the season as a whole.

The cows are free to go wherever they like. Sexually they belong to the bull in whose territory they may happen to be at any one moment. This is unlike the social system of the northern fur seal (*Callorhinus ursinus*) and of the elephant seal (*Macrorhinus angustirostris*) each of which species collects a harem of up to seventy cows, and the harems are herded by the bulls. This type of rutting behaviour which makes for a large surplus of bulls hanging about on the outskirts of the breeding ground, and is associated with great disparity in size between bulls and cows, is biologically wasteful, especially so when the eager bulls reach the extreme of injuring cows. Bertram (1940) has drawn attention to this correlation between the size of bull and size of harem. In the grey seal the harem is up to twenty, though rarely, and the difference in size between the sexes, though marked, is not extraordinary; bulls are 7-8½ feet and cows 6½-7 feet. The bull grey seal is not extraordinarily active among his few cows. They are within an area of possibly ten yards square. The denser the cows and bulls are on the ground the higher are the chances of injury and death to calves from mauling and crushing. This is well seen on Rona; the gulleys of Fianuis seethe with seals and are littered with dead calves.

Matings take place 11-18 days after the birth of the calf. Coition usually occurs ashore, often in a shallow pool but the bare rock or the sea may be used. The bull, then, has been ashore at least three weeks without any cow having been in season. During all this time he has been ready to fight for his territory in which sexual satisfaction has not been obtainable. The north end of Rona is by this time, the latter end of September, completely invaded by the seals and some have climbed a very steep hillside to the top of the ridge. We once saw a pregnant cow seal half-way up the west cliffs of Rona where they were about 150 feet high and at an angle of 45°. Had it not been bare, rough, stable rock she could never have finished her climb. Another calf was successfully reared at the edge of the 300-foot sheer column of the western cliff.

Almost all those cows ashore in the territories of Rona are ones with a calf (plate xiv) or about to calve, and all the bulls are adults with territories or seeking them. Now the social system of the seals is finely adapted for lessening the danger from their association with the land. In 1937 on the Treshnish Isles no maiden cows came ashore for breeding on Lunga, and there were no immature animals to be seen either. Circumstances did not allow a solution to this puzzle, but later when we visited Rona there was no maze of little islets and skerries to obscure the issue; we could see almost everything to do with the seals from the island itself. Maiden cows collect on the large flat skerry, Loba Sgeir (Portuguese for seal is *Lobito*), at the south-west corner of the island, and a few are seen among the reservoir of bulls on Leac Mor Fianuis and on the breeding grounds. There is a large number of bulls on Loba Sgeir, mostly young adults, probably mating with maiden cows. This flat skerry is practically always safe for the seals because it is ringed by a very bad surf, but very few calves are born and no breeding territories are apparent on Loba Sgeir. Bulls and cows lie cheek to jowl and matings occur with comparatively little quarrelling. Frequent mock battles took place on Loba Sgeir between young bulls, sometimes between a cow and a bull, but only rarely between young cows.

It is not desired to draw a teleological conclusion but one must point, all the same, to the value which the fully adult territorial behaviour has for the survival of the calves. It makes for sufficient room for each cow and calf during that fevered fortnight of maternal jealousy after the calf is born. Hewer has drawn attention to the behaviour of the cow in maintaining territory, being centred on the space round the calf and not necessarily on the nature of the ground.

There remains to be described the behaviour of mother and calf. Birth is usually very rapid and the afterbirth is shed within half an hour. The calf when born is clad in a thick coat of fluffy hair, yellow until washed by rain when it becomes cream or ashen in colour. The head appears large and discrete from the body, and the limbs look relatively long, for as yet the calf is thin. If a still-born calf is skinned a dense loofah-like layer of connective tissue is found immediately below the skin. This tissue opens up to accommodate fat in the same way as a new loofah opens up to take water. The new-born calf is about 35 inches long and weighs about 32 lb.

The mother takes very little notice of the calf for the first quarter of an hour after birth, but clears a space around it ; then she offers it her two teats and within half an hour the calf is taking its first meal of milk which is about twelve times as rich in fat as cow's milk ; she does not lick the calf at all though she will smell it, which is her means of individual identification. The first two days of its life the seal calf is more active than it is for the next month. It is possible to tell a new calf at a glance because its two hind flippers tend to spread to the side and it half-uses them in scrambling about in those journeys of a few yards hither and thither. After two days the flippers remain longitudinal and are not used. Here, presumably, is some measure of evolutionary recapitulation, a half-successful use of the hind feet for a few hours. Hewer suggests that the young calf's use of the fore-flippers alternately instead of together may also be recapitulatory. These small adventures of the calf are responsible for much trouble between the cows, and any calf is liable to be severely bitten by a cow not its mother. If a bull finds a calf in his way he will pick it up by the neck or flippers and toss it several yards or shake it vigorously.

The over-anxiety and jealousy of the cows over their calves mean that the bull of a harem is sometimes attacked with great ferocity. The bull backs away quickly from the cow behaving in this fashion and makes as if to defend himself only when the cow is upon him and then he does her but little harm. Why are cow grey seals so jealous? The limited number of suitable nursery sites might impose a density of population causing extreme quarrelsomeness, but in all nurseries in the Hebrides seals could spread out more if they wished. Though North Rona has such a large herd only a very small area of the island, all of which is accessible to seals, is densely colonised. There is probably a biological advantage. On the Treshnish Isles, where cows could get to and from the water easily, they spent much more time out of the water with their calves when they were closely gregarious than when they were isolated.

Immediately after the first feed the cow begins to show maternal affection, which increases in intensity during the following three or four days. She shuffles round in order to touch the calf with her muzzle and then to scratch it lightly with her fore paw. This flippering is a habit sometimes practised before, during and after feeding, and after the first feed the calf comes itself towards the mother's head and is scratched from head to tail down the back. Often the cow's vibrating flipper does not actually touch the calf in these interludes.

The seals on Rona, Gasker and Haskeir have their calves well up from the sea, so the danger from swell and spring tides is small. Locally, however, great danger does exist in these colonies such as on Sceapull and at the Tunnel Cave on Rona. At Treshnish, Shillay, Coppay and Causamul many calves are born within reach of the spring tides and the gale-driven swell, but none more so than at St. Kilda where most of the calves born probably perish by being unable to regain the nursery rocks and are swept exhausted into the open ocean. One is accustomed to seeing carnivorous animals carrying their young in their mouths in the face of danger, but the grey seal cannot do this.

What the cow can do for her calf, then, is limited, but that little she usually does well. We have seen a cow move her newly-born calf 20-30 yards by shuffling it along between her paws. Where there is a heavy surf with ground swell at high tide, the cow lies below her calf at the water's elge and breaks the force of the waves to the calf. She curls herself almost half round it, and the calf is caught against her instead of being sucked back by the swell into the sea. If the calf is perverse, its mother will make as if to snap, and these threats are successful in helping to get the calf above the reach of the surf. We saw a cow on Rona holding her pup against the cliff edge with her paws at a place where it could climb on to a ledge. This type of behaviour is often conducted with perseverence and a cow has been seen to maintain it for the six tides of a three-day onshore gale at the time of a spring tide.

In attendance at all the seal nurseries are flocks of greater black-backed gulls. The size of the flocks are proportionate to the size of the seal colony; on Rona there are usually about 500 scavenging gulls, some of which are herring gulls. The birth of the calf is an occasion for a squabble for the afterbirth, and these are moments of great danger for the newly-born calf. The blood-stained body may still have membranes clinging to it and the blood-red umbilicus is exposed as the calf wriggles free. A cow delivering a calf for the first time may not be prepared for the sudden onslaught of the gulls and her off-spring may receive fatal injuries while she recovers her senses. The numbers of dead and miserable broken-winged black-backs staggering through the seal ground on Rona, however, bear sure testimony to the alertness and speed of most cows in protecting their calves. The treatment meted out to the gulls is in contrast to that to the turnstones which work without harm in the mud between mother and calf.

Young grey seal calves will play happily in the pools of an erosion platform or in the sea if it is quiet and there is an easy beach for them to climb ashore. But their long white coats are unsuitable for much swimming exercise and a calf would not seek escape into the sea. If they get there by accident such as by the lick of the swell at high spring tides, young calves will swim vigorously and make valiant efforts to get ashore. In heavy surf they are dashed against the rocks and quickly killed. Sometimes, when they have been unsuccessful, we have examined the bodies. The claws have been worn away; the chin and palms of the hands have been raw. When hungry or in danger the calf cries like a human child.

The ground on which calves are reared varies from a mud bath to clean spray-washed rock. The mud is strewn with the dead and the manure of seals and birds and is highly septic; consequently the wounds and eyes of calves quickly become infected. The infection also spreads to the lungs through nose and mouth with resulting pneumonia. On Rona there are a few access routes to and from the sea usually up rocky grooves on the slopes of which large numbers of calves are born. In their infant wrigglings some of them roll down into the main path of the traffic of adults, and are soon crushed and killed. Mortality of calves in such an area is usually over twenty per cent. On Shillay, however, the seals do not come on to the island along a few narrow paths, but on the broad front of a boulder and sandy beach; mortality of calves there is probably less than ten per cent.

The calf is fed at about two-hourly intervals during the first few days and then at rather longer periods. Each meal appears to be a good one, for suckling takes from ten minutes to half an hour. The growth rate is very rapid; Coulson found that Farne Island seals reached between 75 and 90 lb. at a fortnight old, i.e. an average of 3 lb. per day confirming our own findings on Treshnish and Rona. The increase has been made on milk alone and wholly at the expense of the mother's body, for when she comes up from the sea before calving she begins a fast which probably lasts until she returns after mating, a period of at least three weeks. An observer is soon able to tell to a day or two how long any cow has been out of the sea by her decrease of fatness. Similarly, the age of a calf can be judged accurately by its increasing degree of fatness and the shedding of its white coat.

The calf begins to shed the fluffy white coat between 13 and 20 days; the date probably varies greatly with the feeding and general condition of the infant. The moult begins on the muzzle, the paws, and a patch on the belly; the female assumes a very beautiful blue coat and the male a dark grey, occasionally black one, in the next fortnight. It is common to see an almost-blue calf lying in the middle of the old hair which it has been several days in casting and rubbing off its back by rolling this way and that. The time when the white coat is shed is usually synchronous with weaning though moulting and fully moulted calves are sometimes suckled. At weaning they are left absolutely by their mothers and have to find their own way to the sea.

The calves have little way to go in most colonies where they may already have had some experience of going in and out of the water in playful fashion, but on Rona, Gasker and Haskeir where many are born high on the island or on the edges of the cliffs, the journey is fraught with danger. Many survive sheer drops of up to 50 feet into the sea. These calves which have fed and prospered so richly on nothing but mother's milk, face a period of complete starvation from a fortnight or three weeks old. They may get to the sea in a week but some take a month to do it and even then to not appear in an urgent hurry. From a telescoped infancy they enter a protracted childhood, for their live-weight increase from then to one year is small.

After entering the sea the young seal travels far in the first few weeks of its life. Two marked with green dye on North Rona were seen at Sule Skerry about 60 miles to the east a few days later. They have been recovered from many points between east Iceland and Donegal in the first three months after being tagged on Rona, and probably within ten weeks of leaving the island. The few recoveries which have been obtained from Shillay show a wide overlap with those of Rona ; they range from Gruinard, Ross-shire to Co. Kerry. Tagged seals seen at St. Kilda could have been from either Shillay or Rona. Young seals from Orkney find their way to the Hebridean and West Highland coasts, and one from the Farne Islands was found at Raasay. E. A. Smith on a recent visit to Pentland Skerries saw three young seals playing together ; two had red Orkney tags and one a pale green from Rona.

The early records of seals in the Hebrides are those of seal hunts, particularly at Haskeir and Causamul. Before 1850 the emphasis was on grey seals as a natural resource for local inhabitants, in days when the islands were much more thickly populated than they are today. Then they bred only at places out of reach of heavy human exploitation: Haskeir, Causamul, Gasker, Sula Sgeir and Sule Skerry (before the lighthouse was built). It is unlikely that large colonies such as we have today on North Rona existed in the Outer Hebrides before the human depopulation. The evidence points to the grey seal moving into breeding localities such as Rona, St. Kilda and the Monachs when man moved out ; the opposite occurred on Sule Skerry when man moved in with his lighthouse.

After 1850 the emphasis shifted to the conservation of the grey seal as a rare species. In 1858 the killings were stopped on Haskeir by the proprietor but clandestine hunts continued. Throughout the latter half of last and the early years of this century, hunts were carried out on all the seal islands including North Rona. The Grey Seals Protection Act of 1914 was based on the supposition that continued hunting by natives in open boats would bring the species to extinction. Had the scale of the hunts been increased to satisfy a wider market there would have been greater cause for anxiety but there is little evidence of large exploitation. Those who promoted the Bill claimed that numbers around Scotland were as low as 500. The figures on which this estimate was based have never been found, and it seems probable that the population in the Outer Hebrides alone was very greatly in excess of 500. Since then, however, the species has been protected in the breeding season by the Act which was renewed annually until 1932, when it was made permanent with a close season from 1st September to 31st December.

Although some of the old records are to be mistrusted they provide a useful time-scale against which to compare the influential factors: human depopulation of potential breeding localities, fall in demand for products of the seal-hunt, and protection by law. It is sometimes inferred that the thriving population of grey seals which we have on our coasts today is the result of protection, but we must not lose sight of the fact that the increase in the population commenced half a century before the Act was passed. This increase was in the face of local seal hunting, but there was no big business involved and the Act has served to prevent this.

In the last decade the grey seal has been brought more and more into the public eye as a threat to the coastal salmon fisheries. Around this there has arisen an argument between the protectionist and "seal-lover" on the one hand and the fishermen on the other. Somewhere between the two there lies the scientist who is responsible for marshalling the facts in a dispassionate way, and the politician who must make the best of interpreting the scientific, economic and social enigma. Out of this in 1959 came a Suspension of Close Season Order, renewed each year, by which the Secretary of State is now able to issue permits, as he thinks fit, for the shooting of grey seals during the breeding season in Scotland, except on National Nature Reserves. Never in the history of the grey seal has there been such widespread concern about its innocence and guilt, its scarcity and abundance, its food and parasites, its beauty and its market value. One thing remains for us to remember—the gathering of these seals to a very few breeding stations, and their comparative helplessness ashore at that time, lay them open to particular danger from commercial exploitation. It is for us all to protect them adequately.

PLATE IX: *Above*, Eastern Pine Forest at Invereshie, Cairngorms National Nature Reserve. *Below*, Western Pine Forest at Coille na Glas Leitire, Beinn Eighe National Nature Reserve, Wester Ross

PLATE X: *Above*, Western Oakwood, Ariundle Forest Nature Reserve, Sunart. *Below*, heavy seas on the exposed rocky coast of Ardnamurchan Point, Argyll

PLATE XI: Birds of the shingle and shore *Above*, Ringed Plover (*Charadrius hiaticula*). *Below*, Oystercatcher (*Haematopus ostralegus*) Allt Mor, Inverness-shire

PLATE XII: *Above*, English Stonecrop (*Sedum anglicum*) on Rhum. *Below*, Moss Campion (*Silene acaulis*) on Ben Macdhui

PLATE XIII: *Left*, Guillemots (*Uria aalge*), one of the bridled form, North Rona

Below, a Fulmar (*Fulmarus glacialis*) among the sea-pinks (*Armeria maritima*), North Rona

PLATE XIV: *Above*, a blind Grey Seal (*Halichoerus grypus*) cow with her calf, 30 yards from the sea on North Rona

Right, a growling Grey Seal bull on North Rona

PLATE XV: *Above*, Red-throated Diver (*Gavia stellata*) nests by small lochans throughout the north-west Highlands. *Below*, Slavonian Grebe (*Podiceps auritus*) cock, a well established breeding species in Inverness-shire

PLATE XVI: The River Lyon in the elegant hill country of Central Perth-shire: a salmon water, the head waters of which have been diverted for hydro-electricity

Chapter Twelve

Freshwater Lochs and Rivers

PHYSICAL AND CHEMICAL FACTORS

Look at a tracing of freshwater lochs and river systems of the Highland area: here alone is material for a whole evening's imaginative thinking. First we can relate it to the orographical map which, in effect, it mirrors. Throughout the Highland region we are struck by the large numbers of large and small lochs, many of which act as reservoirs for the rivers running from them. The great ridge of Scotland, Drum Albyn, is much nearer the west coast than the east; in fact, at one point in Glen Dessary above Loch Nevis, the ridge comes to within four miles of the western sea. The general result of this conformation is that western-running rivers of the Highlands are short, and those going east much longer. The western rivers have far to fall in a short distance and their velocity is on the whole many times greater than that of the eastern rivers. Water is one of the great carriers of this earth, and the power of water to carry is governed by its velocity. A law of hydrodynamics says that a stream able to move shingle of one ounce in weight would, if the velocity were doubled, be able to move boulders four pounds in weight. The scouring power of these western rivers is, therefore, very great.

There is another pleasant hour to be spent comparing the map of the freshwater systems of the Highlands with the geological map. The varying hardness of different rocks, of different complexes within the same formation, and the effects of faults and intrusions referable to cracks, slips and crinkles in the earth's surface, cause both obstacles and opportunities to the Highland rivers. Glaciation in earlier time has had a profound effect on the forms of

L

freshwater systems of today. In gouging, the glaciers have engraved the basic pattern of loch and river and have truncated the flanks of the glens leaving tributary corries high above the floor of the main glen. These are the hanging valleys and there is an outstanding example of such above the Steall Gorge in Glen Nevis, complete with waterfall.

Furthermore, the rain-water which makes the rivers does not fall as pure H_2O. It gathers carbon dioxide and ammonia from the atmosphere which renders it a powerful solvent. When rain-water seeps through a mass of decaying vegetation such as peat, as it often does in Highland country before it reaches the rivers, it has a definite influence on moulding the ultimate courses of rivers. Rock phosphate, for example, is a highly satisfactory fertiliser for hill ground in Scotland because the acidity of the water is sufficient to dissolve the particles, but does not do nearly so well on neutral soils.

Different rocks vary in resistance to the solvent power of river water charged with carbon dioxide. Archean gneiss and granite contribute less to chemicals in freshwater than rain itself, but limestone and Old Red Sandstone contribute much more chemicals than rain. The gneiss is highly resistant; thus, if we look at the courses of rivers and burns in western Sutherland on a large-scale map we find them having to go round the rock and often not managing to get away again until a small loch has been formed and egress found at a new level. The same county of Sutherland will show us the other extreme in its limestone country. There are the caves of Smoo on the north coast, wholly water-worn, and the caves and disappearing burn in the limestone of the country east of the road between Stronchrubie and Inchnadamph. Water laden with carbon dioxide can dissolve limestone so that it will hold one part in ten thousand of limestone. This may not be much, but it must be remembered that there is a lot of time and a lot of water. Limestone-laden water is in turn an extremely potent factor in changing vegetational complexes and the abundance of small invertebrate animals.

The action of carbon dioxide in the micro-habitat of a small loch on the limestone island of Lismore is responsible for summer cloudiness and winter clarity of the water, described by A. Scott in T. Scott (1890). In the vigorous growth of spring and summer the vegetation of the loch will take up, in sunlight, much of the carbon dioxide needed to keep the calcium carbonate in solution. This action is more or less continuous in summer when there is little darkness. The calcium carbonate is therefore precipitated in an extremely fine state of division, giving the milky appearance to the water. The following is the descriptive equation:

$$CaCO_3 + CO_2 + H_2O \rightleftharpoons Ca(HCO_3)_2$$

It is humbling to consider the vast field of action and interaction which a study of the physics, chemistry and dynamics of water reveal. The river, gouging out its base level of erosion, has met obstacles which have changed its course and momentarily its velocity, it has gathered salts and altered landscapes, it may have meandered and deposited alluvial flats, and ultimately it comes to the sea. The chemistry and physics of sea water are a study in themselves, but this much most of us know, that the river and the sea have their lifeless, unconscious, swift-moving rhythmical battle. Not only dynamics are concerned with the deposition of the bar, where the fight is fiercest, but chemistry as well, and forms of life have evolved which can take advantage of these conditions.

The days of the ice sheet are past in Scotland, but its consequences are plain. Frost allowed the immense weight of water to accumulate, but as ice it was still fluid and could move mountains. Ice could also act as a mass, and dam water to make lakes. The Highlands have one fine example of such action which left a mark that puzzled man for centuries. The southward-flowing glacier from Glen More dammed up the entrance to Glen Spean, Glen Roy and Glen Gloy. The waters flowing into these glens from the tops formed lakes over a thousand feet deep. The

different levels of these lakes through time have left beaches
high on the present-day hillsides, and are known as the
Parallel Roads of Glen Roy. Just as the river brings down
detritus to form a bar at its mouth, the glaciers did likewise
and formed some of the lochs near the sea which we know
today, such as Loch Shiel in Moidart and Sunart.

The faster the ice travelled the deeper the glacier
trenched the landscape. Just as with a river, the speed of
flow is fastest in the narrows, and it was in the narrow upper
reaches of the glens that the down-cutting power of the ice
was greatest. As it progressed into the broader lower
reaches of the glen the speed was reduced, resulting in an
upward sweep of the floor at the entrance. The western glens
are mostly of this rock basin type and are filled with
either a sea- or fresh-water loch. Loch Coruisk in Skye is
an outstanding example of such a freshwater loch, and
corrie lochans are also of this type. Harry Slack (1957)
illustrates this well with an echo-sounding transect of Loch
Lomond ; in the narrow fjord-like reach near Tarbet the
depth is about 600 feet, as the loch widens in the Luss basin
it is about 200 feet and near the widest part of the loch beside
Inchmurrin it is less than 100 feet. Loch Morar, which is
but a mile from the sea, reaches a depth of 1,077 feet ; a
depth which the ocean does not reach until the dip of the
Continental Shelf outside St. Kilda. It is the deepest freshwater in Britain. Loch Ness, which forms the largest single
mass of freshwater in the United Kingdom, is 754 feet deep
and not much more than 50 feet above sea level.

These deep lochs are obviously more constant in
temperature than the shallow ones. Their mid-water
temperature does not rise above 60°F in summer or fall
below 42°F in winter. The abyssal water of Loch Ness and
Loch Morar remains at slightly above 42°F permanently,
not varying more than a degree in several years. Such
water is but little aerated. Large lochs are subject to a long
fetch of wind which piles up the water at the downwind
end and when the wind abates leaves the loch in a state of
oscillation. In shallow lochs, or in the shallows of deep
lochs, this oscillation is enough to mix the water from

surface to bed and give a constant temperature throughout the loch. Dr. Slack has shown the isotherm patterns for the deep, middle and shallow reaches of Loch Lomond. In the shallow Inchmurrin basin the water is mixed by the wind all year round except during settled weather in May and June. In the Luss basin mixing of top and bottom water takes place only during periods of high winds from December to April and in the deep Tarbet basin mixing is limited to February, March and April. This means that deep lochs are thermally stratified for most of the year and while in this state there is no passage of water from the upper levels to the bottom or *vice versa*. It also means that the plankton which live in the sunlit upper layers are deprived of the nutrients in which the abyssal water is rich, and that the bottom living organisms are deprived of the oxygen in which the upper layers are rich. The division between the upper and lower layers of water is the thermocline which rises in summer and falls in winter. After the turnover of the loch the thermocline is absent and for a period the plankton have nutrients upswept from the bottom and the abyssal fauna have oxygen in plenty downswept from the surface.

The shores of the lochs vary from those which plummet into the water in vertical cliffs to spacious shingly flats some times with alder scrub, which are covered only in the seasons of high rainfall. In Highland lochs too little light for photosynthesis usually penetrates below a depth of twelve feet and this restricts the coverage of green plants on the bed. The plant zone is usually limited to about ten per cent of the area of the loch, a little more in shallow ones and a little less in deep ones. The nature of the bed changes as one moves outwards from the high-water mark, through gradations of large pebbles, gravels, sands, silts to muds in the deep still water (see also page 352). The pattern of the bed is varied by the presence of outwashed fans from streams entering the loch or erosion cuttings made by rivers issuing from it. As described in the chapter on the sea shore, the nature of the shore is also influenced by its exposure to surf, and though the waves in the lochs are small they are nevertheless incessant and limiting to life. The zonation of plants such as the shore-weed (*Littorella uniflora*), quillwort (*Isoetes lacustris*), the water milfoil (*Myriophyllum alterniflorum*), water lobelia (*Lobelia dortmanna*), the moss *Fontinalis antipyretica* and the green alga *Nitella opaca* are influenced by all those factors and these plants in turn influence the animals which find shelter and feed among them.

The rainfall map shows that the short rivers of the West Highlands north of the Firth of Lorne and Loch Linnhe have to take a particularly heavy rainfall. Many of them provide a spectacle during the heavy spates, but their catchment areas are so small and the slopes so steep that these rivers may rise and fall within a few hours, and quite often in summer they may fall from a raging torrent to little more than a trickle in two or three days. Such quickly-changing conditions much affect the biology of both the river and its banks.

The catchment areas of the many lochs vary enormously, from a few acres for a Sutherland *dubhlochan* to 22.8 square miles for Loch Ness. The character of the waters varies also: first the acidity expressed by the symbol pH. The peaty lochans of the far north are very acid and give a pH value as low as 4.2. Some of the larger lochs strike a stratum of basic rock in their depths and this raises the pH reading to possibly 6.5 or 7.0. The water of a loch on limestone may read pH 8.0-9.0 if there is not a mass of decaying vegetable matter in the shallow parts.

The *machair* lochs of the Hebrides have much of their beds covered with shell-sand and are alkaline. There are biological influences on the pH also, so that the figure for a loch may change both diurnally and seasonally. On Priest Island in the Summer Isles, for example, there are eight lochs within 300 acres and sharp differences in the pH of the water referable to the behaviour of sea-birds, and in one instance to a very slight dilution by sea-spray. One loch is the favoured place for gulls of several species to bathe. The accumulated effects of these birds defecating in the water raises the pH from 6.0 to 7.6 in spring and early summer. This loch alone appeared to have the fresh-water shrimp *Gammarus* plentiful all summer. *Gammarus* is almost always found in alkaline waters, even in isolated limestone lochs. It can occur in mildly acid waters down to pH 6.5 but these are rare occurrences. It is almost absent from the great tracts of the Highlands, for example the north-west. *Gammarus pulex* is generally a stream species in the Highlands and *G. lacustris* more often occurs in lochs.

Loch Lomond, like most other Highland lochs is generally poor in mineral salts, but unlike most, it has a fairly wide variation in salt concentration due to the changing nature of bed and the diverse geological character of the catchment area. The pH range is 6.6 to 7.1 for surface samples, and the mineral content of the water increases from about 22 mg. per litre in the northern fjord-like portion of the loch to about 28 mg. per litre in the wide southern portion. This is to be expected since the inflowing streams from north to south contain an increasing amount of salts. Take calcium for instance; the River Falloch and the Inveruglas Water have respectively 2.2 and 5.8 mg. per litre, while the Rivers Endrick and Fruin have respectively 10.3 and 15.8 mg. per litre of calcium (Slack, 1957).

Generally speaking the more calcium and other alkaline salts there are in Highland streams and lochs the more fertile they are. Fertility is often measured in terms of the fish population, and fish together with the crustacea and molluscs on which they feed require calcium to build their skeletons, integuments and shells. In the Fionn Loch and its neighbouring *dubhlochain* in Wester Ross there is another example of the contrast which we have seen in the two portions of Loch Lomond. The bed of the Fionn Loch has a thin band of basic rock running through it, and this raises the fertility with plenty of fish food, mainly crustacea and insects. The fish of Fionn Loch are numerous and famed for their size and quality, while a few hundred yards away the *dubhlochain* have a few small trout. These lochans usually have a shallow layer of clear water above a deep layer of peat, several feet thick. The roots of water lilies may be in the peat and their flat leaves ornament the surface.

Perhaps the most vivid contrast of chemical conditions in neighbouring waters can be obtained in the Outer Hebrides, and this is nowehere more so than on the Loch Druidibeg National Nature Reserve in South Uist. There the grey lag geese go to islets in the moorland lochs, principally Druidibeg, to nest, but soon after the goslings are hatched the parents usher them from the dark, peaty and comparatively barren waters of Loch Druidibeg itself to the fertile *machair* lochs, Lochs Stilligarry and a' Machair, which are nurseries. These *machair* lochs have excellent stocks of brown trout and are also the feeding grounds for numbers of ducks, geese and swans all year round. The same cannot be said for Loch Druidibeg only half a mile to the east where the bed of the loch is not of shell-sand but rock and peat.

FISH OF RIVER AND LOCH

The brown trout (*Salmo trutta*) is so widely distributed throughout the area that it takes thought to remember which lochans and burns do *not* hold them. Man has been an important influence in this general distribution. He is one of the predators on the fish and in days long past there can be no doubt that he shifted stocks to empty waters and to burns above unscalable falls. Sportsmen of later years have done this on a large scale.

A brown trout may reach no greater length in its life than three inches, or it may reach three feet. Many West Highland rivers subject to rapid rise and fall and carrying peaty water have very small brown trout, and may yet have a good reputation for salmon and sea trout. But the brown trout is the sedentary form dependent on what the river produces and what falls upon its surface and is washed into it in the way of terrestrial invertebrates. Man's destruction of forest and scrub growth has greatly emphasized the swing between spate and trickle. The birch and alder leaves falling into the burns in a good length of their courses provide organic matter for food and cover for those invertebrate forms on which fish live. Putting so much of the gathering ground of Highland rivers to sheep farming, to the virtual exclusion of cattle, has also been a loss to the trout. Cattle are great makers of organic matter —plain muck—which raises the carrying power of both land and water for invertebrate life. The dung that by rights should go back to the grassland enriches the water. In the rivers of the Eastern Highlands where the ground is better wooded than in the west, where spates are less excessive and spawning grounds better, and where the cattle-sheep ratio is narrower, the trout are much larger.

At Lochan an Daim, Dunalastair Reservoir and Loch Moraig all near Pitlochry in Perthshire, T. A. Stuart (1953 and 1957) found that brown trout become ripe at approximately the same time each year; the date of spawning is associated with the seasonal lowering of temperature and adverse climatic conditions may result in failure of the runs. It appears that a proportion of the trout population do not spawn every year. Large fish reach the spawning beds first. Females predominate in the first run of spawning fish, but thereafter males are more numerous on the redds. The large males tend to mate with the large females in the first run, then stay on the redds to meet subsequent arrivals of ripe females and this results in a hierarchical behaviour in the large males. When hatched the alevins live in the gravel and are held there by a negative reaction to light; when they are ready to feed, however, the reaction is reversed and they emerge from the gravel to begin their active life. Young trout feed sporadically unlike salmon parr which feed continuously.

Migrations of young immature trout from streams to lochs occur in autumn and to a certain extent in the opposite direction in spring, when the streams become repopulated. The departure of the immature fish in autumn is usually just before the arrival of the spawning fish and some of the males with ripening testes await the arrival of the spawners and depart with them back to the loch.

Experiments on the stream system of the Dunalastair Reservoir have given little indication that brown trout stray from their own stream when moving in from the loch.

Niall Campbell (1961) has compared the growth rate of trout in a number of acid and alkaline lochs in the Highlands and found that there was no direct relationship between growth rate and alkalinity. Fincastle Loch and Lochan an Daim, both in Perthshire, which have moderate high alkalinities, do not produce better trout than Strathkyle Loch in Easter Ross which is acid. The population density of trout is probably the controlling factor in fish growth. Lochs with good spawning facilities, whether acid or alkaline, produced small fish ; conversely lochs with poor spawning produced large fast-growing fish. Recruitment of young fish to the population is important in determining the size of the fish in a loch. In the Highlands it is exceptional to find a loch which has four-year-old trout attaining a length of about 16 inches.

Alkaline lochs usually have a larger standing crop of invertebrate food for trout than acid ones. The changes in the fauna of the mud on the bottom of low-productivity lochs in the Scourie district of Sutherland were examined by Neville Morgan after the addition of super-phosphate and N.P.K. fertiliser in 1954 and 1955. By the spring of 1960 the weight of the bottom fauna had increased by 25 and 16 times respectively, after which they declined to the prefertilisation levels. In one loch the number of midge larvae per unit area increased by 43 times the prefertilisation level. The chemical and botanical sides of these experiments are described by Brook and Holden (1957) and Holden (1959). The addition of fertilisers did seem to improve the growth rate of trout in the lochs over the first few years at least (Monro 1961).

The dramatic effects of impoundment, initially boosting the growth of resident brown trout, are usually followed in a year or so by an equally dramatic decline. The changes in food, growth and age distributions of the brown trout population of Loch Garry, Inverness-shire during pre- and post-impoundment conditions, until three years after impoundment, were monitored by Campbell (1963). Nearly all age-groups of trout grew exceptionally well, 3 to 5 year-old trout almost doubling their pre-flooding size within two years. However, 1 to 2 year-old trout did not show this response, probably because much of their life was spent in inflowing nursery streams. Unlike most other hydro-electric reservoirs the growth rate of trout did not eventually decline because, says Campbell, of an increased area of shallow water with valuable types of trout food and the tremendous amount of restricted angling reducing the trout population and relatively increasing the food supply. The great increase in growth was due, firstly, to the large numbers of earthworms, spiders, and other soil fauna that were available to trout over the newly flooded land; the ratio of terrestrial to aquatic organisms found in trout stomachs during the early period of flooding was 19:1 by weight and 4.3:1 by numbers. Campbell considered that the larger average size of terrestrial food organisms enabled the trout to obtain more food with less effort and that trout actually eat more food at a time when the food organisms are larger. After the initial increase in growth rate due to the availability of terrestrial food, a second boost to growth was given by the considerable increase in aquatic organisms due to the decomposition of organic material. Normally little or no growth is shown by Highland trout until May or June, but in the first year of impoundment growth commenced in mid-March, indicating that trout grow even at winter temperatures provided an abundance of food is available to them. Campbell (1955 and 1957) also studied a smiliar situation on Loch Tummel after completion of the Clunie Dam.

Studies of the natural food of trout showed that the larval and adult caddis flies (*Trichoptera*) and midges (*Chironomidae*) occurred in 30 to 40 per cent of the stomachs examined; terrestrial wind-blown insects were found in 23 per cent of them and aquatic beetles (*Coleoptera*), stone flies (*Plecoptera*), mayflies (*Ephemeroptera*), crustaceans (*Gammarus, Asellus*), young fish and amphibians in 10 to 20 per cent. From April to

September more surface and mid-water food was taken than bottom food, and *vice versa* from October to March. Large fish between 20 and 24 inches did not seem to have a predominantly piscivorous diet.

There was a large overlap in the food of the trout and the perch, though the former did not scavenge organisms washed into the loch to the same extent as the latter, and the latter did not take surface food. The main overlap occurs in late spring and early summer when the perch have just spawned and both species are feeding vigorously in shallow water on the ascending myriads of midge pupae. At other times of the year competition for food is not so acute with the two species more widely distributed in both deeps and shallows. Worms which are abundant in the littoral and shallow water muds are probably unavailable to most fish except eels (*Anguilla anguilla*).

Salmo trutta not only varies in size ; it varies in form. Where brown trout live in the estuary of a river, they gain the name of slob trout, being neither like river trout nor sea-trout in appearance, but something in between. Whether the slob is a sea-trout become sedentary or a brown trout turned estuarine, we do not know. Lastly, there is the migratory form of *S. trutta*, known as the sea-trout or salmon-trout. This race is hatched in small burns where brown trout may also breed and the young of the two races are indistinguishable. But at the age of from two to four years the migratory form begins to move down-stream towards the sea. These smolts develop a silvery quality of the skin before taking completely to the sea. After a few months at sea, during which the young sea-trout may travel far, it comes back to its own river or a neighbouring one as a finnock or whitling, and it may be a breeder. It may stay a few weeks or a few months, and then go off to sea again for a further period of growth. Brown trout of many *machair* lochs in the Hebrides are large and possess the pink flesh characteristic of the sea-trout. It was once thought that cannibal trout took on a ferocious and carnivorous appearance, with the loss of the red spots. This is not so ; cannibal trout can be well proportioned and have a range of colour markings similar to young trout. Anglers have a habit of calling large "declining" fish "cannibal", but their appearance may have little to do with their food.

The salmon (*Salmo salar*) which run up the Highland rivers (plate XVI) show every sign of being closed societies and though it is very difficult to recognise the differences anglers claim to know them well. This view has been confirmed from marking as a routine procedure in various rivers. Smolts marked in one river are not usually taken as grilse or salmon in any other river but their own. Each river has its own season for the run of its salmon and in some rivers there are two runs, and this may indicate two races using the one river. Adult salmon can return to freshwater in almost any month of the year, but the main runs are (a) grilse, which run in east coast rivers from June and in west coast rivers from July onwards, (b) spring salmon, which enter freshwater in the late winter or early spring and (c) summer salmon, which run during the summer. Grilse have spent one winter in the sea but spring and summer salmon may be divided into "spring", "large spring" and "very large spring" salmon (with a similar series for summer salmon), which have spent two, three or four winters respectively, in the sea before returning to freshwater.

Salmon parr remain in the river of their birth for two or three years before moving down to the sea. The growth and behaviour of parr have been the subject of three interesting papers by K. R. Allen (1940, 1941). He shows that growth takes place from early April until late October, but does not continue through the winter months. There is a decrease in the rate of feeding as the summer passes, though the amount of food available remains constant. The survey of a wide variety of Highland streams mentioned later on page 326 was interesting in respect of food for parr. In summer there were on average three times more invertebrates available than in spring, and the increase in weight of the summer invertebrate crop was one and a half times that of the spring. At temperatures below 7°c or 45°F, parr are less active and lie in deep pools, but when the temperature exceeds this level they move to shallower water of moderate current and feed actively. Allen (1944) has also studied the behaviour of the young salmon when they migrate to the sea as smolts. He trapped at various points on the Thurso River system in Caithness from the middle of April onwards ; migration declined until May 16th, when it was *nil* for a few days, but after May 20th the rate increased to a maximum on May 25th ; then it declined again to *nil* by the end of the first week of June. These dates may, of course, vary from river to river and from year to year. The smolts did not appear to travel in permanent shoals. It appeared that rising water was a definite stimulus to migration, and temperature probably has its influence also. The passing of other actively-migrating smolts seems to act in itself as a stimulus to migration. Scale examination showed that the great majority of smolts migrated after two years and a very small proportion after one or three years. Migration began about eight days earlier in the upper reaches of the river than in the lower.

Male parr which have become sexually mature are found in most Highland rivers. Jones and King (1952) have described how male salmon parr take part in the normal spawning of adult fish. Sperm from the cock salmon may not be effectively disposed over the eggs because of the distance he is from them when discharging in adverse currents. Parr on the other hand deposit sperm much closer to the eggs. Young male brown trout have also a habit of mating or attempting to mate with large female fish.

Once these young salmon reach the sea they may move quickly to the marine feeding ground. Some observations by Vibert (1950) on the recapture of smolts from the River Adour in southern France indicate that the rate of movement may be rapid and the same may apply in Scotland. The marine feeding grounds of the British salmon stock are not yet fully known. The current theory is that the grounds of the British and eastern Canadian salmon lie between Iceland and Greenland, and this has received support by the capture on the south-west coast of Greenland of a kelt tagged at Loch na Croic (River Blackwater, Conon River system), a smolt tagged on the River Usk (Monmouth) and fish tagged as smolts from eastern Canada. It is rather hoped that the precise feeding grounds of the salmon are not found until the conservation of deep-sea stocks of fish is much more highly organised than it is today.

A Pacific salmon (*Onchorhynchus gorbuscha*), sometimes called the pink or hump-backed salmon, was caught off the Kincardineshire coast in July 1960 (Shearer and Trewavas, 1960). This fish is almost certainly derived from a large transplantation experiment which the Russians have been carrying out since 1956. Millions of pink salmon ova have been transferred to the rivers of the Kola peninsula from the Far East. However, no further captures of this salmon have been made in Scottish waters.

The grilse return to their Scottish river in summer after one year at sea. Spates and freshets stimulate movement upstream but are only essential when the river is low. Some grilse return to the sea after spawning and spend a further period of years on the grounds before returning as salmon. Salmon spawn between October and December. How common it is in November in some of the western deer-forest country to find the small rivers far into the hill almost alive with spawning salmon ploughing up the gravel! The early-running fish usually spawn in the upper reaches and the later ones lower down. The intensity of netting is greater on the eastern Scottish coasts than on western ones, and once clear of the nets the eastern salmon has much farther to go to the head waters. The run is an exhausting acitvity, and it is found that on the east coast a salmon rarely runs more than once. West Highland salmon may breed four or five times.

Before the second world war most of the research on sea-trout and salmon was carried out by the naturalists and inspectors of the Fishery Board of Scotland. Since 1948, however, research into many aspects of the biology of brown trout, sea-trout and salmon has been in progress at

Pitlochry. During the first ten years, the work was done under the Supervisory Committee for Brown Trout Research. By 1958, however, the emphasis of the work had shifted from brown trout to sea-trout and salmon and the Committee, feeling that it had achieved its aim in brown trout work, which was to continue as part of a wider research programme, handed over to the Scottish Home Department. In 1960 the Freshwater Fisheries Laboratory was transferred to the Department of Agriculture and Fisheries for Scotland. The work on brown trout and salmon has been directed by Kenneth Pyefinch who has published *Trout in Scotland* (1960), a story of brown trout research at Pitlochry. The detailed results of the work appear in the Freshwater and Salmon Fisheries Research series, many of which are referred to in this chapter. During this post-war period, fishery research in relation to hydro-electric development in the Highlands has been done by the Freshwater Fisheries Laboratory with the advice of W. J. M. Menzies and the co-operation of the North of Scotland Hydro-Electric Board and the District Salmon Fishery Boards. This has involved the design of fish passes and traps, the instrumentation of fish counting and the setting up of hatcheries at Invergarry and Contin.

The migrations of salmon in the River Conon system, Ross-shire, have been closely studied. Traps were erected below the Meig and Luichart dams and an account taken of the fish ascending and descending at each trap. Smolts caught in the traps have been tagged and allowed to proceed to sea. The downstream run of smolts and the upstream run of adult fish has varied greatly from year to year. In 1957 about 14,000 smolts came though the Meig trap and about 600 adult fish passed in the opposite direction, but in 1961 the smolt and adult counts were about 4,600 and 150 respectively. At Luichart about 3,300 smolts came down and 270 adults went up in 1959, but in 1961 the respective counts were about 440 and 90. In some years more adults ascended the Luichart than the Meig, but the run of smolts in the Meig trap has been consistently much heavier. In early experiments about 40 per cent of marked smolts planted above the Meig dam were afterwards recovered in the trap, but only 5 per cent of those at Luichart were retrapped. Some of the marked fish from the Meig system were later caught in the Luichart and these probably descended the tunnel which connects the two. Other investigations on fish movements have been in progress in the Tummel system, at the Clunie, Gaur and Dunalastair dams, and in the Beltie Burn which is a tributary of the Dee. During 1957 the spawning stock of the Beltie Burn was about 160 salmon and 2,100 sea-trout.

The reactions of migrating fish at such obstacles as falls and weirs are being studied by T. A. Stuart (1962) both in the field and in the laboratory. When the river is comparatively low, sea-trout and salmon parr pass up over the falls easily while large numbers of salmon do not move until the river rises.

The stimulus which prepares fish to leap is related to the impact of falling water in the surface of the holding pool; as the force changes different sizes of fish move forward. The actual stimulus to leap is closely related to the standing wave and the distance of the wave from crest of the fall. The leap is made from the surface of the pool at the standing wave and the fish do not run from a distance. If the distance between the standing wave and the crest of the fall is greater than the fish's visual range, it will perform unoriented leaps. Leaping can be inhibited by the introduction of silt into the laboratory test system and this has been observed in the River Endrick where, with nearly equal discharges over falls, fish only leap when the water is relatively clear. It appears that visual examination of the obstacle by the fish may be necessary before it will attempt a leap; it may not be just a matter of orientation and stimulation by the current.

It is natural that man, embroiled in hydro-electricity and uncertain of its effects on fish, should be attracted to unravel the mysteries of the migration of the salmon and trout and the effects which such obstacles as weirs, fish passes, tunnels and turbines have on the errant fish, but the conservation of the stocks cannot be attempted until we know more about their ecology and population structure. In the last few years work has been proceeding on the early stages in the life of salmon and on survival of fry in the Lui Water, Aberdeenshire, by W. R. Munro and W. M. Shearer and at various points on the River Bran and at the Luichart dam in Ross-shire by D. H. Mills. Unfed fry and ova were introduced at different densities at various places and the recapture rates in the traps noted. For example, in three Ross-shire burns, 16,800, 16,900 and 50,900 unfed fry were planted respectively in April. By electrical fishing, it was found that only 0.7 per cent of the fry remained in the first burn in June, and by September only 0.2, 1 and 1.3 per cent, respectively, remained in each burn. Most of the fry which moved out of the burns did so within a few weeks of planting.

Remark was made in the chapter on the summits of the hills that these were biological islands holding a flora of more northern habitat. Plants which we find on our summits are to be found at sea level in Spitsbergen. The summits hold the relics of the last glacial age. Similarly, the depths of sea lochs are islands of converse kind, possibly holding their own types of living arctic relics. The char (genus *Salvelinus*) is an animal of this kind. It is found in many Highland and some Hebridean lochs. It is surprising how few char are taken considering that there must be a fair population of them in some lochs. Zooplankton is commonly supposed to be the food of char, but examination of stomachs has revealed that midge and caddis larvae, corixids and sticklebacks are also included in the diet. Char come to the surface of deep lochs at night, presumably in search of plankton, and may be met with in considerable shoals.

The only migratory char we have is that of Loch Insh, Inverness-shire. Other races of char in the Highlands and Islands are as follows: Willoughby's char (*S. willoughbii*), in Loch Bruiach, Inverness-shire; Loch Maree, Ross-shire; Lochs Borollan, Loyal and Baden, Sutherland; and Loch Fada, North Uist, the Struan char (*S. struanensis*) in Loch

Rannoch, Perthshire is a brilliantly coloured race and has a large, deep-water eye ; the "haddy" (*S. killinensis*) of Loch Killin, Inverness-shire—a very distinct form ; the large-mouthed char (*S. maxillaris*) in a small loch near Ben Hope, Sutherland ; Malloch's char (*S. mallochii*) in Loch Scourie, Sutherland—a short-faced, small-mouthed race (Regan, 1911 ; Jenkins, 1942). There are also char in Loch Builg (Cairngorms) and Loch Lee (Grampians). In 1955 upwards of fifty unusual char were caught in Loch Eck in Cowal, with patches of golden yellow on and around the maxillae, cheeks and opercula (Friend, 1956).

Loch Lomond and Loch Eck have a freshwater herring, the powan (*Coregonus clupeoides*), another relict form which has been studied by Harry Slack and his colleagues (1957) at Glasgow University. This fish ranges over the whole of Loch Lomond and is probably the most numerous fish in the loch. It spawns on the gravelly shallows where the bed consists of stones and gravel usually with vegetation, close to headlands or off-shore banks and reefs. Nearly all the fish have spawned by early January. The eggs are scattered at random on the bed of the loch and are fed upon by the powan themselves and by the larvae of the caddis flies *Phryganea varia* and *P. grandis*. Like the marine herring, the powan occurs in shoals and feeds mainly on plankton. Dr. Slack estimated that 85 per cent of its summer and autumn food consisted of zooplankton (Cladocera, Copepoda and Nematocera), 14 per cent bottom food (snails, insects and crustaceans) and 1 per cent surface food (adult flies). In winter the fish feed more on the bottom, mostly on isopods (*Asellus aquaticus*), snails (*Limnaea peregra*) and its own eggs ; up to 500 powan eggs have been found in a single powan stomach!

The pike (*Esox lucius*) is a common fish of Highland lochs. It was at one time valued as a sporting fish and was in fact introduced into Loch Tulla as such by a previous proprietor ; now the present owner is faced with their making depredations on his salmon and cannot get rid of them. The introduction of pike into some lochs may have been done by Roman Catholics as a regular supply of food for meatless days and there are many lochs in which pike is the only fish. Pike have the reputation of having an entirely piscivorous diet and in some of the lochs in which they are the only fish, they may well have eliminated the others. The poisoning of Loch Choin in October 1955 with rotenone to exterminate the pike provided an excellent opportunity of elucidating the situation (Munro, 1957).

There were 2,078 pike, 9 eels and 1 small trout obtained from the 65-acre loch. One female pike was in its 13th year but all the others were less than 10 years old, and all older than two years were maturing when killed. The Loch Choin pike fed mostly on invertebrates, frogs and small fish of their own species ; they were as large and in as good condition as pike taken from lochs in which there were good stocks of forage fish. A pike of 74 lb. has been caught in Loch Lomond and other big ones in Loch Awe.

Brown trout are found almost everywhere in Highland fresh waters ; they have been caught in Loch Etchachan, 3,100 feet up in the Cairngorms. Sea-trout and salmon are found wherever they have access from the sea and up river systems. The habits and distribution of other fish are, however, less well known. Pike and perch (*Perca fluviatilis*) are locally plentiful but absent from the North-West Highlands ; the minnow (*Phoxinus phoxinus*) is also a local Highland fish but its distribution is becoming more widespread by anglers using it as live bait. Eels are ubiguitous except possibly in very high lochans. The loach (*Nemachilus barbatula*) is not a Highland fish but may occur in the South-East Highlands in the richer rivers only. Grayling (*Thymallus thymallus*) were introduced to the Tay at Kenmore and now occur throughout running water in the Tay and Earn systems below the lochs. Roach (*Rutilus rutilus*) is unheard of except in the lower Tay and Earn, and tench (*Tinca tinca*) is restricted to a few ponds and reservoirs along the Highland Boundary Fault. Gudgeon (*Gobio gobio*) are also released by anglers and are found in the Don and North Esk. The three-spined stickleback (*Gasterosteus aculeatus*) is locally numerous both on the mainland and in the islands but the ten-spined one is unknown in the area. Rainbow trout (*Salmo irideus*) and brook-trout (*Salvelinus fontinalis*) have been widely introduced to Highland lochs and rivers without becoming firmly established. The sea-lamprey (*Petromyzon marinus*), lampern (*Lampetra fluviatilis*) and the brook lamprey (*L. planeri*) are not common except in lower reaches, major rivers and in Loch Lomond. Hydro-electric schemes have caused fish to be transported into new localities through connecting tunnels and aqueducts, e.g. pike to the Lochy system by way of Loch Pattack-Loch Laggan-Loch Treig ; perch from Luichart to Meig ; char from Garry to Errachty.

Loch Lomond lies astride the Highland Line and its fauna is a characteristic mixture of Highland and Lowland conditions. A survey of the lower vertebrates of the loch and district revealed 3 lampreys, 13 freshwater fish, 8 brackish water fish including flounder (*Platichthys flesus*) and plaice (*Pleuronectes platessa*), 3 newts (warty, smooth and palmate), common frog (*Rana temporaria*), common toad (*Bufo bufo*), slow-worm (*Anguis fragilis*), viviparous lizard (*Lacerta vivipara*) and adder (*Vipera berus*) (Hunter, Slack and Hunter, 1959).

INVERTEBRATES

The biology of freshwater invertebrates, like the biology of terrestrial invertebrates already mentioned, is beyond the scope of this book. As a result of the investigations into the food of game fish a great deal has been learned of the invertebrate fauna of river, hill stream and loch ; much of this has already been mentioned.

Morgan and Waddell (1961) have described the emergence of insects in Loch Dunmore near Pitlochry and relate this to the food supply of brown trout. The mayflies *Leptophlebia marginata*, *L. vespertina* and *Cloeon simile*, the alder fly *Sialis lutaria*, the caddis flies *Limnophilus lunatus*, *Anabolia nervosa*, *Mystacides azurea*, *Polycentropus flavomaculatus* and *Cyrnus flavidus*, midge larvae and the snail *Limnaea peregra* were the most widespread. The surface and mid-water food taken by trout in early summer consisted mostly of emerging nymphs and pupae of these aquatic insects and later in summer the diet became more variable with terrestrial insects and zooplankton included. The bottom fauna which provided the food in winter consisted mostly of snails, shrimps, slaters, caddis and midge larvae. Morgan (1957) has shown that the caddis fly *Athripsodes* (*Leptocerus*) *aterrimus* is an important constituent of the diet of loch trout in May and June.

The maximum emergence rate of insects in Loch Dunmore is in early June from deep water and in late June and early August from shallow water. In one loch the following number of species of various groups were caught in emergence traps: one species of stone fly, 5 mayflies, 2 dragonflies, 18 caddis flies, one moth or butterfly, 5 Culicidae, 100 midges and 4 Ceratopogonidae. The midges formed from 70 to 90 per cent of the emerging insect population per annum. Annual catches varied from 6,580 to 22,502 insects per square yard, and the weight of annual catches above vegetation was three times that taken from bare mud at similar depth. Mayflies and dragonflies emerge in the middle of the day, caddis flies after dusk, and midges chiefly at midday or immediately following sunset.

Collections of the bottom fauna from fifty Highland streams showed that the most abundant and widespread species of stoneflies in order of importance were: *Leuctra inermis, Amphinemura sulcicollis, Brachyptera risi, Chloroperla torrentium* and *Isoperla grammatica*. The most important mayflies were *Baetis rhodani* and *Rhithrogena* spp.; *Baetis pumilus, B. vernus-tenax, B. scambus-bioculatus* and *Ephemerella ignita* were also numerous. Streams which were chemically poorest were also those with the poorest fauna.

A great deal of work on the natural history of invertebrates without special reference to fisheries has been done by Slack and his colleagues at the Loch Lomondside Field Station of the Department of Zoology, Glasgow University.

The exploration of the loch and the biology of the powan already mentioned were accompanied by an inventory of the invertebrates and special studies on the adaptations and ecology of the freshwater snails, on midges and the parasites of fish. W. Russell Hunter (1957) in summing up his work on the snails says that they are relatively poor in numbers of species, though parts of the loch are relatively rich in numbers of individuals. Thirteen species were recorded but in only four of these were high population densities achieved: *Limnaea peregra, Physa fontinalis, Planorbis albus* and *Valvata piscinalis*. There is evidence that the lack of calcium brings about an exclusion of several species. Hunter compares the snail communities in the Lowlands and Highlands and lists 25 species; besides the 13 for Loch Lomond, there are 7 for smaller Highland lochs in Dunbartonshire and Argyll and 14 for lochs on the island of Lismore.

Many aquatic insects in their larval stages inhabit off-shore reefs and banks which are never exposed and which are so deeply submerged as to be unnoticed on the surface. Between these banks and the shore there are considerable stretches of deep water in which these particular shallow water larvae are not found, but they occur again on the littoral zone of the loch shore. These larval populations are isolated communities, and the puzzling thing is the departure of the adult insects after hatching to a terrestrial or semi-terrestrial habitat on the shore line. How are these populations maintained? This question is discussed by Weerekoon (1957) with reference to the re-population of the McDougall Bank in Loch Lomond by a mayfly (*Caenis horaria*), and several species of caddis flies and midges. It was concluded that such isolated shallows owe their populations of insect larvae to a temporary planktonic phase in the insect immediately after hatching from eggs deposited on the littoral zone of the loch shore. This planktonic larva, borne into deep water by currents, may be able to select a favourable substratum on which to settle, swimming upwards from unsuitable ones until exhausted or until a favourable one is reached.

The most common of the "biting midges" in the Highlands is *Culicoides impunctatus*. *C. heliophilus* bites in bright sunlight early in the season and *C. pulicaris* is widespread but not common. Most species have eggs hatching in summer within a few days of being laid and within a few weeks the larvae have grown rapidly through two moults. They spend the winter as third or fourth stage larvae and in spring and summer pupate; it is on the ascending pupae that the fish gorge themselves, but those that do reach the surface—and millions of them do as we know to our cost—emerge as flying adults. The dominance of *C. impunctatus* over all other midges was well demonstrated by D. S. Kettle on Loch Lomondside; out of more than 67,000 midges caught and identified, more than 65,000 were of this species (Lawson, 1957).

The term "dancing midges" is sometimes used to discriminate between those that do not bite and those that do. The biting midges however, also become involved in flying swarms similar to those characteristic of the chironomids. The larvae live mostly in bottom muds and on moss and algae and are a staple food of fish. The larvae of some species are known as blood-worms because of their haemoglobin which gives them a blood-red colour. In the well aerated waters the genus *Tanytarsus* is most numerous and in poorly aerated waters *Tendipes* (*Chironomus*). Dancing swarms of *Tendipes anthracinus* occur on the loch margins in sunlight, and male swarms react very strongly to sound. Morgan and Waddell (1960) provide a list of 100 species of Chironomid collected mostly at Loch Dunmore in Perthshire and give some idea of just what a diverse and numerous group this is.

In Highland river systems the brown trout acts as the host for the early stage, the glochidia, of the fresh-water pearl mussel (*Unio margaritifer*) which is a common resident on the bottom of quick-flowing, clear rivers about three feet deep which have a bottom strewn with stones. The glochidia attach themselves to the gills of the fish. Boycott (1936) says this mussel may reach the great age of 70 years and remarks that destruction by man seriously diminishes mussel populations. This is precisely what has happened in the Highlands. Certain rivers such as the Kerry and Polly in Wester Ross, the Laxford in Sutherland, and the Gruline in Mull, fulfil conditions for the mussel and also for their being easily fished. Pearl fishing is still done by tinkers on the Rivers Laxford and Polly, but the fishery is unprofitable and is dying out. The tinkers stand in the bed of the river, slowly quartering it, bent double, a glass box in one hand in order to see the bottom clearly, and in the other a long stick notched at the end. They press this stick over the mussel, pull it from the bed of the river and put it in a sack slung over the shoulder. It is said that only those which bear the "pearl scar" are opened, but this is probably not so and the fishery may be very wasteful.

BIRDS OF RIVER AND LOCH

The Highlands are relatively poor in many forms of living things common in England and lowland Scotland, but in the birds whose habitats call for reaches of fresh water the Highlands are rich. First there is the osprey (*Pandion haliaetus*). From being relatively common at the beginning of the 19th century it became extinct by the beginning of the 20th, but a pair returned to breed in Speyside in the early 1950's. Exactly when the first successful nesting took place in re-establishment is uncertain. Philip Brown and George Waterston (1962) have given the detailed story of the osprey's return and only a sketch of events need be repeated here.

In 1952 there were unconfirmed reports of a nesting pair not far from Rothiemurchus where they nested on the ruins of the old castle in Loch an Eilean until 1899. Since 1902 none had visited the old castle site, and it was not to it that the new birds were attracted. The new site was in the Loch Garten area near Boat of Garten. At another Inverness-shire site, at Loch Arkaig in the Locheil country, ospreys nested until 1908 and a single bird returned to the site on the island at the east end of the loch until 1913. A pair is reported by a stalker to have bred at Loch Loyne as late as 1916.

The ospreys in Sutherland fared even worse. At the old castle of Ardvreck on the shore of Loch Assynt the nest was deserted in 1848, and in Wester Ross the nests on Eilean Subhainn and Loch an Iasgair (Loch of the Fishing Eagle) were deserted in 1857; the one at Loch Luichart failed in 1892. It is not known when the last osprey nested in the north-west, but Brown and Waterston (*op. cit.*) quote an interesting letter in Harvie-Brown's correspondence which states that there was a collector visiting Scourie about 1910 with an *unblown* egg of an osprey in a collection which he had then in his possession. There is said to have been an eyrie in Glen Sheil in 1836 and a clutch was obtained at Loch Awe in 1850, but the bird had long since ceased to breed there in 1871. In 1840 Macgillivray recorded ospreys at Loch Tay and they probably bred in Rannoch. The odds have been heavy against the osprey not only during last century when the Scottish population was fighting for its life against collectors but also in the 1950's when the species was fighting again to re-establish itself.

Charles St. John in his *Tour in Sutherland* (1884) says, "I walked on to look at the osprey's nest on the old castle (Ardvreck), and an interesting sight it is, though I lamented the absence of the birds. Why the poor osprey should be persecuted I know not, as it is quite harmless, living wholly on fish, of which everyone knows there is too great an abundance in this country for the most rigid preserver to grudge this picturesque bird his share." A few pages further on the same Charles St. John is shooting a hen osprey and taking her two eggs from the nest. He sentimentally describes the calling of the distraught male bird and finishes up, "I was really sorry I had shot her." Later he is trying to get another shot at an osprey while his friend William Dunbar swims out to the nest and comes back with a half-grown young and an addled egg. On this same after-noon they find another nest, the friend has another swim, and comes back with three young ospreys. To cap it all, St. John stalks the male bird: "I am sorry to say that I shot him deliberately in cold blood as he sat." It is no surprise to find William Dunbar writing in a letter to the egg-collector John Wolley, dated 29th June 1850: "I believe, at this moment, there is only one osprey's nest in this country (Sutherland) and that has been taken by Lord Grosvenor's keeper. I am afraid that Mr. St. John, yourself and your humble servant, have finally done for the ospreys." The stories of how the Loch an Eilean nest was robbed five times by Lewis Dunbar, a brother of the marauder of the Sutherland eyries, between 1848 and 1852 are like something out of Robert Louis Stevenson. At that time ospreys also nested at Loch Morlich and a similar treatment was meted out to the birds there. On 29th April 1851, Dunbar walked the twenty miles to Loch an Eilean from Grantown and robbed the eyrie for the fourth year running. He arrived at 3 a.m. in a snow-storm and after swimming to the island was slipping about in the snow. He swam on his back holding an egg in each hand, blew the eggs there and then rinsed them out with whisky.

On 4th June 1958 after returning ospreys had nested unsuccessfully for several years in the Loch Garten area, and after the Royal Society for the Protection of Birds had set up a round-the-clock watch near the eyrie, the nest was robbed before the eyes of the watchers. At 2.30 a.m. a cunning intruder climbed the tree, snatched the eggs, left two hen's eggs daubed with boot polish and vanished into the forest. This man, whoever he was, made the climax to Brown's story *The Return of the Osprey*. The robber's guile was much more than would ever have been demanded of the Dunbar brothers a hundred years before.

Since that disastrous year the ospreys have bred successfully in a neighbouring tree at Loch Garten though in 1963 both this eyrie and another at Inshriach failed to produce young. The Loch Garten area has been made a statutory bird sanctuary and the R.S.P.B. have organised a base from which the eyrie can be viewed by the public. In the nesting seasons between 1959 and 1964 a total of about 106,000 people saw the eyrie. From the continuous watch which has been kept and the instruments which have been placed in the vicinity of the nest a detailed study of the behaviour of the birds has been possible.

The birds arrive at the nest site singly about mid-April; in most years the male has appeared before the female. Waterston describes the magnificent display flight of the cock over the eyrie at a height of between 500 and 1,000 feet. "He climbs upwards with rapid wing beats—hovers for a moment with tail fanned out—then dives very quickly downwards with flexed wings. All the time he is moving in a wide sweeping circle, calling a high-pitched 'chee-chee-chee' note almost continuously. He usually carries a fish in his talons during the display. . . ." Mating takes place on the eyrie within a few hours of arrival and up to the laying of the first egg the birds may mate as often as 50 times.

The first eggs are probably laid in the last days of April or first in May, three are usually laid and the incubation takes about 37 days. The male is responsible for supplying the food, bringing in on average 1.7 fish per day during the incubation period. After the young are hatched he requires to step up the food supply and in 1959 he brought on the average 4.6 fish per day for three chicks and in 1960, 3.8 fish per day for two chicks. Feeding intensity is greatest between 5 a.m. and 7 a.m., about 3 p.m. and between 8 p.m. and 10 p.m. On windy days when the water surfaces of river and loch were ruffled he took noticeably longer to return with fish. The food included trout varying in size from 4 lb. to $\frac{3}{4}$ lb. and pike from $\frac{1}{4}$ lb. to $\frac{1}{2}$ lb. In 1960 the cock apparently took more pike than other fish which was a matter of some considerable satisfaction to local anglers who have to contend with pike in the Spey and neighbouring trout lochs.

The young are weak and helpless at first but after three weeks are well feathered and aggressive. At the end of a month they are exercising their wings and towards the end of the fledgling period spend long periods wing-flapping on the edge of the eyrie. At about six weeks old they begin to tear up food with the help of the hen. As the young become fledged the hen takes less and less part in feeding them. With three chicks in the nest in 1959 the first flights of young took place on 2nd August, 59 days after the hatch ; in 1960, however, with two chicks the first flight occurred 52 days after the hatch.

The birds of the freshwater systems of the Highlands divide up fairly easily into those of lochs and those of rivers. Dippers (*Cinclus cinclus*) are regular inhabitants in most rivers and burns and ascend to considerable elevations. It is common to hear the song of the dipper in deep winter at Loch Toll an Lochan, 1,750 feet, in Dundonnell Forest. Kingfishers (*Alcedo atthis*) are very rare but have bred in the West Highlands, including Islay, also on the Tay, and the North Esk. They have also been seen on the Dee and Don in Aberdeenshire.

Goosanders (*Mergus merganser*) again are birds of the rivers rather than the lochs, occurring on the upper reaches of river systems in spring and summer and on the lower reaches and lochs in autumn and winter. The spread of the species in Scotland since about 1870 has been rapid. Baxter and Rintoul believed that Scotland was invaded simultaneously in several localities from overseas. Within twenty years of the great immigration which took place in the winter of 1875-76 the goosander had become common in most part of the Highlands as a breeding bird, and it has since remained so. D. H. Mills (1962) states that the population density in the Rivers Bran and Meig in summer is two to three birds per ten miles of river. The species colonised the interior of the Highlands first, the earliest records of breeding coming from the watersheds and subsequent ones from farther down-stream. Except in the north-west, where there may have been a considerable breeding population before the invasion of the 1870's, the goosander is still an inland species (fig. 15). It has been subject to considerable persecution and bounties have been paid on the heads by certain District Fishery Boards. It seems well able to hold its own, however, nesting in hollow trees and cavities among the rocks, in holes on river banks

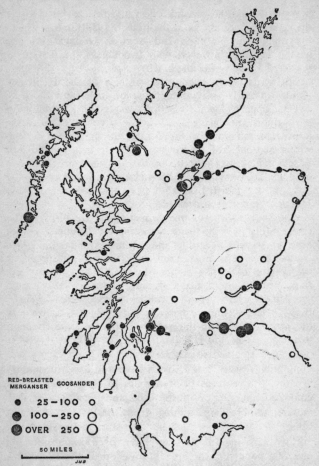

FIG. 15. The main concentrations of the red-breasted merganser
(dots) and goosander (circles) in the Highlands and Islands
*From 'Wildfowl in Britain', Wildfowl Trust, 1963 by permission
of the Controller of H.M. Stationery Office*

and occasionally in such unusual places as open ledges and deserted huts.

The red-breasted merganser (*Mergus serrator*) is much more a bird of the loch and sea-loch than the goosander (fig. 15). It is probably the most common sea duck on West Highland coasts in autumn and winter. In those seasons we have seen flocks of well over 100 birds in the sea-lochs and the sandy bays of Tiree and other islands, and flocks of up to 400 have been seen on the Beauly Firth. The birds go up the river to breed, sometimes far inland on the Tay, Dee, Spey and Ness, nesting very close to the water, usually in the undergrowth on the bank. Beside lochs they will nest in solitary clumps of gorse or similar cover or on open moorland ; nests are sometimes found several hundred yards from the water. The merganser is a common breeder in all the Highland counties. There is a fast-moving communal display in April which includes tilting, hydroplaning and opening wide the beak, ending in what appears to be promiscuous coition.

The food of these saw-billed ducks has been investigated at the Freshwater Fisheries Laboratory (Mills, 1962). The stomachs of 147 goosanders from 16 rivers, two lochs and four firths or bays have been examined for contents. The results show that salmon parr and smolts are by far the most important food of both the goosander and red-breasted merganser. The percentage of goosander stomachs containing various foods was: salmon 57.2, salmonid fish remains 19.2, eels 14.5, brown trout 9.7, perch 9.7. Other food included insects, minnows, pike, gudgeon (*Gobio gobio*), miller's thumb (*Cottus gobio*), unidentified birds and water shrews (*Neomys fodiens*). There were also 122 tags from salmon smolts, three fish hooks and stones of bird cherry (*Prunus padus*). In the case of the red-breasted merganser stomachs the percentages were: salmon 42.5, fish remains (mostly salmonid) 31.8, brook lampreys 6.2, crustaceans 6.2. Other food here included insects, eels, perch, flounders, three-spined sticklebacks, minnows, common gobies (*Gobius minutus*), brown trout, butter fish (*Pholis gunnellus*), and coal-fish (*Gadus virens*) ; 36 tags from salmon smolts, tagging wire and one fish hook were also found. It is not thought that the tags attract the attention of the bird to the fish, for experiments have shown that internally tagged fish are taken in numbers comparable to externally tagged. Most of the lampreys taken were from the spring spawning migration in the River Conon and were distended with ova. It will be noticed that though the goosander takes many salmon, it also takes many salmon predators: eels, brown trout, pike, miller's thumbs and water shrews.

The mallard (*Anas platyrhynchos*) is found on freshwater systems throughout the Highlands but is much commoner as an estuarine species. Nowhere in the Highlands proper does it occur in the large concentrations found on the east coast and in the south-west of Scotland. The teal (*Anas crecca*) is widely distributed, but in our experience is a rare duck over the whole west coast region, especially in the north. Teal are common on moors in the eastern Highlands breeding in fair numbers by moorland pools up to 2,000 feet, along with mallard. The gadwall (*Anas strepera*) is a comparatively rare breeding duck in the Highlands and Islands. It bred in Inverness-shire in the late 1930's and in 1952 in the Spey Valley. It has also bred in Easter Ross and pairs have been seen in May and June in Wester Ross and on Loch Awe. It bred recently in Caithness, at Crianlarich in 1947, and a duck with three young was seen in Tiree in late July 1955. The wigeon (*Anas penelope*) is now a well-established breeder in the Highland region. This bird is one of our gains in the last 150 years, having spread from an original colonisation in Sutherland. It breeds in all the mainland Highland counties, in the Uists and in Coll, but not in other islands. The wigeon becomes estuarine in winter and its numbers are augmented by birds presumed to come from Iceland. The pintail (*Anas acuta*) has bred in Aberdeenshire, Angus, Moray, Inverness, Easter Ross and Sutherland but it is not common. In 1951 a nest was found in Tiree. The shoveler (*Spatula clypeata*) is resident in the southern part of the Outer Hebrides, the Inner Hebrides and in the Moray

Basin. Its possible habitats are limited by its need for shallow muddy conditions. Baxter and Rintoul (1953) state that the pochard (*Aythya ferina*) now breeds regularly in Aberdeenshire, Moray, Nairn, Ross and Caithness. It has bred in Perthshire, Angus, Argyll, Inverness and Inner Hebrides. The colonisation of Scotland by the tufted duck (*Aythya fuligula*) has taken place within the last ninety years. With the possible exception of the north-west, most of the Highland lochs have become populated and it is now a common breeding species in most of the mainland counties and also in both Inner and Outer Hebrides. Scaup (*Aythya marila*) are indeed rare breeders and most of the records are old. A nest was found in Wester Ross in 1946. The goldeneye (*Bucephala clangula*) has not yet been recorded breeding in the Highlands and Islands. Birds have been seen on many occasions in summer but no nests have been found. The common scoter (*Melanitta nigra*) breeds in Inverness-shire, Ross-shire, Sutherland, Caithness, Argyll (Islay), Perthshire (1939), Orkney and Shetland.

The cormorant (*Phalacrocorax carbo*) is a marine cliff-nesting species, though in Wigtownshire it nests in a fresh-water loch. It has attempted unsuccessfully in the past to nest on Tower Island in Loch Rannoch. Many lochs have cormorant rocks. We have seen them roosting on the old osprey island in Loch Arkaig, on Loch Insh in Speyside, on Loch Clair and Loch Coulin and on Loch Maree. The appearance of the shag (*Phalacrocorax aristotelis*) in fresh-water is rare.

The little grebe (*Podiceps ruficollis*) is a common breeder in all the Highland counties with the possible exception of Nairn, and in both Inner and Outer Hebrides. The great crested grebe (*P. cristatus*) is not generally a Highland species, although it nests in low-lying lochs in Perthshire, Angus, Aberdeen, Moray and Ross. It has bred in Moray and up to 700 feet on lochs in Perth and Angus. The Slavonian grebe (*P. auritus*) (plate xv) is a well established breeding species in Inverness-shire and has recently bred in Aberdeenshire. The black-necked grebe (*P. nigricollis*) breeds in Perthshire and Angus, and we

M

have seen a pair on a sedgy, waterlilied *dubhlochan* in Sutherland, but were not able to confirm that the birds were breeding. The coot (*Fulica atra*) is of far more general distribution in the Highlands and Islands than the moorhen (*Gallinula chloropus*), though both breed in all counties and in the larger islands. The moorhens are much commoner in the wider valleys of the Central and Eastern Highlands than in the west. The coot is a dabbler in shallow water as well as a shore feeder and can become adapted to most lochside situations. The skulking habits of the water rail (*Rallus aquaticus*) make its nest difficult to find. Breeding is known from Perthshire, Angus, Kincardine, Aberdeenshire, Moray, east Inverness-shire and Easter Ross. It bred in North and South Uist in 1919-20.

The whooper swan (*Cygnus cygnus*) is a common winter visitor and has bred in scattered localities in Perthshire, Inverness-shire, Orkney and Shetland (once at an altitude of 1,500 feet). During hard winter weather in the West Highlands large numbers are found on the *machair* lochs of the Hebrides and also in the wet stubbles in the croftlands. Favourite haunts also are Loch Park near Keith in Banffshire, Loch Strathbeg in Aberdeenshire, Loch Insh in Inverness-shire, and they are common in most estuaries and firths in winter. The mute swan (*Cygnus olor*) is common all the year round but the two species usually keep apart.

The native grey lag goose (*Anser anser*) is still a rare breeding species on the mainland in Caithness, Sutherland and Wester Ross but is more numerous in the Outer Hebrides where its headquarters is the Loch Druidibeg Nature Reserve. There, in the crofting townships of Stilligarry and Dremisdale, and with the co-operation of the crofters and the South Uist estates, the Nature Conservancy is successfully conserving the large breeding group of the native grey lag remaining in Britain. When all the young are on the wing, flying in large noisy flocks back and forth from Druidibeg to the *machair* in autumn, one would hardly believe that this was so rare an animal. In winter many of the native birds seem to move out of the breeding locality and are replaced on the *machair* by white-fronted geese (*Anser albifrons*) and barnacle geese (*Branta leucopsis*, fig. 12, p. 236). The grey lag breeds also in South Uist, Benbecula, North Uist and Lewis; it may still breed in Coll, Islay and the Summer Isles. At Anancaun Field Station, Kinlochewe, the Nature Conservancy have augmented the local breeding stock in the Loch Maree district with birds of native stock raised and allowed to fly free from a compound.

The black-throated diver (*Gavia arctica*) breeds in Sutherland, Ross-shire, Inverness-shire, Perthshire, Argyll and the Outer Hebrides. The distribution of the red-throated diver (*Gavia stellata*) (plate XXIIIa) is very similar as are its habitat preferences. The red-throat will inhabit a small lochan but the black-throat is shyer and probably takes more food from its nesting loch than does the other. The red-throat resorts much more to the sea for food while nesting and will fly several miles to the sea from its nesting loch. The observer may see or hear it coming down to the sea with its wild calling in the evenings and early mornings of summer. It flies high, the wing movement is quicker than that of a duck and its looks more streamlined; the call is also sharper than a duck's quack and uttered in rapid staccato succession. Both species breed in the Outer and Inner Hebrides where they are both much more marine in their habit than on the mainland. The black-throat lays its eggs (normally two) in May on a scrape close to the water on an islet in a freshwater loch. The red-throat usually makes more of a nest with grass and moss and the nests of both species are so placed to enable the duty bird to survey the whole loch and slip into the water unseen. They usually incubate for about a month and the young are flying in two months.

The heron (*Ardea cinerea*) finds so much more food on the innumerable miles of sheltered coasts that in the West Highlands it is not the freshwater species of inland districts. The late Elizabeth Garden (1958) has compared the data obtained in the national censuses of heronries in Scotland in 1928-29 and 1954. In the Highlands and Islands there were almost twice as many heronries and breeding pairs in 1954 as 25 years earlier.

The following table, compiled from Miss Garden's paper, summarises the information for the main Highland counties:

	OCCUPIED HERONRIES		BREEDING PAIRS		AVERAGE SIZE OF HERONRY (Approx. pairs)	
	1928–29	1954	1928–29	1954	1928–29	1954
Aberdeen	6	10	15–16	48–49	3	5
Argyll	22	33	101–127	270–289	5	8
Inverness	11	28	111	176–181	10	6
Perth	3	7	49	28	16	4
Ross and Cromarty	4	14	34	55–58	9	4
Sutherland	2	7	19	33–34	–	5
TOTALS	48	99	329–356	610–639	9	5

Herons usually nest in trees. The census revealed that in 140 heronries with 903 nests, 51 per cent were on coniferous trees, mostly Scots pine; 33 per cent on deciduous trees, mostly beech, oak and birch; 7 per cent on cliffs and 9 per cent on other sites. At Poltalloch, Argyll, they nested in ruins, on the Monach Islands we have seen the nests in the rafters of a deserted house and in the Outer Hebrides they nest on the ground (plate vib).

The common gull (*Larus canus*) breeds on the banks of shallow lochans in the hill country. It is very common and its daily passage from the hill lochs down to the estuaries and sea lochs is a characteristic sight. It breeds far inland on high lochs up to 2,800 feet in the Lochnagar-Cairngorms area. Black-headed gulls (*L. ridibundus*) do likewise, but do not go so high, up to 2,000 feet, and lesser black-backed gulls (*L. fuscus*) nest fairly commonly on moors well into Banffshire, Moray and Aberdeenshire. Common terns (*Sterna hirundo*) breed far inland in the Eastern Highlands. There are old records from Glen Livet, the River Deveron above Beldormie, Ashie Moss in Inverness-shire, and on the Tay above Ballinluig. Nethersole-Thompson has found them at over 1,000 feet on river-side shingle. In the Hebrides both common and arctic terns (*S. macrura*) nest on the islands in freshwater lochs but resort to the sea for food.

In summer the common sandpiper (*Tringa hypoleucos*) is one of the commonest river and loch-side birds in the Highlands. It arrives during the last week in April, and after much show and sound during the breeding season it becomes a silent and almost furtive bird of the coasts in August and until it leaves. The oystercatcher (*Haematopus ostralegus*) (plate xviib) is almost entirely a shore-nesting and shore-living bird in the West Highlands, but in the Central Highlands it is a river bird. Oystercatchers are a feature of the shingle beds and barely-grassed islands and banks of the Spey, and are sometimes found on upland grazings at over 1,500 feet. They also occur widely on lower moors, breeding well away from streams and rivers, and the nest is commonly found on fields and in-bye land in the hills and glens. Lapwings (*Vanellus vanellus*) are common in the Eastern and Central Highlands on arable land and in grassy places in the hills up to 2,000 feet and occasionally 2,800 feet. Temminck's stint (*Calidris temminckii*) bred in the 1930's and 1956 and the green sandpiper (*Tringa ocrophus*) bred in 1959, both in Strath-spey. The wood sandpiper (*Tringa glareola*) has also bred recently in Sutherland and west Inverness.

This section on freshwater birds closes with the red-necked phalarope (*Phalaropus lobatus*), the daintiest of them all. In recent years a few pairs have nested at a wide scatter of stations in the Outer Hebrides and in Tiree. Nesting is irregular in some of these places, the birds appearing in some years for a week or so in June but apparently failing to breed. It is amazing that such small colonies which fare so badly over the years should have power to keep the sites occupied.

We cannot begin here to describe the large winter immigration of ducks and waders to the estuarine end of the freshwater systems. It is not a striking phenomenon in the Highland area proper. Had we included the Beauly

Firth and the coasts of the Moray Firth, then another chapter could well have been devoted to them.

One is left with the impression that the birds of the fresh-water systems are too few in total numbers to have any considerable effect on the remainder of the eco-system. The goosander and merganser do make some sort of impression on the stocks of salmon, but what a vastly greater impression on the stocks is made by the decline of the invertebrate crop on which the fry and parr feed, or by the open jaws of trout, eels and pike. Birds are at the summit of a pyramid of fresh-water life and as a group they are subject to human persecution. We are now beginning to appreciate that to animals such as the fish-eating birds of freshwaters, there is a great danger from poisoning by toxic chemicals used in industry and agriculture. The spiritual loss to humanity of the water birds would be crushing.

MAMMALS OF RIVER AND LOCH

The otter (*Lutra lutra*) and common seal (*Phoca vitulina*) both occur in freshwaters, but as has been mentioned in chapter 9 both are mainly marine. The otter is widespread throughout the interior of the Highlands in all the river systems, following the spawning fish into the hill-streams. No comprehensive account is at present available on the otter in the Highlands and Islands, though much valuable work on the animals has been done by Gavin Maxwell. In *Ring of Bright Water* (1959) he tells the story of his life with Persian otters at his home on the shores of the Sound of Sleat. We believe that he has now succeeded in rearing and studying, in the same intimate way, the native otter of the West Highlands. Otters are mercilessly hunted by District Fishery Boards and gamekeepers, as are common seals in the firths and estuaries. Both otter and seal, however, seem capable of maintaining their numbers and this may result from their great mobility. The otter population of the Highlands has in the marine part of its range a great reservoir of breeding animals, comparatively untouched by man, from which to replenish the stocks in rivers and lochs. Though numbers of common seals are shot locally this does little to reduce the vast British population and depleted areas are quickly recolonised.

One can only guess what the food of the otter is in Highland freshwaters. Fish, of course, will be staple, particularly salmon, trout, eels and pike. Ducks, geese, grebes, divers, waders and song birds all fall prey in the breeding season. On Loch Iosal an Duin in North Uist there is a small island which was completely paved with about 30 nests of herons in 1954. The loch and surrounding country has a population of otters; yet the heronry is unscathed. We have observed two otters swimming close to this heron island and although each paused and looked, neither attempted to come ashore to kill the newly hatched young; this is not to be wondered at for the island was bristling with long necks and sharp beaks poised to deal deadly thrusts. Frogs, toads, newts, water voles and water shrews the otter will take in its sweeping strides through the water, and when in transit over fields and moorlands it will eat terrestrial small mammals. Rabbits, hares and moles (*Talpa europaea*) are taken locally. All these animals are recorded as food of the otter in the south (Stephens, 1957).

Probably the bulk of the Scottish otter population occurs on the much indented coasts of the West Highlands and Hebrides, and many otters frequenting freshwater lochs and rivers may return to the coast to raise their young. Holts are commonly found, however, among the roots of trees on the river bank, in recesses under fallen rocks, in rabbit burrows and on nests on top of islets in freshwater lochs. On the coast the holt may be in a dry recess under fallen rocks, or in a deep cleft or sea-cave often on an off-shore island. A white otter has been recorded from such an island in Argyll (Fletcher, 1956) and we have seen fine holts on Eilean Ghaoideamal off Oronsay.

There is no comprehensive picture of the distribution of the water vole (*Arvicola amphibius*) and the water shrew in the Highlands and Islands. Water voles are present in the limestone country of Sutherland at Durness and Assynt. Harrison Matthews (1952) states that those inhabiting the Scottish Highlands are a sub-species *reta* which is slightly smaller and distinctly darker in colour than *amphibius* which occurs south of the Tay and Clyde. It probably does not occur commonly on the shores of large lochs except where they offer fen conditions. In reedy lochans and ponds, especially in lime-rich situations with luscious vegetation, and along the slow-flowing mature reaches of the rivers, particularly those flowing east across fertile land, the water vole is probably common.

Water shrews require a rich crop of invertebrates and occur in rich still waters. They have been reported in the food of the goosander and are probably much more common in the east than the west where they are recorded from Applecross, Inchnadamph, Skye, Raasay and Mull. In his survey of small mammals in north-west Scotland, Michael Delany (1961) caught water shrews in both deciduous and coniferous scrub woodland close to burns or lochs, on farmland, and on moorland over 100 yards from the nearest water.

VEGETATION OF RIVER AND LOCH

The botanical complexes of the freshwaters of the Highlands are of great interest when studying distribution in relation to rate of flow, altitude, depths of lochs and chemical nature of the water and the soil of the floor. The burns carry plants and seeds from their sources, so that alpine plants may be found far below their normal habitat on the banks of rivers. The very presence of water is necessary to the fertilization of many mosses and ferns.

The banks of burns in the Highlands are often sanctuaries for a woodland flora. Immediately the banks come within reach of the questing, destructive muzzles of sheep, they are woodland no longer, but sedgy moor and bracken. In these cleughs are birches, hazels, rowans, oaks, willows, ashes and hollies—and sometimes pines—with an undergrowth of shade-loving grasses, garlic, wood sorrel, stitchwort, hawkweeds, wild hyacinth, primroses, polypody and scented fern. The trees of the cleughs are acting as mother trees and the situation in their vicinity shows how soon there could be regeneration of (and reversion to) forest to an altitude of almost 2,000 feet. Wherever it gets a chance on the lower reaches, alder is common. This tree has considerable value for the surrounding country where it is allowed to grow, for it has considerable leaf-fall and its roots fix nitrogen by means of symbiotic nodule bacteria.

The lochs of the Highlands vary so much in size, slope of shore, depth, type of shore and floor, acidity, exposure and so on that a whole body of ecological work is waiting to be linked up with a similar study of the rivers. The vegetation of islands in lochs has already been mentioned in chapter 7 on the point of tree regeneration, and tends to be similar to that of the cleughs just described. In the Outer Hebrides where the terrain is wind-blasted, the islands possess thickets of willow, rowan, juniper and introduced conifers and rhododendrons, while on the mainland a full forest growth is found (McVean, 1958, Spence, 1960). In the *Flora of Uig* in Lewis (Campbell, 1945), A. J. Wilmott gives an account of the flora of the rivers and lochs there.

We have already described the vegetation of bogs and flushes on moorland, forest and mountain top in their appropriate chapters but we deal now with the truly lacustrine flora. Access to light is a very important factor in the life of submerged plants. If a loch is shallow and clear, a rich variety of water-growing plants is possible, to an ultimate depth of between 40 and 50 feet, but the peaty lochs of the Highlands allow sufficient light only to much shallower depths. West (1910) has suggested the following rough rule: "The extreme depth to which such plants as *Nitella opaca* (a charad) and *Fontinalis antipyretica* (a moss) will flourish in peaty water may roughly be estimated by multiplying by four the greatest depth at which one can see the gravel at the bottom, when looking over the shaded side of a boat about midday in the summer, when the sun is shining brilliantly, the water being perfectly calm and the boat still. Such a depth in Loch Ness and others is from 7 to 8 feet; in many peaty lochs, however, the depth is considerably less."

It is possible to divide any loch except very shallow ones, like those of the *machair,* into (a) a photic zone, in which the higher flora can develop; (b) a dysphotic zone, normally carrying a cryptogamic flora of charads and mosses, but still supporting a few attenuated members of the higher zone; and (c) an aphotic zone, where organisms needing light cannot exist. Taking Loch Ness as an example of the deep, narrow loch with water of intermediate clarity and with gravelly shores, it is obvious how poor the flora must be. The great mass of the floor of the loch is without plants because the fall from the shore is steep. Furthermore, wave action can be so great on the loose gravelly shore that no vascular plants can get a hold. It is on fixed rocks that the algae and mosses grow without being eroded away. The action of the wind and a deep range of water level are inimical to the growth of a waterside vegetation. Urquhart Bay is one of the few places with a considerable water-edge flora, including an alder swamp.

In the fairly typical zonation of the Highland freshwater loch from the shore-line into deep water, there is a marginal band of diatoms, especially *Gomphonema* spp., filamentous algae and clumps of moss (*Fontinalis anti-pyretica*) which lies between the seasonal high and low water marks. Outwardly to a depth of about one metre the shore weed and water lobelia grow in mosaic to a uniform height of about 5 cms. Beyond this there is a similar band of quillwort growing to a height of up to 10 cms. depending on exposure to wave-action. This zone extends to depths of two to three metres with the quillwort interspersed with water milfoil on the gravelly bed and the green algae *Nitella opaca* on finer sediments. Where the bed reaches a depth of three metres and is of organically rich mud, the pondweed (*Potamogeton perfoliatus*) grows luxuriantly, sending up its shoots to the surface. Finally beyond a depth of four metres there are only diatoms. This zonation is modified by the topography of the shore, the exposure to wave-action, the nature of the bottom deposits, the chemistry of the rocks and the water, the clarity of the water and the presence of animals. In sheltered places the accumulation of the rich organic muds favours the growth of reeds (*Phragmites communis*), bullrush (*Scirpus lacustris*), yellow water-lily (*Nuphar lutea*), and the pondweeds (*Potamogeton lucens* and *P. praelongus*). These conditions are sometimes found in the disused meanders of rivers, in the *dubhlochain* and the *machair* lochs, and in artificial lochs and ponds. Sometimes *Phragmites* finds the *dubhlochain* too sour, and in such the bottle sedge (*Carex rostrata*) takes its place.

J. W. Heslop-Harrison (1947) has studied the pondweeds of the Western Isles since 1934 and has followed their distribution from the *machair* lochs were the pH is high through the transition zone to the moorland lochs where the pH is low (see p. 316). In this transect, pondweeds fall off in numbers and there is some degree of correlation between the distribution of the weeds and the pH of the waters. *Potamogeton natans* and *P. polygonifolius* are ubiquitous. *P. pectinatus* and *P. filiformis* are found in brackish lochs of very high pH. The *machair* lochs are rich in species and in stands of plants but it must not be supposed that the rare pondweeds are restricted to the *machair*; *P. praelongus* and *P. lucens* were taken from Loch na Meilich (pH: 7.3) on Raasay.

Studies of the phyto-plankton of numerous freshwater lochs have been carried out by E. M. Lind (1950 and 1952) and A. J. Brook (1957 and 1958). Loch Laidon, Loch Ossian and lesser lochs on the Moor of Rannoch have been examined as have freshwaters elsewhere in Perthshire, on Rhum and in Sutherland.

The shallow limestone lochs of Lismore have the richest flora of any Highland lochs and they are not large enough for wave action to have any considerable effect on their shores. The following list is typical for these lochs:

Water crowfoot   *Ranunculus aquatilis*

Lesser spearwort	*R. flammula*
Marsh marigold	*Caltha palustris*
White water-lily	*Nymphaea alba*
Yellow water-lily	*Nuphar lutea*
Cuckoo flower	*Cardamine pratensis*
Bog stitchwort	*Stellaria alsine*
Meadowsweet	*Filipendula ulmaria*
Purple loosestrife	*Lythrum salicaria*
Mare's tail	*Hippuris vulgaris*
Spiked milfoil	*Myriophyllum spicatum*
Water chickweed	*Montia fontana*
Marsh pennywort	*Hydrocotyle vulgaris*
Marsh ragwort	*Senecio aquaticus*
Buckbean	*Menyanthes trifoliata*
Common bladderwort	*Utricularia vulgaris*
Shoreweed	*Littorella uniflora*
Amphibious polygonum	*Polygonum amphibium*
Iris	*Iris pseudacorus*
Soft rush	*Juncus effusus*
Branched burweed	*Sparganium erectum*
Broad pondweed	*Potamogeton natans*
Shining pondweed	*P. lucens*
Perfoliate pondweed	*P. perfoliatus*
Lesser pondweed	*P. pusillus*
Obtuse pondweed	*P. obtusifolius*
Slender pondweed	*P. filiformis*
Cotton-grass	*Eriophorum angustifolium*
Tufted sedge	*Carex aquatilis*
Bottle sedge	*C. rostrata*
Common reed	*Phragmites communis*
Flote-grass	*Glyceria fluitans*
Horsetail	*Equisetum limosum*

A typical *dubhlochan* of the Sutherland gneiss country has the following group of plants:

Lesser spearwort	*Ranunculus flammula*
White water-lily	*Nymphaea alba*
Water milfoil	*Myriophyllum alterniflorum*

Water starwort	*Callitriche hammulata*
Water lobelia	*Lobelia dortmanna*
Buckbean	*Menyanthes trifoliata*
Shoreweed	*Littorella uniflora*
Small burweed	*Sparganium affine*
Broad pondweed	*Potamogeton natans*
Cotton-grass	*Eriophorum vaginatum*
Bottle sedge	*Carex rostrata*
Reed	*Phragmites communis*
Horsetail	*Equisetum limosum*
Quillwort	*Isoetes lacustris*

Conclusion

As senior author of this book, the sole author of the first edition, I sought and chose the junior author and thereafter left most of the work of revision and expansion to him. I feel this second edition is a happy result of collaboration, in that the work hangs together and has a quality of singleness of effort and yet, coming to write this final chapter, I feel I must write it myself and not involve Morton Boyd in what I have to say. Naturally, I read through the earlier Conclusion again and saw it was written in 1947 with a depth of feeling. To some extent the inadequacy of my own knowledge which troubled me then, despite the sustained enthusiasm of writing the first edition, has been corrected now, though we both admit the book as a whole to be inadequate because it attempts to give a conspectus of natural history which could fill several volumes. I must write this myself because I have more years of perspective than my colleague and I was concerned with the formulation of the Nature Conservancy which Morton Boyd now serves as Assistant Director (Conservation) for Scotland.

The establishment of the Nature Conservancy is the most considerable event in British natural history in this century; especially so, I think, for Scotland, and most especially for the Highlands and Islands. When a few of us sat as a Scottish National Parks Survey Committee in St. Andrew's House under Sir Douglas Ramsay in 1943-4, we were rather an innocent bunch of idealists. I personally was soon aware that notions of what national parks should be differed widely, yet each idea was good and right. How far could these be reconciled in a series of national parks for Scotland? We were far too modest to think different categories of wild open spaces would be acceptable to the Government. After our Report had been digested, Mr. Tom Johnson, then Secretary of State for Scotland, appointed a larger and more widely expert Committee under Sir Douglas Ramsay to develop a blue-print of the structure of a national parks system for Scotland. The necessity was obvious almost immediately to appoint a Sub-Committee on wildlife conservation, under the late Professor James Ritchie and subsequently Professor J. R. Matthews. I was on both committees. The sub-committee gradually developed the idea of a separate governmental agency for the administration of National Nature Reserves (this being greatly helped by the deliberations of the sister English committee under Arthur Tansley). I remember a remark by a member of the Scottish National Parks Committee when the sub-committee reported to them, "If we are not careful, the tail will wag the dog!" This was prophetic of the sardonic decision of the government of the late '40's to establish no national parks in Scotland. The short-sightedness and wrongheadedness of that decision becomes the more apparent as the recreational function of wild lands is realized as an urgent and proper form of land use.

The Nature Conservancy established under Royal Charter in 1949, operates throughout Great Britain ; Scotland having a headquarters in Edinburgh but with little autonomous power. Nevertheless, Scotland can have no valid grumble over the degree to which her wildlife, vegetation and land is being conserved and brought into the areas of National Nature Reserves. Over 173,000 acres are so styled, of which about a quarter is owned by the Conservancy. Indeed, over 75 per cent of the National Nature Reserve land of Great Britain is in Scotland, and a great part of it in the Highlands and Islands. Quite frankly, the present situation is beyond my rosiest expectations of sixteen years ago. The National Nature Reserves in the Highlands and Islands are well dispersed and demonstrate an admirable flexibility in conception, administration and management.

Beinn Eighe, over 10,000 acres, was the first Scottish

acquisition, in which we had to learn what we could and could not do on land owned outright. The Cairngorms including Glenfeshie totalling 64,000 acres is a different kind of Reserve, only partly owned by the Conservancy and the rest regulated under agreements with the owners. Established land uses of forestry and sheep-farming continue and long-established recreational uses are admitted. This kind of Reserve has provided a situation which has educated Conservancy and private owners alike in co-operative efficacy in a mutual ideal. Then there have been National Nature Reserves established with particular species of animals or plants in mind, such as North Rona for the Atlantic grey seal and Leach's fork-tailed petrel, and Loch Druidibeg in South Uist for the grey lag goose. St. Kilda is owned by the National Trust for Scotland but is also a National Nature Reserve. The collaboration of the two bodies is smooth and effective. The attempt has been made—and as I think the end largely achieved—to bring into National Nature Reserve status as large a variety of natural habitats as may be found in the Highlands. Even the small limestone cliff-face at Knockan in west Sutherland has not been forgotten and the reserve there was in some measure made possible by co-operation with the County Council over the reconstruction of the road. A management plan is set out for each Reserve on its acquisition, a good exercise in clarifying intentions, but it is well realized that management plans must have elasticity in the light of experience. Their formulation and general pattern stems from the original work of W. J. Eggeling, Director for Scotland of the Nature Conservancy in Scotland. There is no point in repeating here a large amount of factual information which can be extracted from the Reports of the Nature Conservancy, publications from H.M.S.O. which are neither dry nor heavily statistical but topical running commentaries on the progress of conservation of habitats and wildlife in Britain as a whole (*Progress 1964-68*).

Research is a primary aim of the Nature Conservancy and the second edition of this book has mentioned a good deal of what has been going on in these fruitful years. I could not have imagined in 1947 that so much would have been found out by 1963. The monograph by McVean and Ratcliffe on *Plant Communities of the Scottish Highlands* is an example of the depth of investigation which has been attempted: the work is definitive. Research on the larger and smaller mammals has been done in Scotland: Lowe and Mitchell have worked on the red deer, following up with population studies the census work I did for the Conservancy between 1953 and 1958. The acquisition of the island of Rhum, about 26,400 acres of varied habitat, as a National Nature Reserve, has given a finite area with a finite population of deer for study. I wished myself to do my original study of red deer on Rhum 30 years ago, but that was not to be. The fact that the whole island is now a Research Reserve is an indication of how far we have travelled. E. A. Smith has done much work on the Atlantic grey seal, mainly in Orkney, carrying through marking of calves to ascertain dispersals from breeding stations and how far each breeding island becomes a closed society. This is work I was about to start in the autumn of 1939 when war put it all in abeyance. My colleague Morton Boyd not only worked intensively on the grey seals on remote Hebridean stations even before he joined the Conservancy, but has since written a thesis on the measurement and conservation of populations of sea-birds, St. Kilda mice, seals and Soay sheep in the Hebrides contained in some twenty papers. Lockie and Charles have studied the pine marten and vole populations.

Research on invertebrate life has not been followed up as a principal field of research by the Conservancy in Scotland because the Merlewood Research Station in Lancashire is particularly active in this field, but there has always been co-operation with University departments, to the extent that the Reserve system plays a steady educational role beyond the immediate staff of the Conservancy.

The distance of many of the Highland Reserves from well-equipped research laboratories of the usual type and the shyness of most of the mammals and birds of primary importance in the Reserves have perhaps called forth field-work of pioneer order in the Highlands and Islands of Scotland. The study by James Lockie and David Stephen of the food habits of the golden eagle in the Hebrides was necessary in view of certain local objections to protection of this bird; the study was undertaken *ad hoc* and was an effective reply to exaggerated and uncritical attitudes. All carnivorous and raptorial creatures need careful research to help them achieve their proper status in our semi-natural world. The fox in the Highlands is now the subject of a similar study by Lockie. We live in an era of admitted over-population of the Highlands by deer, yet the persecuted fox is the surest check on the calf crop, failing adequate stalking of the adult stock by man. Highland sheep farming has been one of the depressants of biological condition of the wild lands and the intense persecution of the fox which does so much else than eat lambs has been one of the factors putting the Highlands out of ecological equilibrium.

When this book was first written the only scientific investigation of grouse populations in Scotland was already forty years old. Happily, the Grouse Report was a fine piece of work standing the test of time, but new concepts called for a fresh study. A Unit of Grouse and Moorland Ecology was set up by the Nature Conservancy in conjunction with Aberdeen University in Deeside following an enquiry financed by the Scottish Landowners' Federation on the estates of the Earl of Dalhousie in Glen Esk. David Jenkins, Adam Watson and Gordon Miller have conducted this investigation with Professor V. C. Wynne-Edwards of Aberdeen as Hon. Director of the Unit. The Conservancy Report of 1962 says:

... it has become apparent that the numbers of grouse on the study areas are not directly governed by external accidents of recruitment and mortality, although these are of course secondarily involved, but rather that they depend upon and reflect underlying changes in the primary productivity of moorland vegetation. A possible implication therefore is that the long-term decline in

numbers has resulted from a gradual overall impoverish-
ment of the moorland itself. For this there seems to be
three likely causes: first, that the climate is changing,
making the environment less suitable for heather growth;
second, that moor management is less efficient than
formerly; third, that existing methods of moor management are causing a run-down on moorland fertility.
Research on grouse is consequently bringing the
ecologists involved into close contact with practical
problems of land management, and with primary aspects
of moorland ecology including the study of micro-
climate, and of the mineral content of soils.

This is a beautiful example of the place of fundamental
disinterested research bearing on larger fields and on the
whole usage of Scotland's wild lands. It is also an affirmation
of the attitude I took up originally in this book in the
chapter, *The Human Factor*. At that time I was also con-
ducting the West Highland Survey, an essay in human
ecology, the foundations of which were a synthetical study
of soil, sea and the natural resources on which the human
population was finally dependent.

The West Highland Survey studied man as part of the
natural history, and natural history as a large part of man's
environment. It was concerned also with history or process
which is an essential part of ecological investigation. The
section of the volume of the *West Highland Survey*, on
The Ecology of Land Use emphasized the ecological
influence of the political factor and showed the cumulative
sociological effects of continued wrong land-use practices.
My view was that the original Highlands were in effect a
balanced complex organism, wherein the biotic communities, though apparently competing among their
individual component species, were really co-operating
towards that climax state or biological continuum which
showed a richness beyond what the geology and climate
would have promised. We know through studies in recent
years that the more complex biotic communities represent-
ing the upper stages of ecological succession have a higher
rate of energy flow than have simple communities, and that

the functions of individual species in the intricate process of conversion of organic matter are important in maintaining the efficiency of energy flow in the eco-system. I am not quite prepared to explain in detail what is meant by primary productivity because the world's ecologists are still debating the term, but its general nature is sufficiently clear to give comprehension in the notion that when species are lost from an eco-system, functions of conversion are lost, bottlenecks occur in energy flow and biological productivity of the environment declines.

The Highlands had become a devastated terrain which political and administrative acts could not restore without biological understanding. This was a most unpopular opinion to express in the 1940's and even now has scarcely penetrated St. Andrew's House. In my view the Nature Conservancy has done more than any other public body concerned with land use in the Highlands to arrive at truth through scientific investigation which, by definition, should be impartial and without bias. The ecological continuum nobly bearing its character of wilderness, can yield greater and more subtle returns to the nation than the subsidized devastation accepted as normality when this book was first written. Research is substantiating this view.

Unfortunately, there is still evident administratively the neat notion of efficiency which leads away from maintenance of the varied habitat towards a simple monocultural one. As I saw it myself, one of the most important demonstrations in the West Highland Survey was ascertaining and plotting on maps the cattle-sheep ratios in West Highland parishes, their widening in time—meaning more sheep and less cattle—and the close correlation between this tendency towards monoculture and depopulation and bad condition of the land. I had pointed to this need for more cattle in the Highlands in the early 1930's but the economic situation not linked to any long term view of improvement caused this opinion to fall on deaf ears. It was not until the second war had begun that a few large owners began to stock Highland sheep farms with cattle. The Ben Callum Company, of which that fine Shorthorn judge and breeder, D. M. Stewart of Millhills, was a director, was foremost in the change. Monetarily, they probably lost in the early years but they could see a greenness coming back on the hill. Lord Lovat followed the same idea on his large estates crossing Inverness-shire and by the late '40's the movement was general and even the Department of Agriculture saw that hill cows were more worthy of subsidy than ewes. Too few of the early enthusiasts for cattle on the hills, as distinct from men of knowledge, foresaw the concomitant necessity of cultivating the limited arable land at the highest level of husbandry to provide winter keep for the animals. It also needed money in the pocket to buy protein feed from January to March if the cows were to calve annually. This is the way for organic fertilizers to get back up on to the hill ground.

The Hill Farming Research Organisation was established in Scotland in 1950 under the Directorship of A. R. Wannop, with lands in the Borders and Argyll. A much more scientific attitude has been adopted in its investigations than we are accustomed to find in agricultural research [*sic*] and it has been possible for the Nature Conservancy and the Hill Farming Research Organisation to co-operate in the general aim of conservation in Scotland, and in the precise study of the wild sheep of St. Kilda.

But as I said, the notion of efficient monoculture dies hard. Several Forestry Commission biologists see quite well that variety is a necessary part of the creation of new forests, but what will be the effect of the noisily-heralded pulp mill at Fort William? The fact is that a pulp mill is efficient as an industrial unit when there are 100,000 acres of spruce, preferably Sitka, in the neighbourhood. Then, would say the technologists, grow the 100,000 acres of spruce and make the mill efficient. Soon we would be in the monocultural whirlpool which involves epidemic outbreaks of pests, followed by aerial spraying of lethal chemicals which do so much more than kill the pests. Waters are polluted, fish food is destroyed, fish are killed and there is a fair possibility, it seems, that raptorial birds, accumulating the poison at second-hand in their tissues, become sterile. The greatest strength of any countryside is preservation of maximum natural variety. Charles Elton has well explained this in his lucid book on *The Ecology of Invasions by Animals*.

Mention of fish and chemicals brings me to another splendid development of natural history in the Highlands, the establishment of the Brown Trout Research Laboratory, now the Freshwater Fisheries Laboratory at Pitlochry. Just as Mr. Thomas Johnston as Secretary of State for Scotland encouraged establishment of the West Highland Survey, so as Chairman of the North of Scotland Hydro-Electric Board did he set about initiating research on the brown trout in the waters over which his Board was exercising control and disturbance. He had heard of increasing fish size and overall crop in fresh waters by adding chemical fertilizers, and thought immediately that this might be a wonderful development in Highland waters. That such a fisherman's dream was not quite so simply realized is not the point: Mr. Johnston had a belief in science and at such a moment he called together a committee of scientists of varied disciplines under Professor C. M. Young, F.R.S., of Glasgow University. I had the honour to be a member of this Committee throughout its ten years of existence and I can never be grateful enough for its educative influence on me. When the first edition of this book was written I alluded to my own limnological ignorance and also to the paucity of work done in the Highlands since the days of Murray and Pullar in the early years of this century. The best mine of subsequent work was the little-read Annual Reports of the Fishery Board for Scotland.

The situation has been changed almost dramatically by the Brown Trout Research Laboratory under the direction of Kenneth Pyefinch. Administrative duties on the Brown Trout Research Supervisory Committee were kept down to a minimum and most of the time in our meetings was devoted to listening to the progress reports of the zoologists, botanists and chemists who were discovering the nature of Highland lochs and freshwaters. We were privileged to be able to discuss their work with these men in the relaxed atmosphere of a pleasantly sited country laboratory where there was yet a high enthusiasm characteristic of pioneer work of discovery. We could also go to the lochs and outstations where field work was being done. Not least we enjoyed the ingenious working models of streams which T. A. Stuart constructed to demonstrate the behaviour of trout and the way water and air acted in passing over the redds. The Scottish Home Department published the results of many of the investigations as separate papers in their Freshwater Series. It is possible this form of publication from Her Majesty's Stationery Office may have limited the circulation of this knowledge abroad, though it may have been greater within Scotland than if the medium of the scientific journals had been used. The highly specialized work of botanists and chemists often did appear in such journals. A strikingly large contribution has emerged in these years towards our understanding of the biological circulation in Highland fresh waters and the ecology of brown trout, sea-trout and salmon.

Birdwatching has always been popular in Scotland, but the enthusiasm displayed since the war has been extra-ordinary. George Waterston did a great service to Scottish natural history when he courageously bought Fair Isle as a personal burden. The island is now owned by the National Trust for Scotland. The traps at Fair Isle and Isle of May have added much to our knowledge of the migration of birds; Williamson's studies of the initiating factors in flights and the influences of meteorological conditions have made possible a remarkable degree of accuracy in forecast-ing migration trends. The Scottish Ornithologists' Club is well organized so that amateurs' observations can swell our knowledge of distribution and of the flux of distribution.

Lastly, I would like to mention the strong natural-history interest and influence of the Forestry School of Aberdeen University under Professor H. M. Steven. *The Native Pine-woods of Scotland* by Steven and Carlisle was published in 1959, the result of painstaking work over many years to which a volume like ours, which gropes and synthesizes, owes a special debt. Here is no "sticks of timber" approach but that of scientists touched by wonder and a sense of history which inspires stewardship. They say, speaking of these remnants of the Great Caledonian Forest first recorded by Pliny:

> Even to walk through the larger of them gives one a better idea of what a primeval forest was like than can be got from any other woodland scene in Britain. The trees range in age up to 300 years in some instances, and there are thus not many generations between their earliest predecessors about 9,000 years ago and those growing today ; to stand in them is to feel the past.

It seems to me that, despite increasing pressures on land in the Highlands as well as elsewhere and the fact that we have lost some examples of natural habitats in these sixteen years, we have made great advances, not only in securing Reserves in which we shall try to perpetuate forest, moor, sand dunes, salt marsh, Hebridean islands, and the denizens and vegetation of all these places, but the national ethos has changed toward greater respect for the natural world around us and a sense of our trusteeship. Our human dominance now gives us less of the view that the natural world was made for man, as that we are fellow organisms with other living things, our intellectual superiority placing upon us the responsibility of wise care. The Highlands and Islands of Scotland are still one of the great wildlife areas of the world, and having now travelled farther than I had in tropic and arctic, I am aware that habitats do not necessarily go into irreversible decline when damaged or reduced to semi-desert. Time is the great healer, indeed, and ecological knowledge backed by spiritual conviction in what one is doing can help to shorten the period of repair. The task is to our hand and our duty plain.

Bibliography

AINSLIE, J. A. and ATKINSON, R. (1937). On the breeding habits of Leach's fork-tailed petrel. *Brit. Birds*, 30: 234-48.

ALLAN, R. M. (1955). Observations on the fauna of Heisker or Monach Isles, Outer Hebrides. *Scot. Nat.* 67: 3-8.

ALLEN, K. R. (1940-4). Studies on the biology of the early stages of the salmon (*Salmo salar*).
1. Growth in the River Eden. *J. Anim. Ecol.* 9: 1-23.
2. Feeding habits. *Ibid.* 10: 47-76.
3. Growth in the Thurso River system, Caithness, *Ibid.* 10: 273-95.
4. The smolt migration in the Thurso River in 1938. *Ibid.* 13: 63-85.

ANDERSEN, J. (1953). Analysis of a Danish roe-deer population (*Capreolus capreolus* (L)): based upon the extermination of the total stock. *Dan. Rev. Game Biol.* 2: 127-55.

ANDERSON, A. (1957). A census of fulmars on Hirta, St. Kilda, in July 1956. *Scot. Nat.* 69: 113-16.

(1962). A count of fulmars on Hirta, St. Kilda, in July 1961. *Scot. Nat.* 70: 120-5.

ANDERSON, A., BAGENAL, T. B., BAIRD, D. E., and EGGELING, W. J. (1961). A description of the Flannan Isles and their birds. *Bird Study*, 8: 71-88.

ANDERSON, M. L. (1967). *A History of Scottish Forestry.* 2 vols. (Ed.: C. J. Taylor). London, Nelson.

ANON. (1949). Golden eagle *versus* wild cat. *Scot. Nat.* 61: 121-2.

ATKINSON, R. (1940). Notes on the botany of North Rona and Sula Sgeir. *Trans. Proc. Bot. Soc. Edinb.* 30: 52-60.

(1949). *Island Going.* London, Collins.

ATKINSON, R., and ROBERTS, B. (1952). Notes on the islet of Gasker. *Scot. Nat.* 64: 129-37.

BADEN-POWELL, D. and ELTON, C. (1936-7). On the relation between a raised beach and an Iron Age midden on the Island of Lewis, Outer Hebrides. *Proc. Soc. Antiq. Scot.* 71: 347-65.

BAGENAL, T. B. (1957). The vertical range of some littoral animals on St. Kilda. *Scot. Nat.* 69: 50-1.

(1958). The feeding of nestling St. Kilda wrens. *Bird Study*, 5: 83-7.

BAGENAL, T. B., and BAIRD, D. E. (1959). The birds of North Rona in 1958, with notes on Sula Sgeir. *Bird Study*, 6: 153-74.

BAIRD, P. D. (1957). Weather and snow on Ben Macdhui. *Cairngorm Club J.* 17: 147-49.

BANNERMAN, D. A. (1953 *et seq.*). *Birds of the British Isles*. Edinburgh, Oliver and Boyd.

BARRINGTON, R. M. (1886). (In "Further notes on North Rona", J. A. Harvie-Brown). *Proc. R. Phys. Soc.* 9: 289.

BAXTER, E. V., and RINTOUL, L. J. (1953). *The Birds of Scotland*. Edinburgh, Oliver and Boyd.

BENNET, A. (1907). The plants of the Flannan Islands. *Ann. Scot. Nat. Hist.* 1907: 187.

BERTRAM, D. S. (Ed.) (1939). The natural history of Canna and Sanday, Inner Hebrides: a report upon the Glasgow University Canna Expeditions, 1936 and 1937. *Proc. Roy. Phys. Soc. Edinb.* 23: 1-71.

BOURNE, W. R. P. (1957). The birds of the island of Rhum. *Scot. Nat.* 69: 21-31.

BOYCOTT, A. E. (1936). The habitats of freshwater mollusca in Britain. *J. Anim. Ecol.* 5: 116-86.

BOYD, J. M. (1956). The Lumbricidae in the Hebrides. II, Geographical distribution. *Scot. Nat.* 68: 165-72.

(1956). The Lumbricidae of Hirta, St. Kilda. *Ann. Mag. Nat. Hist.* (12), 9: 129-33.

(1956). Fluctuations of common snipe, Jack snipe and golden plover in Tiree, Argyllshire. *Bird Study*, 3: 105-18.

(1957). Ecological distribution of the Lumbricidae in the Hebrides. *Proc. Roy. Soc. Edinb.* B. 66: 311-38.

(1958). The birds of Tiree and Coll. *Brit. Birds*, 51: 41-56, 103-18.

(1959). Observations on the St. Kilda field-mouse *Apodemus sylvaticus hirtensis* Barrett-Hamilton. *Proc. Zool. Soc. Lond.*, 133: 47-65.

(1960). Studies of the differences between the fauna of grazed and ungrazed grassland in Tiree, Argyll. *Proc. Zool. Soc. Lond.*, 135: 33-54.

(1960). The distribution and numbers of kittiwakes and guillemots at St. Kilda. *Brit. Birds*, 53: 252-64.

(1961). The gannetry of St. Kilda. *J. Anim. Ecol.* 30: 117-36.

(1961). Home range and homing experiments with the St. Kilda fieldmouse. *Proc. Zool. Soc. Lond.*, 140: 1-14.

(1962). Seasonal occurrence and movements of seals in North-West Britain. *Proc. Zool. Soc. Lond.*, 138: 385-404.

(1964). The grey seal (*Halichoerus grypus* Fab.) in the Outer Hebrides in October 1961. *Proc. Zool. Soc. Lond.*, 141: 635-61.

(1967). Grey seal studies on North Rona. *Oryx*. 9: 19-24.

(1967). Experimental wildlife management in the Scottish Highlands. *Symp. Zool. Soc. Lond.* In press.

BOYD, J. M., HEWER, H. R., and LOCKIE, J. D. (1962). The breeding colony of grey seals on North Rona, 1959. *Proc. Zool. Soc. Lond.* 138: 257-77.

BOYD, J. M., and LAWS, R. M. (1962). Observations on the grey seal (*Halichoerus grypus*) at North Rona in 1960. *Proc. Zool. Soc. Lond.* 139: 249-60.

BOYD, J. M., DONEY, J. M., GUNN, R. G., and JEWELL, P. A. (1964). The Soay sheep of the island of Hirta, St. Kilda. A study of a feral population. *Proc. Zool. Soc. Lond.* 142: 129-63.

BRISTOWE, W. S. (1927). The spider fauna of the Western Islands of Scotland. *Scot. Nat.* 1927: 88-94, 117-22.

BRÖGGER, A. W. (1929). *Ancient Emigrants: A History of the North Settlements of Scotland.* Oxford, Clarendon Press.

BROOK, A. J. (1957). Notes on freshwater algae, mainly from lochs in Perthshire and Sutherland. *Trans. Bot. Soc. Edinb.* 37: 114-22.

(1958). Notes on the algae from the plankton of some Scottish freshwater lochs. *Ibid.* 37: 174-81.

BROOK, A. J., and HOLDEN, A. V. (1957). Fertilization experiments in Scottish freshwater lochs. 1. Loch Kinardochy, *Sci. Invest. Freshwat. Fish. Scot.* 17. Edinburgh, H.M.S.O.

BROWN, L. H. and WATSON, A. (1964). The golden eagle in relation to its food supply. *Ibis* 106: 78-100.

BROWN, P., and WATERSTON, G. (1962). *The Return of the Osprey.* London, Collins.

BUCKLEY, T. E., and EVANS, A. H. (1899). *A Vertebrate Fauna of the Shetland Islands.* Edinburgh, Douglas.

BUDGE, D. (1960). *Jura, an Island of Argyll: its history, people and story.* Bristol, John Smith.

CAMERON, A. E., DOWNES, J. A., MORISON, G. D., and PEACOCK, A. D. (1946). *A Survey of Scottish Midges.* Dept. of Health for Scotland. Edinburgh, H.M.S.O.

(1948). *The Second Report on Control of Midges.* Dept. of Health for Scotland. Edinburgh, H.M.S.O.

CAMERON, A. E. HARDY, J. W., and BENNETT, A. H. (1951). *The Heather Beetle. Its Biology and Control.* Petworth, British Field Sports Society.

CAMERON, A. G. (1923). *The Wild Red Deer of Scotland.* Edinburgh, Blackwood.

CAMPBELL, B. (1963). Crossbills. *Forestry Commission Leaflet,* 36. London, H.M.S.O.

(1958). The crested tit. *Forestry Commission Leaflet,* 41. London, H.M.S.O.

CAMPBELL, J. L. (1938). The macrolepidoptera of the parish of Barra. *Scot. Nat.* 1938: 153-63.

CAMPBELL, J. L. (1954). The macrolepidoptera of the Isle of Canna. *Scot. Nat.* 66: 101-121.

CAMPBELL, M. S. (1945). *The Flora of Uig (Lewis)*. Arbroath, Buncle.

CAMPBELL, R. N. (1955). Food and feeding habits of brown trout, perch and other fish in Loch Tummell. *Scot. Nat.* 67: 23-8.

(1957). The effect of flooding on the growth rate of brown trout in Loch Tummel. *Sci. Invest. Freshwat. Fish. Scot.* 14. Edinburgh, H.M.S.O.

(1961). The growth of brown trout in acid and alkaline waters. *Salm. Trout Mag.*, 161: 47-52.

(1963). Some effects of impoundment on the environment and growth of brown trout (*Salmo trutta* L.) in Loch Garry (Inverness-shire). *Sci. Invest. Freshwat. Fish. Scot.* 30. *Edinburgh*, H.M.S.O.

(1966). Grey seal marking at North Rona. *Scot. Fish. Bull.* 25: 14-16.

CARRICK, R., and WATERSTON, G. (1939). The birds of Canna. *Scot. Nat.* 1939: 5-22.

CHARLES, W. N. (1956). The effect of a vole plague in the Carron Valley, Stirlingshire. *Scot. For.* 10: 201-4.

CHARLESWORTH, J. K. (1955). The late-glacial history of the Highlands and Islands of Scotland. *Trans. Roy. Soc. Edinb.* 62: 769-927.

CLARK, W. A. (1939). Noteworthy plants from North Uist, Baleshare, Monach Islands, Harris, Taransay, Mingulay and Berneray. *Proc. Univ. Durham Phil. Soc.* 10: 124-29.

CLARKE, W. E. (1912). *Studies in Bird Migration*, Vol. 2. London Gurney and Jackson.

COCKBURN, A. M. (1935). The geology of St. Kilda. *Trans. Roy. Soc. Edinb.* 58: 511-47.

COLLINGWOOD, C. A. (1951). The distribution of ants in North-West Scotland. *Scot. Nat.* 63: 45-9.

(1961). Ants in the Scottish Highlands. *Scot. Nat.* 70: 12-21.

COUSENS, J. E. (1962). Notes on the status of the sessile and pedunculate oaks in Scotland and their identification. *Scot. For.* 16: 170-9.

CRAIG, G. Y. ed. (1965). *The geology of Scotland*. Edinburgh, Oliver and Boyd.

DAHL, E. (1951). On the relation between summer temperature and the distribution of alpine vascular plants in the lowlands of Fennoscandia. *Oikos*, 3: 22-52.

DARLING, F. F. (1937). *A Herd of Red Deer.* Oxford, University Press.

— (1938). *Bird Flocks and the Breeding Cycle.* Cambridge, University Press.

— (1939). *A Naturalist on Rona.* Oxford, Clarendon Press.

— (1947). *Natural History in the Highlands and Islands.* London, Collins.

— (1955). *West Highland Survey: An Essay in Human Ecology.* Oxford, University Press.

DAVIES, J. L. (1957). The geography of the grey seal. *J. Mammal.* 38: 297-310.

DICKSON, J. A., and INNES, R. A. (1959). Forestry in North Scotland. *Forestry*, 32, 65-109.

DELANY, M. J. (1961). The ecological distribution of small mammals in North-West Scotland. *Proc. Zool. Soc. Lond.* 137: 107-26.

DELANY, M. J., and BISHOP, I. R. (1960). The systematics, life history and evolution of the bank-vole *Clethrionomys* Tilesius in North-West Scotland. *Proc. Zool. Soc. Lond.* 135: 409-22.

DELANY, M. J. and COPLAND, W. O. (1964). The effects of depopulation on the island of South Rona. *Glas. Nat.* 18: 351-362.

DENNIS, R. W. G. (1964). The fungi of the Isle of Rhum. *Kew Bull.* 19: 77-131.

DIAMOND, A. W., DOUTHWAITE, R. J. and INDGE, W. J. E. (1965). Notes on the birds of Berneray, Mingulay and Pabbay. *Scot. Birds.* 3: 397-404.

DONISTHORPE, H. ST. J. K. (1927). *British Ants.* London, Routledge.

DUFFEY, E. A. G. (1959). Spiders taken on St. Kilda, with extracts from field notebook. *St. Kilda Nature Reserve Record.* The Nature Conservancy, unpublished.

DUNNET, G. M., ANDERSON, A., and CORMACK, R. M. (1963). A study of survival of adult fulmars with observations on the pre-laying exodus. *Brit. Birds*, 56: 2-18.

EGGELING, W. J. (1965). Check list of the plants of Rhum, Inner Hebrides. *Trans. Bot. Soc. Edinb.* 40: 20-99.

EGGLISHAW, H. J. and MORGAN, N. C. (1965). A survey of the bottom fauna of streams in the Scottish Highlands. II. The relationship of the fauna to the chemical and geological conditions. *Hydrobiologia.* 26: 173-183.

ELLIOT, R. J. (1959). Moor management. *Enquiry into the Decline of Red Grouse.* Fifth progress report, 19-25. Scottish Landowners' Federation, limited circulation.

ELTON, C. S. (1938). Notes on the ecological and natural history of Pabbay. *J. Ecol.* 26: 275-97.

— (1942). *Voles, Mice and Lemmings.* Oxford, Clarendon Press.

EVANS, P. R., and FLOWER, W. U. (1967). The birds of the Small Isles. *Scot. Birds.* 4: 404-445.

FENTON, E. W. (1937). The influence of sheep on the vegetation of hill grazings in Scotland. *J. Ecol.* 25: 424-30.

FISHER, J. (1952). *The Fulmar.* London, Collins.
(1956). *Rockall.* London, Bles.

FISHER, J., FERGUSON-LEES, I. J., and CAMPBELL, H. (1949). Breeding of the northern golden plover on St. Kilda. *Brit. Birds*, 42: 379-82.

FISHER, J., and VEVERS, H. G. (1943-44). The breeding distribution, history and population of the North Atlantic gannet (*Sula bassana*). *J. Anim. Ecol.* 12: 173-213; 13: 49-62.

FLETCHER, J. M. (1956). A white otter. *Scot. Nat.* 68: 59.

FOOKS, H. A. (1960). The Roe Deer. *Forestry Commission Leaflet,* 45. London, H.M.S.O.

FORD, E. B. (1945). *Butterflies.* London, Collins.

FRIEND, G. F. (1956). A new sub-species of char from Loch Eck. *Glas. Nat.* 17: 219-20.

GARDEN, E. A. (1958). The national census of heronries in Scotland 1954 with a summary of the 1928-29 census. *Bird Study*, 5: 90-109.

GAULD, D. T., BAGENAL, T. B. and CONNELL, J. H. (1953). The marine fauna and flora of St. Kilda, 1952. *Scot. Nat.* 65: 29-49.

GEDDES, T. (1960). *Hebridean Sharker.* London, Jenkins.

GEIKIE, A. (1887). *The Scenery of Scotland* (2nd Edition). London, Macmillan.

GIMINGHAM, C. H. (1959). The maintenance of good heather. *Enquiry into the Decline of Red Grouse.* Fifth progress report, 24-8. Scottish Landowners' Federation, limited circulation.

GIMINGHAM, C. H., GEMMELL, A. R., and GRIEG-SMITH, P. (1948). The vegetation of a sand-dune system in the Outer Hebrides. *Trans. Bot. Soc. Edinb.* 35: 82-96.

GIMINGHAM, C. H., MILLER, G. R., SLEIGH, L. M., and MILNE, L. M. (1961). The ecology of a small bog in Kinlochewe Forest, Wester Ross. *Trans. Bot. Soc. Edinb.* 39: 125-47.

GOTO, H. E. (1955). Collembola from Shillay, Outer Hebrides, including new British and local records. *Scot. Nat.* 67: 29-33.
(1957). Some further Collembola from Shillay, Outer Hebrides. *Ibid.* 69: 1-10.

GRAHAM, H. D. (1890). *The Birds of Iona and Mull.* Edinburgh. Douglas.

GREEN, F. H. W. (1959). Rainfall, evaporation and land-use. *J. Inst. Water Eng.* 13: 575-80.

(1959). Some observations of potential evaporation 1955-57. *Q. J. Roy. Met. Soc.* 85: 152-8.

GRUBB, P., and JEWELL, P. A. (1966). Social grouping and home range in feral Soay sheep. *Play exploration and territory in mammals.* P. A. Jewel ed. *Symp. Zool. Soc. Lond.* No. 18: 179-210.

HALDANE, A. R. B. (1952). *The Drove Roads of Scotland.* London, Nelson.

HALDANE, R. C. (1905). Notes on whaling in Shetland, 1904. *Ann. Scot. Nat. Hist.* 1905: 65-72.

(1906-10). Whaling in Scotland. *Ibid.* 1906: 130-7; 1907: 10-15; 1908: 65-72; 1909: 65-9; 1910: 1-2.

HAMILTON, J. D. (1963). The freshwater fauna of Hirta, St. Kilda. *Glas. Nat.* 18: 233-41.

HARRISON, J. W. H. (1937). The natural history of the Isle of Raasay, etc. *Proc. Univ. Durham Phil. Soc.* 9: 246-351.

(1938). A contribution to our knowledge of the Lepidoptera of the islands of Coll, Canna, Sanday, Rhum, Eigg, Soay and Pabbay (Inner Hebrides), and of Barra, Mingulay and Berneray (Outer Hebrides). *Ibid.* 10: 10-23.

(1948). Potamogetons in the Scottish Western Isles, with some remarks on the general natural history of the species. *Trans. Bot. Soc. Edinb.* 35: 1-25.

HARRISSON, T. H., and BUCHAN, J. N. S. (1934). A field study of St. Kilda wren (*Troglodytes t. hirtensis*), with especial reference to its numbers, territory and food habits. *J. Anim. Ecol.* 3: 133-45.

HARRISSON, T. H., and LACK, D. (1934). The breeding birds of St. Kilda. *Scot. Nat.* 1934: 59-60, 61-9.

HARRISSON, T. H., and MOY-THOMAS, J. A. (1933). The mice of St. Kilda, with especial reference to their prospects of extinction and present status. *J. Anim. Ecol.* 2: 109-15.

HARTING, J. E. (1880). *British Animals Extinct within Modern Times.* London, Trübner.

HARVIE-BROWN, J. A. (1880-1). The history of the squirrel in Great Britain. *Proc. Roy. Phys. Soc.* 5: 343-48. *Ibid.* 6: 31-63, 115-83.

(1892). The great spotted woodpecker (*Picus major* L.) in Scotland. *Ann. Scot. Nat. Hist.* 1892: 4-17.

(1895). The starling in Scotland, its increase and distribution. *Ann. Scot. Nat. Hist.* 1895: 2-22.

(1906). *A Vertebrate Fauna of the Tay Basin and Strathmore.* Edinburgh, Douglas.

HARVIE-BROWN, J. A., and BARRINGTON, R. M (1897). On the ornithology of Rockall. *Trans. Roy. Irish Acad.* 31: 66-75.

HARVIE-BROWN, J. A., and BUCKLEY, T. E. (1887). *A Vertebrate Fauna of Sutherland, Caithness and West Cromarty.* Edinburgh, Douglas.

— (1888). *A Vertebrate Fauna of the Outer Hebrides.* Edinburgh, Douglas.

— (1892). *A Vertebrate Fauna of Argyll and the Inner Hebrides.* Edinburgh, Douglas.

— (1895). *A Vertebrate Fauna of the Moray Basin,* 2 vols. Edinburgh, Douglas.

HARVIE-BROWN, J. A., and MACPHERSON, H. A. (1904). *A Vertebrate Fauna of the North-West Highlands and Skye.* Edinburgh, Douglas.

HEWER, H. R. (1957). A Hebridean colony of the grey seal *Halichoerus grypus* (Fab.) with comparative notes on the grey seals of Ramsey Island, Pembrokeshire. *Proc. Zool. Soc. Lond.* 128: 23-66.

— (1960). Behaviour of the grey seal *Halichoerus grypus* (Fab.) in the breeding season. *Mammalia,* 24: 400-21.

— (1963). Provisional grey seal life table. Grey Seals and Fisheries, *Rep. Consult. Committee on Grey Seals and Fisheries:* 27-8.

HEWER, H. R., and BACKHOUSE, K. M. (1960). A preliminary account of a colony of grey seal *Halichoerus grypus* (Fab.) in the southern Inner Hebrides. *Proc. Zool. Soc. Lond.* 134: 157-95.

HEWSON, R. (1954). The mountain hare in Scotland in 1951. *Scot. Nat.* 66: 70-88.

— (1955). The mountain hare in the Scottish Islands. *Ibid.* 67: 52-60.

— (1962). Food and feeding habits of the mountain hare *Lepus timidus scoticus,* Hilz. *Proc. Zool. Soc. Lond.* 139: 415-26.

— (1963). Moults and pelages in the brown hare (*Lepus europaeus occidentalis*). De Winton. *Proc. Zool. Soc. Lond.* 141: 677-688.

— (1964). Reproduction in the brown hare and the mountain hare in north-east Scotland. *Scot. Nat.* 71: 81-89.

— (1965). Population changes in the mountain hare (*Lepus timidus* L.). *J. Anim. Ecol.* 34: 587-600.

— (1967). The rock dove in Scotland in 1965. *Scot. Birds.* 4: 359-371.

HICKLING, G. (1962). *Grey Seals and the Farne Islands.* London, Routledge & Kegan Paul.

HOLDEN, A. V. (1959). Fertilization experiments in Scottish freshwater lochs. II. Sutherland, 1954. Pt. I. Chemical and botanical observations. *Sci. Invest. Freshwat. Fish. Scot.* 24. Edinburgh, H.M.S.O.

— (1966). A chemical study of rain and stream waters in the Scottish

Highlands. *Sci. Invest. Freshwat. Fish. Scot.* 37. Edinburgh, H.M.S.O.

HUNTER, R. F. (1962) Hill sheep and their pasture: a study of sheep grazing. in south-east Scotland *J. Ecol.* 50: 651-80.

— (1963). Home range behaviour in hill sheep. *Grazing in terrestial and marine environments.* D. J. Crisp ed. *Symp. Brit. Ecol. Soc. Symp. Grazing,* April 1962: 155-71.

HUNTER, R. F., and DAVIES, G. E. (1963). The effect of method of rearing on the social behaviour of Scottish blackface hoggets. *Anim. Product.* 5: 183-94.

HUNTER, W. R. (1957). Studies on freshwater snails at Loch Lomond. *Glas. Univ. Publ. Stud. Loch. Lomond.* 1:56-95.

HUNTER, W. R., and HAMILTON, J. D. (1958). *Pisidium* from Hirta, St. Kilda. *J. Conch.* 24: 247-8.

HUNTER, W. R., SLACK, H. D., and HUNTER, M. R. (1959J. The lower vertebrates of the Loch Lomond district. *Glas. Nat.* 18: 84-90.

HUXLEY, J. S. (1931). The relative size of antlers in deer. *Proc. Zool. Soc. Lond.* 1931: 819-64.

— (1939). Species formation and geographical isolation. *Proc. Linn. Soc. Lond.* 150: 253-64.

JACKSON, A. R. (1914). A contribution to the spider fauna of Scotland. *Proc. Roy. Phys. Soc. Edinb.* 19: 108-28, 177-90.

JENKINS, D. (1962). The present status of the wild cat (*Felis silvestris*) in Scotland, *Scot. Nat.* 70: 126-38.

JENKINS, D., WATSON, A., and MILLER, G. R. (1963). Population studies on red grouse *Lagopus lagopus scoticus* (Lath.) in northeast Scotland. *J. Anim. Ecol.* 32: 317-76.

— (1964). Predation and red grouse populations. *J. Appl. Ecol.* 1: 183-95.

— (1967). Population fluctuations in the red grouse (*Lagopus lagopus scoticus*). *J. Anim. Ecol.* 36: 97-122.

JENKINS, J. T. (1936). *The Fishes of the British Isles.* London, Warne.

KENNETH, A. G. (1964). The flora of Canna. *Trans. Bot. Soc. Edinb.* 39: 489-501.

KING, C. A. M., and WHEELER, P. T. (1963). The raised beaches of the north coast of Sutherland, Scotland. *Geol. Mag.* 100: 299-320.

KNIGHT, J. E., and SUTTON, F. R. (1966). Lepidoptera of the Beinn Eighe Nature Reserve. *Entomologists' Gaz.* 17: 125-8.

KNOX, R. B. (1958). Flora of the Isle of Jura. 1. Flowering plants and ferns. *Trans. Bot. Soc. Edinb.* 37: 251-6.

LACK, D. (1931). Coleoptera on St. Kilda in 1931. *Ent. Mon. Mag.* 67: 276-9.

(1932). Further notes on insects from St. Kilda in 1931; etc. *Ibid*. 68: 139-45.

(1932). Notes on the Diptera of St. Kilda. *Ibid*. 68: 262-6.

(1939). The display of the blackcock. *Brit. Birds*, 32: 290-303.

LAWSON, J. W. H. (1957). The biting midges and the dancing midges. *Glas. Univ. Publ., Stud. Loch Lomond*, 1: 96-112.

LEOPOLD, A. S., and DARLING, F. F. (1953). *Wildlife in Alaska*. New York, Ronald Press.

LEWIS, J. R. (1956). Intertidal communities of the northern and western coasts of Scotland. *Trans. Roy. Soc. Edinb*. 63: 185-220.

(1964). *The Ecology of Rocky Shores*. London, English Universities Press.

LIND, E. M. (1951). The plankton of some lakes and pools in the neighbourhood of the Moor of Rannoch. *Trans. Bot. Soc. Edinb*. 35: 362-9.

(1952). The phytoplankton of some lochs in South Uist and Rhum. *Ibid*. 36: 37-47.

LOCKIE, J. D. (1955). The breeding habits and food of short-eared owls after a vole plague. *Bird Study*, 2: 53-69.

(1961). The food of the pine marten *Martes martes* in West Ross-shire, Scotland. *Proc. Zool. Soc. Lond*. 136: 187-95.

(1962). Grey seals as competitors with man for salmon. The Exploitation of Natural Animal Populations. *Proc. Brit. Ecol. Soc. Symp*. March 1960. (Edit. Le Cren, E. D., and Holdgate, M. W.): 316-22.

(1964). The distribution and fluctuations of the pine marten in Scotland. *J. Anim. Ecol*. 33: 349-56.

(1964). The breeding density of the golden eagle and fox in relation to food supply in Wester Ross, Scotland. *Scot. Nat*. 71: 67-77.

LOCKIE, J. D., and RATCLIFFE, D. A. (1964). Insecticides and Scottish golden eagles. *Brit. Birds*, 57: 89-102.

LOCKIE, J. D., and STEPHEN, D. (1959). Eagles, lambs and land management on Lewis. *J. Anim. Ecol*. 28: 43-50.

LOCKLEY, R. M. (1942). *Shearwaters*. London, Dent.

LODER, J. de V. (1935). *Colonsay and Oronsay in the Isles of Argyll*. Edinburgh, Oliver and Boyd.

LODGE, E. (1963). The bryophytes of the Small Isles parish of Inverness-shire. I. Sphagnaceae and Musci. *Nova Hedwigia*. 5: 117-48; II. Hepaticae. *Ibid*. 6: 57-65.

MACAULAY, K. (1764). *The History of St. Kilda*. London.

MACCULLOCH, J. (1824). *The Highlands and Western Isles of Scotland*. London, Longman.

MCINTOSH, W. C. (1866). Observations on the marine zoology of North Uist. *Proc. Roy. Soc. Edinb*. 5: 600-14.

MCINTYRE, A. D. (1961). Quantitative differences in the fauna of boreal mud associations. *J. Mar. Biol. Ass. U.K.* 41: 599-616.

(1961). New records of polychaetes from Scottish coastal and off-shore waters. *Proc. Roy. Soc. Edinb.* 67: 351-62.

MACKENZIE, Rev. J. B. (1904). Antiquities and old customs in St. Kilda, compiled from notes by Rev. Neil MacKenzie, 1829-43. *Proc. Soc. Antiq. Scot.* 38: 397-402. Episode in the life of Rev. Neil MacKenzie at St. Kilda from 1829 to 1843, privately printed.

(1905). Notes on the birds of St. Kilda. *Ann. Scot. Nat. Hist.* 1905: 75-80; 141-53.

MACKENZIE, O. H. (1924). *A Hundred Years in the Highlands.* (Popular Edition.) London, Edward Arnold.

MACLEAN, H. I. C. (1961). Four observations of killer whales with an account of the mating of these animals. *Scot. Nat.* 70: 75-8.

MACLEOD, A. M. (1948). Some aspects of the plant ecology of the Island of Barra. *Trans. Bot. Soc. Edinb.* 35: 67-81..

MACLEOD, J. (1932). The bionomics of *Ixodes ricinus* L., the "sheep tick" of Scotland. *Paristology*, 24: 382-400.

(1932). Preliminary studies in the tick transmission of louping-ill. *Vet. J.* 88: 276-84.

MCVEAN, D. N. (1958). Island vegetation of some West Highland freshwater lochs. *Trans. Bot. Soc. Edinb.* 37: 200-8.

(1961). Flora and vegetation of the islands of St. Kilda and North Rona in 1958. *J. Ecol.* 49: 39-54.

(1961). Experiments on the direct sowing of Scots pine. *Emp. For. Rev.* 40: 217-27.

MCVEAN, D. N., and RATCLIFFE, D. A. (1962). Plant communities of the Scottish Highlands: a study of Scottish mountain, moorland and forest vegetation. *Monographs of the Nature Conservancy.* 1. London, H.M.S.O.

MANLEY, G. (1952). *Climate and the British Scene.* London, Collins.

MARTIN, M. (1698). *A Late Voyage to St. Kilda.* London.

(1703). *A Description of the Western Islands of Scotland, etc.* London.

MATTHEWS, L. H. (1952). *British Mammals.* London, Collins.

MAXWELL, G. (1952). *Harpoon at a Venture.* London, Hart-Davis.

(1960). *Ring of Bright Water.* London, Longmans.

MILLAIS, J. G. (1904-6). *The Mammals of Great Britain and Ireland.* London, Longmans.

MILLAR, R. H. (1952). The littoral Ascidians of Argyll. *Scot. Nat.* 64: 19-25.

(1961). Scottish oyster investigations 1946-1958. *Mar. Res. Scot.* 1961, 3. Edinburgh, H.M.S.O.

MILLER, G. R., JENKINS, D., and WATSON, A. (1966). Heather performance and red grouse populations. I. Visual estimates of heather performance. *J. Appl. Ecol.* 3: 313-26.

MILLS, D. H. (1962). The goosander and red-breasted merganser in Scotland. *Wildfowl Trust, Thirteenth Ann. Rep.* 79-92.

(1962). The goosander and red-breasted merganser as predators of salmon in Scottish waters. *Sci. Invest. Freshwat. Fish. Scot.* 29. Edinburgh, H.M.S.O.

(1964). The ecology of the young stages of the Atlantic Salmon in the River Bran, Ross-shire. *Sci. Invest. Freshwat Fish. Scot.* 32

(1964) The distribution and food of the cormorant in Scottish inland waters. *Sci. Invest Freshwat. Fish. Scot.* 35.

MONRO, D. (1884). *Description of the Western Isles of Scotland* (*Circa* 1549). Glasgow, Thomas D. Morrison.

MORGAN, N. C. (1956). The biology of *Leptocerus aterrimus* Steph. with reference to its availability as a food for trout. *J. Anim. Ecol.* 25: 349-65.

(1966). Fertilization experiments in Scottish freshwater lochs I. Sutherland, 1954. 2. Effects on the bottom fauna. *Sci. Invest. Freshwat. Fish. Scot.* 36.

MORGAN, N. C., and EGGLISHAW, H. J. (1965). A survey of the bottom fauna of streams in the Scottish Highlands. I. Composition of the fauna. *Hydrobiologia*. 25: 181-211.

MORGAN, N. C., and WADDELL, A. B. (1960). Chironomidae (Diptera) new to Perthshire including some species new to Britain. *Entomologist*, 93: 62-9.

(1961). Diurnal variation in the emergence of some aquatic insetcs. *Trans. R. Ent. Soc. Lond.* 113: 123-37.

MORLEY, A. (1943). Sexual behaviour in British birds from October to January. *Ibis* 85: 132-58.

MUNRO, W. R. (1957). The Pike of Loch Choin. *Sci. Invest. Freshwat. Fish. Scot.* 16. Edinburgh, H.M.S.O.

(1961). The effect of mineral fertilisers on the growth of trout in some Scottish lochs. *Verh. int. Ver. Limnol.* 14: 718-21.

MURRAY, J. (1905). Microscopic life of St. Kilda. *Ann. Scot. Nat. Hist.* 1905: 94-6.

(1906). Scottish alpine Tardigrada. *Ibid.* 1906: 25-30.

MURRAY, W. H. (1962). *Highland Landscape*. Edinburgh, The National Trust for Scotland.

NAIRN, D. (1890). Notes on Highland woods, ancient and modern. *Trans. Gaelic. Soc. Inverness*, 18: 170-221.

NATURE CONSERVANCY (1952-68). *Reports of the Nature Conservancy*. London, H.M.S.O.

NETHERSOLE-THOMPSON, D. (1951). *The Greenshank*. London, Collins.

(1967). *The Snow Bunting*. Edinburgh, Oliver and Boyd.

NICOL, E. A. T. (1936). The brackish-water lochs of North Uist. *Proc. Roy. Soc. Edinb.* 56: 169-95.

(1939). Three rare crabs from the Inner Hebrides. *Scot. Nat.* 1939: 1-4.

O'DELL, A. C., and WALTON, K. (1962). *The Highlands and Islands of Scotland*. London and Edinburgh, Nelson.

PALMAR, C. E. (1965). The capercailzie. *Forestry Commission Leaflet*, 37. London, H.M.S.O.

(1965). Woodpeckers in woodlands. *Ibid.* 42. London, H.M.S.O.

PENNANT, T. (1774). *A Tour of Scotland and Voyage to Hebrides*. Chester.

PENNIE, I. D. (1950-1). The history and distribution of the capercailzie in Scotland. *Scot. Nat.* 62: 65-87; 157-78; 63: 4-17; 135.

(1951). The Clo Mor Bird Cliffs. *Ibid.* 63: 26-32.

PETCH, C. P. (1933). The vegetation of St. Kilda. *J. Ecol.* 21: 92-100.

PICKARD-CAMBRIDGE, O. (1905). Spiders of St. Kilda. *Ann. Scot. Nat. Hist.* 1905: 220-3.

POORE, M. E. D. (1954). Phytosociology of the Breadalbane district of Perthshire. *University of Cambridge Thesis*.

POORE, M. E. D., and MCVEAN, D. N. (1957). A new approach to Scottish mountain vegetation. *J. Ecol.* 45: 401-39.

POORE, M. E. D., and ROBERTSON, V. C. (1949). The vegetation of St. Kilda in 1948. *J. Ecol.* 37: 82-99.

PYEFINCH, K. A. (1960). *Trout in Scotland. A story of brown trout research at Pitlochry*. Edinburgh, H.M.S.O.

PYEFINCH, K. A., and MILLS, D. H. (1963). Observations on the movements of Atlantic salmon (*Salmo salar* L.) in the River Conon and the River Meig, Ross-shire. *Sci. Invest. Freshwat. Fish. Scot.* 31.

RAE, B. B. (1960). Seals and Scottish Fisheries. *Mar. Res. Scot.* 1960, 2. Edinburgh, H.M.S.O.

(1963). The food of grey seals. Grey seals and Fisheries. *Rep. Consult. Committee on Grey Seals and Fisheries*, 28-33.

RAE, B. B., and WILSON, E. (1953-61). Rare and exotic fishes recorded in Scotland, etc. *Scot. Nat.* 65: 141-53; 66: 170-85; 68: 23-38; 92-109; 70: 22-33.

RAE, B. B., and LAMONT, J. M. (1961-4). Rare and exotic fishes recorded in Scotland, etc. *Scot. Nat.* 70: 34-42, 102-19; 71: 29-36; 71:39-46.

RATCLIFFE, D. A. (1958). A limestone flora in the Ben Alder group. *Trans. Bot. Soc. Edinb.* 37: 217-19.

(1960). Montane plants in Ross-shire and Sutherland. *Ibid.* 39: 107-13.

(1962). Breeding density in the peregrine *Falco peregrinus* and raven *Corvus corax. Ibis,* 104: 13-39.

(1963). The status of the peregrine in Great Britain. *Bird Study,* 10: 56-90.

(1965). The peregrine situation in Great Britain 1963-64. *Bird Study,* 12: 66-82.

REGAN, C. T. (1911). *The Freshwater Fishes of the British Isles.* London, Methuen.

REID, D. M. (1935). The range of the sea-urchin *Echinus esculentus. J. Anim. Ecol.* 4: 7-16.

RITCHIE, J. (1920). *The Influence of Man on Animal Life in Scotland.* Cambridge, University Press.

ROBERTS, B., and ATKINSON, R. (1955). The Haskeir rocks, North Uist. *Scot. Nat.* 67: 9-18.

ROBSON, M., and WILLS, P. (1963). Notes on the birds of Bearasay, Lewis. *Scot. Birds,* 2: 410-14.

ROGER, J. G. (1954). The flora of Caenlochan. *Trans. Proc. Bot. Soc. Edinb.* 36: 189-94.

(1956). Flowering plants of the Cairngorms. *Cairngorms Club J.* 17: 57-72.

(1959). The conservation of the Scottish flora. *Trans. Bot. Soc. Edinb.* 38: 168-79.

ROY, A. B. (1955). The spiders of the Black Wood of Rannoch. *Scot. Nat.* 67: 19-22.

(1962). The spiders of the Cairngorm region. *Ibid.* 70: 96-101.

ST. JOHN, C. (1884). *A Tour in Sutherlandshire.* 2 vols (2nd Edition). Edinburgh, Douglas.

(1893). *Short Sketches of the Wild Sports and Natural History of the Highlands* (9th Edition). London, J. Murray.

SANDEMAN, P. W. (1957). The breeding success of golden eagles in the Southern Grampians. *Scot. Nat.* 69: 148-52, 182, 192.

SCHEFFER, V. B. (1958). *Seals, Sea Lions and Walruses.* London, Oxford University Press.

SCOTT, T. (1890). The invertebrate fauna of the inland waters of Scotland. *Rep. Fish. Bd. Scot.* 8: 269-96.

(1895). Invertebrate fauna of the freshwater lochs of the Outer Hebrides. *Ibid.* 1895: 237-57.

SCOTTISH PEAT COMMITTEE (1954). *Report of Scottish Peat Committee.* Edinburgh, H.M.S.O.

(1962). *Scottish Peat.* Edinburgh, H.M.S.O.

SCROPE, W. (1894). *Days of Deer Stalking.* London, Hamilton.

SEEBOHM, H. (1884). On a new species of British wren. *Zoologist,* 8: 333-5.

SERGEANT, D. E., and WHIDBORNE, R. F. (1951). Birds on Mingulay in the summer of 1949. *Scot. Nat.* 63: 18-25.

SHEARER, W. M., and TREWAVAS, E. (1960). A Pacific salmon, *Onchorhynchus gorbuscha*, in Scottish waters. *Nature, Lond.* 188:868.

SLACK, H. D. (1957). The topography of the lake. Physical and chemical data. The lake bed in relation to its flora and fauna. The fauna of the lake. *Glas. Univ. Publ. Stud., Loch Lomond*, 1: 4-48.

SLACK, H. D., GERVERS, F. W. K., and HAMILTON, J. D. (1957). The biology of the powan. *Glas. Univ. Publ. Stud., Loch Lomond*, 1: 113-27.

SMITH, E. A. (1963). The population of grey seals. Grey Seals and Fisheries, *Rep. Consult. Committee on Grey Seals and Fisheries:* 15-18.

—— (1963). Results of marking-recovery experiments on grey seals 1951-1961. *Ibid.:* 18-22.

—— (1963). A review of studies of grey seal reproduction. *Ibid.:* 23-6.

—— (1963). Utilisation of seal products. *Ibid.:* 49-52.

SMITH, F. W. (1948). On our knowledge of the distribution of Macrolepidoptera in Scotland. *Scot. Nat.* 60: 192-4.

SMITH, M. (1963). A collection of invertebrates from St. Kilda between 1961 and 1963. The Nature Conservancy, unpublished.

SMITH, W. G. (1911). Grass moor association. In *Types of British Vegetation* ed. Tansley. Cambridge, University Press.

—— (1918). The distribution of *Nardus stricta* in relation to peat. *J. Ecol.* 6: 1-13.

SOUTHERN, H. N. (1938). Distribution of the bridled form of the common guillemot (*Uria aalge*). *Nature, Lond.* 142: 951-3.

—— (1951). Change in status of the bridled guillemot after ten years. *Proc. Zool. Soc. Lond.* 121: 657-71.

—— (1962). Survey of bridled guillemots, 1959-60. *Proc. Zool. Soc. Lond.* 138: 455-72.

SPENCE, D. H. N. (1960). Studies on the vegetation of Shetland. III Scrub in Shetland and in South Uist, Outer Hebrides. *J. Ecol.* 48: 73-95.

SPINK, P. C. (1967). Scottish snowbeds in summer 1966. *Weather.* 22: 298-299.

STATISTICAL ACCOUNT OF SCOTLAND (OLD) (1791-99). Edinburgh.

STATISTICAL ACCOUNT OF SCOTLAND (NEW) (1845). Edinburgh, Blackwoods.

STATISTICAL ACCOUNT OF SCOTLAND (THIRD) Vols. on Argyll (1961) and Banff (1961). Glasgow, Collins.

STEPHEN, A. C. (1935). Notes on the intertidal fauna of North Uist. *Scot. Nat.* 1935: 137-42.

STEPHENS, M. N. (1957). *The Natural History of the Otter.* A report to the Otter Committee. London, U.F.A.W.

STEVEN, H. M., and CARLISLE, A. (1959). *The Native Pinewoods of Scotland.* Edinburgh, Oliver and Boyd.

STEWART, M. (1933). *Ronay.* Oxford, University Press.

STUART, T. A. (1953). Spawning migration, reproduction and young stages of loch trout (*Salmo trutta* L.). *Sci. Invest. Freshwat. Fish. Scot.* 5. Edinburgh, H.M.S.O.

(1957). The migrations and homing behaviour of brown trout (*Salmo trutta* L.). *Ibid.* 18. Edinburgh, H.M.S.O.

(1962). The leaping behaviour of salmon and trout at falls and obstructions. *Ibid.* 28. Edinburgh, H.M.S.O.

TANSLEY, A. G. (1949). *The British Islands and their Vegetation.* 2 vols. Cambridge, University Press.

THOMAS, H. J. (1958). The lobster and crab fisheries in Scotland. *Mar. Res. Scot.* 1958, 8. Edinburgh, H.M.S.O.

TOD, K. (1953). The distribution of the northern dart, *Agrotis hyperborea* (Zetterstedt) on the mainland of Scotland. *Scot. Nat.* 65: 11-18.

TRAIL, J. W. H. (1905). The plants of the Flannan Islands. *Ann. Scot. Nat. Hist.* 1905: 187.

VIBERT, R. (1950). Recherches sur le saumon de l'Adour (*Salmo salar*, Linné). *Ann. Sta. Cent. Hydrobiol. Appl.* 3: 27: 149.

VOSE, P. B., POWELL, H. G., and SPENCE, J. B. (1957). The machair grazings of Tiree, Inner Hebrides. *Trans. Bot. Soc. Edinb.* 37: 89-110.

WATERS, W. E. (1963). Observations on the fulmar at St. Kilda. *Scot. Birds.* 2: 459-68.

(1964). Observations on the St. Kilda wren. *Brit. Birds.* 57: 49-64.

(1964). Observations on small petrels at St. Kilda, 1961-62. *Scot. Birds.* 3: 73-81.

WATERSTON, A. R. (1936). Land and freshwater mollusca. In "The natural history of Barra, Outer Hebrides", by Forrest, J. E. *et al. Proc. Roy. Phys. Soc. Edinb.* 22: 290-4.

WATERSTON, J., and TAYLOR, J. W. (1905). Land and freshwater molluscs of St. Kilda. *Ann. Scot. Nat. Hist.* 1905: 21-4.

WATSON, A. (1955). The winter food of six Highland foxes. *Scot. Nat.* 67: 123-4.

(1957). The breeding success of golden eagles in the north-east Highlands. *Ibid.* 69: 153-69, 184.

(1963). The effect of climate on the colour changes of mountain hares in Scotland. *Proc. Zool. Soc. Lond.* 141: 823-35.

(1964). The food of ptarmigan (*Lagopus mutus*) in Scotland. *Scot. Nat.* 71: 60-6.

(1965*a*). Research on Scottish ptarmigan. *Scot. Birds*. 3: 331-349.

(1965*b*). A population study of ptarmigan (*Lagopus mutus*) in Scotland. *J. Anim. Ecol.* 34: 135-72.

(1966). Hill birds of the Cairngorms. *Scot. Birds*. 4: 179-203.

WATSON, A., and HEWSON, R. (1963). *Mountain Hares*. London, *Sunday Times*.

WATSON, A. and JENKINS, D. (1964). Notes on the behaviour of the red grouse. *Brit. Birds*. 57: 137-70.

WATSON, A. and MORGAN, N. C. (1964). Residues of organo-chlorine insecticides in a golden eagle. *Brit. Birds*. 57: 341-44.

WATSON, A., MILLER, G. R., and GREEN, F. H. W. (1966). Winter browning of heather (*Calluna vulgaris*) and other moorland plants. *Trans. Bot. Soc. Edinb.* 40: 195-203.

WATSON, E. V., and BARLOW, H. W. B. (1936). The vegetation. In "The natural history of Barra, Outer Hebrides", by Forrest, J. E. *et al. Proc. Roy. Phys. Soc.* 22: 244-54.

WATT, A. S., and JONES, E. W. (1948). The ecology of the Cairn-gorms. 1. The environment and the altitudinal zonation of the vegetation. *J. Ecol.* 36: 283-304.

WATT, H. B. (1937). On the wild goat in Scotland. With supple-ment: Habits of wild goats in Scotland, by F. F. Darling. *J. Anim. Ecol.* 6: 15-22.

WEEREKOON, A. C. J. (1957). The maintenance of isolated faunas. *Glas. Univ. Publ. Stud. Loch Lomond*, 1: 49-55.

WEST, G. (1910). An epitome of a comparative study of the domin-ant Phanerogamic and higher Cryptogamic flora of aquatic habit, in seven lake areas of Scotland. *Bathymetrical Survey of the Scottish Freshwater Lochs*, 1: 156-260.

WHITEHEAD, G. K. (1960). *The Deer Stalking Grounds of Great Britain and Ireland*. London, Hollis and Carter.

(1962). Future of Scotland's wild deer. *Country Life*, 23 August 1962, 132: 424-6.

WILDFOWL TRUST (1963). Wildfowl in Great Britain (Edit. ATKINSON-WILLES, G. L.). *Monographs of the Nature Con-servancy*, 3. London, H.M.S.O.

WILLIAMSON, K. (1964). A census of breeding land-birds on Hirta, St. Kilda, in summer, 1963. *Bird Study*. 11: 153-167.

WILLIAMSON, K., and BOYD, J. M. (1960). *St. Kilda Summer*, London, Hutchinson.

(1963). *A Mosaic of Islands*. Edinburgh, Oliver and Boyd.

WISLOCKI, G. B. (1942). Studies on the growth of antlers. 1. On the structure and histogenesis of the antlers of the Virginia deer

(*Odocoileus virginiamus borealis*). *Amer. Jour. Anat.* 71: 371-415.

(1943). Studies on the growth of deer antlers. II. Seasonal changes in the male reproductive tract of the Virginia deer with a discussion of the factors controlling the antler-gonad periodicity. *Essays in biology in honor of Herbert M. Evans.* Univ. Calif. Press, Berkeley and Los Angeles: 629-53.

WISLOCKI, G. B., AUB, J. C., and WALDO, C. M. (1947). The effects of gonadectomy and the administration of testosterone propionate on the growth of antlers in male and female deer. *Endocrinology* 40: 202-24.

WORMELL, P. (1962). Notes on the Lepidoptera of the Isle of Rhum. *Entomologist.* 95: 94-6.

WYNNE-EDWARDS, V. C. (1957). The so-called 'northern golden plover'. *Scot. Nat.* 69: 89-93.

(1962). *Animal Dispersion in Relation to Social Behaviour.* Edinburgh, Oliver and Boyd.

YAPP, W. B. (1953). The high-level woodlands of the English Lake District. *Northwestern Nat.* 24: 190-207, 370-83.

(1961). Oaks in Scotland. *Scot. Nat.* 70: 2-6.

(1962). *Birds and Woods.* London, Oxford Unversity Press.

ZWICKEL, F. C. (1966). Sex and age ratios and weights of capercaillie from the 1965-66 shooting season in Scotland. *Scot. Birds.* 4: 209-213.

Index